Remote Sensing
for the Control of
Marine Pollution

NATO • Challenges of Modern Society

A series of edited volumes comprising multifaceted studies of contemporary problems facing our society, assembled in cooperation with NATO Committee on the Challenges of Modern Society.

Remote Sensing for the Control of Marine Pollution

Edited by
Jean-Marie Massin

Department of Pollution Prevention
Ministry of Environment
Neuilly-sur-Seine, France

Published in cooperation with
NATO Committee on the Challenges of Modern Society

PLENUM PRESS • NEW YORK AND LONDON

Library of Congress Cataloging in Publication Data

Main entry under title:

Remote sensing for the control of marine pollution.

(NATO challenges of modern society; v. 6)
Includes bibliographical references and index.
1. Marine pollution—Remote sensing—Addresses, essays, lectures. I.
Massin, Jean-Marie. II. Series.
GC1085.R46   1984   628.1′6833′0287   84-9843

ISBN-13: 978-1-4612-9719-2   e-ISBN-13: 978-1-4613-2787-5
DOI: 10.1007/978-1-4613-2787-5

©1984 Plenum Press, New York
A Division of Plenum Publishing Corporation
233 Spring Street, New York, N.Y. 10013

Softcover reprint of the hardcover 1st edition 1984

PREFACE

    This volume represents the findings of the six-year NATO
CCMS pilot study on the use of remote sensing for the control
of marine pollution, a joint study conducted by countries
confronted by the problems that arise from the prevention of,
and the fight against, deliberate and accidental oil spills.

    In 1976, when I submitted to the Committee on the
Challenges of Modern Society the draft of this pilot study,
the use of remote sensing in the area covered by the project
was still at the experimental research stage. Two years later,
the Amoco Cadiz disaster occurring on the Brittany Coast gave
the opportunity to demonstrate the important role that remote
sensing could play in the fight against major oil spills. At
the same time, many countries engaged in the fight against the
deliberate discharge of oil from ships became aware of the potential of remote sensing to help combat this type of illegal
activity.

    Compiling this volume has afforded me the opportunity to
re-read the papers that were presented at the workshops in
Washington, D.C. and Paris, which were held April 1979 and
October 1982, respectively. Not only is the material still of
current interest, but also some recommendations expressed by
the experts have already received international recognition.
This applies, in particular, to the Isowake experiments, which
were originated by the United States (U.S. Coast Guard) and
the United Kingdom (Warren Spring Laboratory).

    In retrospect, the NATO-CCMS pilot study on remote sensing
applied to marine pollution was highly successful and has,
hopefully, maintained the standard of excellence shown many
times previously by the Committee on the Challenges of Modern
Society. This success can be attributed essentially to the
amount and quality of work carried out by the experts who
assisted the different meetings and actively participated in
the workshops. I would like to thank, in particular, the United
States Coast Guard for their collaboration. To no one, however,

do I owe more sincere thanks than Russ Vollmers, chairman of
Working Group I. Russ has been the keystone of the pilot study
and without him, no doubt, it would not have given the same
final impressions. I personally regret the absence of Russ from
the Paris Workshop and would like to emphasize how his absence
was felt by all those at this meeting.

I would like to express my sincere thanks to each of the
participants who assisted at the different meetings. It would
be tedious for the reader to mention them all but I would like
to express my appreciation to Bob O'Neil (Canada), Louis Laidet
(France), Friedhelm Günneberg (F.R.G.), Hugo Brunsveld van Hulten
and Rudi Spanhoff (The Netherlands), Peter Morris (United Kingdom), and, finally, Charles Corbett (United States Coast Guard)
for many stimulating and clarifying discussions.

A special thanks also to all those who organized the
meetings in Washington, Paris, The Hague, Stevenage, Toulouse
and Brussels for making the meetings so welcoming and memorable.

I am deeply indebted to the NATO-CCMS chairmen who chaired
the CCMS meetings during these past seven years: Mr Ozdas and
Mr Chabbal. Their encouragement and advice have been very
valuable.

Finally, I wish especially to thank Thierry Chambolle, the
Director, and Pierre Woltner, the Deputy Director of the Prevention Department in the French Ministry of the Environment. Their
interest in remote sensing when it appeared only as a "promising
tool" has been very encouraging and without their help and
understanding my participation would have been less complete.

The reader is asked to excuse the black and white reproduction of the imagery, which for technical reasons could not be
presented in their original colour. I hope this does not detract
seriously from gaining an understanding of the material presented
but where necessary colour prints of the imagery can be obtained
directly from the relevant authors.

Jean-Marie Massin
Décembre 1983

CONTENTS

TECHNICAL RESULTS

CONTENTS

CONCLUSIONS AND RECOMMENDATIONS

CONCLUSIONS AND RECOMMENDATIONS

In 1976 France submitted to the Committee on the Challenges of Modern Society (CCMS) a draft pilot study on the use of remote sensing for the control of marine pollution. Its purpose was to review progress achieved with techniques for detecting pollution of the marine environment and to identify the research policies to be adopted by NATO countries in this area.

Following the Council's approval of the draft, France was joined by the United States, Greece and Turkey as co-pilots.

In September 1978, after an exhaustive list of commercially available resources and techniques had been drawn up and issued, the countries participating in the pilot study* decided, at the suggestion of the United States, to assign to two working groups:

* the study of the applications of remote sensing to the detection, surveillance and monitoring of oil spills and other dangerous substances at sea (Working group I);

* the study of the opportunities offered by remote sensing for tracking and monitoring the diffusion and coastal movements of pollutants (Working group II).

Because of the feeling that Working group II could not reasonably embark on work until Working group I had submitted its conclusions and, consequently, that Working group II's activities should be deferred, Working group I, chaired by Mr. Russel Vollmers (United States Coast Guard Headquarters), drew up its terms of reference as follows :

----

* France, The United States, Greece, Turkey, Canada, Federal Republic of Germany, Belgium, Italy, The Netherlands, United Kingdom as members of the Steering Committee. Later on, Denmark, Norway, Portugal, Sweden and the Commission of the European Communities joined the participating countries.

* To obtain and disseminate state-of-the-art information
  on oil pollution surveillance in marine waters;

* To list operational techniques in the area under consideration and to describe them in detail;

* To prepare a non-technical document indicating the advantages, disadvantages and operational limitations of
  the various remote sensing systems for oil pollution
  surveillance;

* To recommend new lines of research for the short and
  long terms aimed at developing new sensors or at improving the performance of existing ones to be used for
  surveillance by aircraft or satellites;

* To encourage and participate in joint exercises;

* To collect and disseminate all information on remote
  sensing experiments concerning oil pollution surveillance;

* To provide technical advice to agencies and public authorities dealing with the legal aspects of oil pollution, in particular in the case of deliberate oil spills
  caused by ships in contravention of national and international regulations in force.

To address these goals, five meetings, including two workshops (in Washington, D.C., from 18th to 20th April 1979 and
in Paris from 12th to 14th October 1982) were held on the initiative of the United States Coast Guard.

During the Washington workshop, two panel discussions were
held to consider the presentations made to the workshop and to
try to summarize international requirements concerning the detection, tracking and monitoring oil spills, to identify the
present capabilities of existing systems and to define lines
of research where the development of new remote sensing systems
could be fruitful.

The first panel was to examine the national requirements
and existing programmes of the various participating nations
in an attempt to identify the common goals and areas where the
objectives of ocean surveillance were widely divergent.

Panel I consisted of:

M. Strome (co-chairman)
Cdr. J. T. Leigh (co-chairman)

R. O'Neil
L. Backlund
H. Parker
J. M. Massin
R. Spanhoff
F. Günneberg
R. Burkhalter
R. Landers

The second panel was to study the technical results presented and to evaluate the need for further research. The panel consisted of:

R. O'Neil (chairman)
W. Hovis
J. Knoll
A. Fontanel
R. Neville
V. Thomson
L. Buja-Bijunas
B. M. Sørensen
R. Hawkins
R. Burkhalter
D. Nicholson
U. Alvarado

In both panels, the discussions were very wide ranging and in some cases overlapped. This was undoubtedly because of the diverse experience of the participants in the operational and scientific aspects of oil pollution surveillance.

The conclusions reached are a consensus of the opinions of the experts assembled in the two panel discussions and should not be taken as the policy of any one nation particular. These conclusions were in part revised during the Paris workshop, in the view of the more recent development of the remote sensing techniques.

Thereby, the following conclusions and recommendations may be seen as a summary of the pilot study activities as a whole. They are based on the proceedings of the Washington workshop (which were published in 1978 under the aegis of both the French Ministry of the Environment and the United States Coast Guard) and on the reports submitted at the Paris meeting.

THE IDEAL SYSTEM

Table 1 was drawn up to show how well user requirements can be met by each of the sensors discussed. These requirements expressed by the experts responsible for controlling oil

Table 1. A summary of sensors capabilities

		REQUIREMENTS												
		REAL TIME	DAY	NIGHT	NEAR ALL-WEATHER	CONFIRM OIL	QUANTITY (THICKNESS)	SUBMERGED OIL	GOOD RESOLUTION	MAPPING	LONG RANGE (LARGE AREA)	DATA LINK TO GROUND	VESSEL IDENTIFICATION	DOCUMENTATION INCL. POSITION
SENSORS — CORE PACKAGE	OBSERVER	▨	▨						▨	▨			▨	▨
	PHOTOGRAPHIC CAMERA		▨						▨				▨	▨
	PASSIVE TV LOW LIGHT LEVEL	▨	▨						▨	▨		▨		▨
	INFRARED L.S. (Thermal)	▨	▨	▨					▨			▨		▨
	ULTRAVIOLET LINE SCANNER	▨	▨						▨	▨		▨		▨
	SIDE LOOKING AIRBORNE RADAR	▨	▨	▨	▨					▨	▨	▨		▨
	ACTIVE GATED TELEVISION	▨	▨	▨		▨			▨				▨	▨
	PASSIVE MICROWAVE RADIOMETER	▨	▨	▨	▨		▨			▨				▨
	PASSIVE MICROWAVE IMAGER	▨	▨	▨	▨					▨				▨
	LASER FLUOROSENSOR	▨	▨			▨		▨	▨					▨
APPLICATIONS	POLICING	▨	▨	▨	▨	▨							▨	▨
	OIL SPILL MANAGEMENT	▨	▨			▨	▨		▨	▨	▨	▨		▨
	LONG TERM MONITORING		▨		▨	▨	▨	▨	▨	▨	▨			▨

pollution of the sea in terms of unreported intentional oil spills and the reported accidental oil spills reflect three different marine surveillance concepts:

(1) so-called "routine surveillance" which is particularly aimed at the detection of illicit oil spills and the

prosecution of the offenders (Table 1: Policing);

(2) wide-area surveillance in the event of a major accident causing massive pollution of the marine environment, such action being designed to locate, quantify and monitor the movements of usually extensive oil spills (Table 1: Oil spill management);

(3) baseline surveillance for a variety of purposes which include data cataloguing for pre-spill and post-spill management decision facilitation and scientific research (Table 1: Long term monitoring).

To fulfil all these missions, the ideal remote sensing system should:

* Provide continuous day night, all-weather, wide area, real-time surveillance;

* Detect any oil spill on (or below) the surface;

* Unambiguously confirm the presence of an oil spill;

* Assess the thickness (and thickness variations) of the pollutants and, hence, the amount of oil spilled;

* Identify the source of pollution and specify the type of pollutant being discharged;

* Provide precise navigational accuracy for spill and source location and positioning clean-up vessels;

* Assemble all the collected data in the form of hard-copies and recordings for subsequent prosecution purposes and legal evidence.

The requirement which can be met by each sensor is shaded; the shading is chosen to indicate the specific applications for which the requirements were identified.

It should be appreciated that this method of presentation is a simplification of the actual situation but serves to demonstrate benefits of each of the sensors listed. Many of the terms used to describe the requirements are used relative to the demands of the application. As an example, "good resolution" for oil spill management purposes is resolution sufficiently high to show the structure of the slick, i.e. windrows within the overall area affected by oil.

It is apparent from Table 1 that to meet the requirements

of any one of the applications, an integrated system of overall sensors is called for. A laser fluorosensor, for example,
can confirm the presence of oil and also detect submerged oil
but cannot map the spill or present it as an overall picture.
A passive microwave instrument may be developed to measure the
thickness of oil but could not unambiguously confirm the presence of oil. A side-looking airborne radar provides large area
coverage for close to all-weather conditions. The spatial resolution of real aperture systems is felt to be adequate for
the long range detection of anomalies caused by oil spills. If
higher spatial resolution is necessary, for other tasks which
might be carried out with the same aircraft, i.e. synthetic
aperture systems can also be used.

On the basis of experience gained so far, only a multi-
sensor approach exploiting various portions of the electromagnetic spectrum could theoretically meet all the requirements.
(Table 1).

From this table, the user is able to draw a first list of
sensors to fulfil the needs of his particular application. It
is then necessary for a more detailed understanding of the conditions under which the sensor performance can be improved
beyond that claimed in the Table.

STATE-OF-THE-ART OF SENSORS

Core Sensor Package

Not all the sensors which might be employed operationally
to detect, identify, classify, map and track oil spills are unfortunately at the same stage of development.

Broadly speaking, two categories may be distinguished, the
difference lying in the degree of perfection.

The first category included sensors that have the following
features in common:

* their development is sufficiently advanced for them to be
  more or less immediately available on the market in a
  form that can provide the services required;

* they use a comparatively advanced data processing and interpretation method, so that they can provide useful data
  without a great deal of further research;

* they have already been extensively used to tackle the
  missions involved.

Besides the human observer, the sensors are:

* Photographic cameras,
* Side looking airborne radar (or other radar types)
* Ultraviolet line scanner
* Infrared line scanner
* Visible line scanner
* Passive television

Because of their performances, these sensors have been considered as the core of any integrated system to meet the basic requirements of most countries not only as regards oil pollution surveillance and monitoring but also as regards sea search and rescue, maritime traffic control and fishery surveillance.

The systems currently used by the United States Coast Guard (AIREYE), the Swedish Coast Guard and the Danish authorities reflect this approach.

It is believed that this sensor package could be used operationally without a great deal of supporting research work; nevertheless, a number of topics for further investigations were suggested.

The contrast between oil slicks and the surrounding water might be enhanced if the observers were to wear eye glasses fitted with suitable filters. Similarly the optimum combination of film type and filters may not have been found for use in photographic cameras under various sea states and illumination conditions.

As a scientific question, it would be interesting to understand more completely the mechanism for the detection of oil slicks on the water surface. Although all sensor systems should benefit, the need was particularly noted for microwave sensors. As an example, there are many hypotheses for the apparent temperature reversal often visible in thermal imagery of oil slicks at varying thicknesses. The response of the various sensors to oil/water emulsions should be investigated. In studies of the detection mechanism, the effect of sea state, current and wind velocities, and weather conditions should be among the topics considered.

Radars and side-looking airborne radars in particular are most often configured for the detection of military targets. In frequency, it is difficult to deviate from the bands conventionally used because many of the specialized components are not available, though as many frequencies as possible should be tried. Future work should be undertaken to determine the optimum

polarization, and viewing angle for any given radar system. The
question of the resolution required to detect and identify an
oil slick also merits study.

Developmental Sensors

The second category included sensors which, also catering
for certain aspects of oil pollution surveillance and monitoring, are still at the initial development stages and have therefore not yet proved their effective capability to meet user requirements.

While the optimum configurations of some of these sensors
are known, tests are still necessary in order to have a better
idea of their performances and to determine optimum conditions
for success.

The sensors particularly concerned are the ones employing:

* Fluorescence (for example, the laser fluorosensor) to
  identify and classify the oil;

* Passive microwaves (for example, profilers and imagers)
  to measure the thickness of slicks;

* Active optics (range gated laser illuminated television)
  to identify polluting ships at night.

These sensors have yet to produce a great deal of data
over oil slicks largely because of the limited availability of
the hardware. As additional experience is gained with these
systems and users become more confident in their interpretation
of the data (and its limitations), the sensors could be integrated into an operational system.

These sensors still require considerable research, in particular as regards the spectral signature of oil, owing mainly
to the limited capabilities of current hardware.

The fluorescence detectors, such as the active laser
fluorosensor or the passive Fraunhofer line in-filling devices,
must be developed to provide real-time identification and classification of oil spills. A laser fluorosensor also promises
to be able to detect and identify oil lying under the water
surface. In this field, the active research conducted by the
Joint Research Centre at Ispra (Italy) (Commission of the European Communities) and the parallel field trials carried out by
Canada and Denmark have to be taken into account. Fluorescence
imagers were likely be the next stage of development in this
class of sensors.

Passive microwave sensors were thought to require work in the same areas as those mentioned for the active microwave sensors in the previous section. The spatial resolution of the passive sensors was recognized as being limited by the antenna dimensions which can be conveniently carried by aircraft.

Passive microwave imagers may evolve to the point where they can display oil slick thickness maps in real time.

An active gated television system is being developed by the United States Coast Guard for use in the Aireye sensor package. The main purpose of this sensor will be to identify oil pollution violators at night. Active optical imaging systems will also be useful in mapping and tracking the oil and monitoring clean-up operations. The sensor will be of particular value in other ocean surveillance missions: vessel traffic management, fisheries patrols, search and rescue, etc.

THE COST-PERFORMANCE PERSPECTIVE

A cost effective approach to sensor development and acquisition was recommended by the experts assisting the Washington workshop.

The development of airborne remote sensing systems for maritime oil pollution surveillance must compete for funds with various other national requirements. As such, the discussion of national requirements for oil pollution surveillance systems must be sensitive to the type of cost/performance curve in Figure 1 which is typical for large integrated systems.

For more coastal and offshore applications, the minimum useful system as shown in Figure 1 was·defined as an aircraft with an observer using visual sightings to detect oil spills on the sea surface.

The ideal system was defined as a state-of-the-art remote sensing aircraft instrumented with an integrated, multi-sensor package and equipped with a real-time data link to a data processing, interpretation and distribution centre. The ideal system may also include satellite systems for a synoptic presentation of very large oil spills.

DATA PROCESSING AND UTILIZATION

The view of the various national authorities responsible for the surveillance of the marine environment (and for controlling oil pollution) is that it is essential that data gathered in flight be presented on real-time displays in the surveillance aircraft itself. On these displays the data could

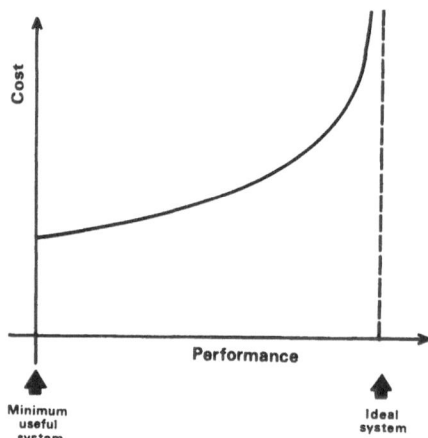

Figure 1. Cost/performance for oil pollution
           surveillance systems

be enhanced interactively even to the point of contouring the
imagery prior to its transmission to operation centres.

   In order to enhance the effectiveness of anti-pollution
facilities, it is necessary that the imagery obtained be clearly
identified in terms of time and space (for example, by trans-
ferring space-time data from a navigational data generator onto
the edge of the images) or that the position of ships, buoys,
and shore lines be clearly marked thereon.

   Data processing algorithms are often fairly complex and it
was suggested that the software language in which they are
written should be standardized to allow joint software development through the exchange of programs.

   However, not all countries agreed on the subject of the
utilization of such data. Some feel that they should be handled
on the aircraft by experts, while others consider that this
should be the responsibility of the operational centres, on-
board display serving simply to guide the aircraft during its
mission. In the latter case, real-time transmission to the
ground of sensor-provided imagery is of fundamental interest.

   This being, there are numerous operational advantages in
the near real-time transmission of oil spill imagery to operation centres. In many locations, where it is impossible to set
up a receiving station at the operation centres or where the
distances were too great to obtain good transmission, data
could be relayed reliably from the aircraft to shore by means
of satellite links.

In order to facilitate the exchange of imagery between experimentors or comparisons from experiment to experiment, particularly from different countries, it may be wise to establish guidelines for annotation of the data, for magnetic tape formats and for the ancillary data required. Ancillary data would include such things as sea state, surface wind, oil type, and sky irradiances. Similar guidelines would prove valuable whenever remotely sensed data was exchanged on pollution incidents involving two or more jurisdictions.

A number of panel members commented on the past and continuing experience in the development of sensors for satellites in which the software required was unavailable to process and interprete data at the time the sensor began to produce data. Not only had the development of the software started later than the design and fabrication of the sensor but it was not ready when the satellite was flown. In the newer sensors, such as the SEASAT synthetic aperture radar, the data processing facilities were unable to keep up with even the modest duty cycles available onboard the satellite for the acquisition of the imagery.

Thereby, it was recognized by a number of experts that, for both airborne and satellite sensors, software was an integral part of the sensors. It was suggested that for both airborne and satellite sensors, the design of instrumentation and data processing should be started simultaneously. The cost of implementing the two facets (hardware and software) of the system is expected to be approximately equal.

INTERPRETATION OF INFORMATION

Interpretation keys for each of the sensors will have to be prepared before remote sensing techniques can be truly operational tools.

Reliable and consistent interpretation of imagery, as would be provided by a set of keys, will be particularly important if remotely sensed imagery is to be used as legal evidence for the prosecution of offenders having deliberately caused oil spills not in conformity with international agreements or national regulations in force. In the field of combatting accidental pollution, accurate and timely interpretation is, of course, essential in all aspects of oil pollution control, from the initial detection and assessment of the incident to the final phases of the clean-up activities. Keys have to be matched to the data as available in the aircraft so that the initial interpretation can be arrived at quickly and easily. Supplementary keys could be used at operational centres if the data has been subjected to more sophisticated processing or if the format has been altered.

Emphasis has to be placed on the integration of the data from all sensors in order to improve the reliability of the interpretation and, in particular, to reduce false alarms. The same integration will likely allow additional information about the spill to be extracted from the data, beyond that available from consideration of the data from individual sensors.

SATELLITES

The general view in 1979 was that in the near future, satellite-borne sensors were not expected to play a major role in oil pollution surveillance.

The reasons were: atmospheric conditions limit the usefulness of optical (visible and infrared) line scanners, particularly in northern and equatorial latitudes; the spatial resolution of passive microwave imagers is too course to be of value from satellites altitudes; the operational centres could not have real-time information since neither the data processing nor the distribution of interpretation centres has been envisaged for this purpose.

In fact, all the sensors considered to be of value in the detection and mapping of oil at sea, only the synthetic aperture radar (SAR) holds much promise. It is clear that a great deal of effort must be placed on the interpretation of SAR imagery of the sea surface to eliminate the ambiguities now seen. To be useful, the SAR imagery would have to be processed in near real time. Because of cost, the National Ocean Surveillance Satellite (NOSS), to be launched by NOAA, will not have a SAR on board.

Several satellites may have to be placed in special orbits to obtain frequent overpasses over limited coastal areas. Daily coverage is likely the minimum acceptable for oil pollution control activities.

Satellites were seen, however, to be very useful in determining the fate of oil spills by monitoring weather conditions, and specifically, surface winds and currents. Weather information is available from existing meteorological satellites, the surface winds and sea state are not.

Satellites will undoubtly provide reliable wide band data links from the aircraft to data processing centres and to operational headquarters. The on-scene co-ordinator can assess the data arriving from the aircraft in real time and deploy the resources at hand effectively. This same facility might allow more complex sensor packages to be placed in the aircraft because the data processing and recording can then all take place on the ground.

This was the opinion of the experts in 1979.

Since 1979, there have been swift and encouraging developments in the following areas:

Concerning sensors, optical performances (IR and visible) permit (or will permit in a near future) attainment of ground resolution of some 10 to 30 meters as well as of high radiometric resolutions as a result of the development of new detector technologies. In addition, microwave sensors (such as SEASAT's synthetic aperture radar) have demonstrated their ability to provide all-weather high resolution (25 m) on land and sea.

Concerning algorithms, the sustained efforts of laboratories, universities and institutes have made it possible to develop, test and validate algorithms permitting quantitative translation of the measurements of the various sensors into geophysical parameters and/or production of the corresponding imagery.

The distribution of data and processed information will be improved: the time between acquisition on the ground of raw data from a satellite and the supply to the appropriate authorities of processed information will soon be reduced to between three and six hours (ESA's ERS-1 satellite).

At least, the launching of pre-operational and operational systems (MOS-1, ERS-1, SPOT, LANDSAT 4) accompanied by co-ordination efforts with respect to performance harmonization orbits, data formatting and the information provided by these different satellite systems.

NEED FOR INTERNATIONAL CO-OPERATION

The need for international co-operation in the area covered by the pilot study stems basically from the following observations:

(1) As mentioned previously, the funding requirements for remote sensing of oil pollution must compete with the funding requirements for other national priorities. In order to justify the expense of the ideal system, it was felt that the extent of the oil pollution problem in each participating country must be quantified.

The various parameters to be taken into account (the time,

day, weather conditions, frequency of occurrence, location
source, type of pollutant, areal extent and quantity of
pollutant spilled into the marine environment) should be
documented sufficiently to establish a statistical base for
each participating country.

The importance of collecting baseline data was emphasized
since it is the key for identifying realistic, affordable
requirements. Baseline data could also be used effectively for
cataloguing pre-spill and post-spill conditions and for law
enforcement and regulation development.

Once the problem was quantified in the above terms, the
expense of developing similar sensor systems at the national
level may prove to be warranted and therefore receive a higher
priority when competing with other national needs. On the other
hand, the data may prove that a system with fewer capabilities
is adequate, within the spirit and intent of the law, and in
the best interest of the general public.

To do this, member nations should pool their knowledge and
combine their resources to develop a close approximation of the
ideal system. This system would be used internationally to
collect the necessary baseline data which would quantify the
seriousness of the oil pollution problem.

It was out the terms of reference of the experts group to
fully explore the details of creating an international program
to develop and operate a dedicated oil pollution surveillance
system. However, it was generally thought that the system could
fly scheduled surveillance patrols in each participating
country until sufficient data was collected to establish the
baseline requirements. It could also respond to major oil spill
disasters such as the AMOCO CADIZ or the IXTOC I oil spills. An
international committee could plan the development and allocate
the development cost equitably to each participating nation.
Operational use of the system could also be controlled by the
same international committee.

(2) When the risk probabilities are known and, hence, the
amount of "reasonable" investment established, the experience
or data already acquired by third countries in the field of
sensors (and, in particular, with regard to the capabilities
and employment limitations of the latter) are essential in
order to make a strict choice taking account of requirements,
the expected outlay and the specificity of the asker country.

(3) With regard to equipment and, in particular, equipment
that is new or under development, trials under so-called
"operational" conditions of necessity require effective

discharge of more or less considerable quantities of oil in
the marine environment. Given to somewhat restrictive national
regulations concerning spills, this type of trial may be impossible in some countries and thus require the collaboration of
others that are less stringent in this field.

Thus a means of informing a selected group of any spills
which might be used for remote sensing experiments would be
beneficial. A simple newsletter was suggested as a medium for
this exchange of plans. This recommendation could be implemented easily within the structure of the Maritime Environment
Protection Committee (International Maritime Organisation)
(See Recommendadtion (12)). It was recognized that any co-operative effort beyond this becomes exceedingly difficult to implement due to the very nature of formal multilateral international agreements.

(4) A comparison, on the basis of a single experiment, of
the various sensors available to countries provides a unique
source of information for each country participating in the
trial as well as for all the other countries.

It was felt that joint exercises would enable a wide
variety of sensors to be brought to bear on the same oil slick
and the different analytical techniques available among the
participants could be applied to the data. It was again pointed
out that large scale experiments were expensive, particularly
if aircraft and equipment must be deployed from one continent
to another. A source of sufficient funds was not immediately
apparent.

The observation was made that airborne experiments with
adequate ground truth are difficult to manage and a lot can
go wrong particularly if weather and sea state are important
considerations. Thus one experienced hand requested (or recommended) that the scope and complexity of each particular experiment be limited.

There is an interaction between the atmospheric conditions,
the sea state and the nature of the hydrocarbon slick. One of
the main goals of future research using experimental spills
could be to establish the limiting conditions under which the
various sensors can detect, map and identify oil. Because a
large number of test spills would be required to arrive at a
complete understanding of the sensor performance, some theoretical models may have to be created to extend predictions from
the few existing experimental points.

Routine patrols can be used to gather simple statistical
information on the nature of oil spill target. Little is
presently published concerning the number of slicks, their

location, area, thickness and the relationship of the slicks
to vessels in the vicinity. This would enable the effectiveness
of the overall system to be measured, in terms of: the number
of slicks observed, the number of ships observed and the number
of ships that could be associated with a particular slick
(ideally one). If the object of oil pollution surveillance is,
among other things, to reduce number of deliberate discharges
by acting as a deterrent such a study may be the only way to
judge its success.

(5) Finally, in the context of measures against the
deliberate discharge of oil from ships, the development by the
International Maritime Organization (IMO) of guidelines
concerning the content and layout of the infringement reports
to be sent to the authorities of the country of registration
leads, with regard to data provided by remote sensing to the
search for some degree of standardization in the submission
of the data and to the provision of reference material (imagery
and recordings) to the tribunals enabling them to reach a
decision with full knowledge of the facts.

This approach is reflected in the ISOWAKE project (International Standardized Oil Wake Experiment), conducted, on the
initiative of the United States and the United Kingdom, with a
view to studying the appearance of the wakes of ships, as seen
by different sensors, when such wakes are polluted by
quantities of oil whether or not conforming to the provisions
of the MARPOL 73/78 International Convention.

RECOMMENDATIONS

With respect to the use of remote sensing to combat
pollution caused by accidental or deliberate oil spills in the
marine environment, IT IS RECOMMENDED:

(1) that countries should concentrate efforts on
integrated multisensor systems, which alone are capable of
meeting the basic requirements of most countries in the area
concerned and of performing a number of other marine
surveillance functions such as search and rescue, maritime
traffic control, fishery surveillance, etc.

This can be met by using an integrated sensor package
consisting of the following :

        * Side-looking radar
        * Infrared/ultraviolet line scanner
        * Passive television system
        * Photographic cameras

and, naturally, human observer.

A second set of sensors including the passive microwave imager, the laser fluorosensor and the active (range gated laser illuminated) television is being developed for the measurement of slick thickness, identification of oil and the identification of ships respectively. The optimum configuration of few sensors have been established, though the sensors themselves have proved to be effective as implemented. Continued testing of the sensors over oil spills is required to improve sensor performance and to determine the limiting conditions for successful operation of the sensors, the latter of which should be thoroughly defined.

(2) That the surveillance package should be equipped with real-time displays capable of integrating the data from the various sensor systems; such integration possibly resulting in better system performance than those of each sensor employed and considered separately.

With the exception of the photographic cameras, all the sensors referred to above can be fitted with real-time displays individually; however, a significant effort will be required to carry out an effective integration to meet the operational requirements. A data link carrying real-time annotated imagery from the aircraft to the on-scene co-cordinator will greatly enhance the utilization of the remotely sensed imagery.

(3) That for all sensors, whether on aircraft or satellite, the data processing software be considered as an integral part of the sensors and that, consequently, sensor and software development be started simultaneously and be allocated the same credits.

(4) That for all integrated systems and sensors considered individually the operational phase be preced by the development of interpretation keys taking into account, in particular, the provisions laid down by national regulations and international agreements. These keys will allow a consistent and acceptable interpretation of remotely sensed oil spill data. The keys would probably deal with the data from each sensor individually and also in an integrated fashion. The keys would be useful for situation assessment by the on-scene co-ordinator, and the keys must reflect both the national and international conventions defining oil pollution for use in courts of law.

(5) That countries planning to acquire such a marine surveillance system first undertake a statistically based evaluation of the pollution risks they run, such evaluation being aimed at identifying their real requirements and, hence, at determining and justifying their choices, in particular as regards the sensors.

(6) That, with a view to facilitate the exchange of data, particularly in connection with the legal aspects of oil pollution, <u>standardization efforts</u> be made in the following areas:

* The computer language used to develop data processing software in order to facilitate if appropriate the joint development of software;

* The acquisition and presentation of data derived from national trials in order to facilitate, in particular, the comparison of experiments of the same type conducted in different countries;

* The content and layout of reports on the observation of deliberate or accidental oil spills to ensure in particular that national experts on receiving such reports from third countries can come to a conclusion without having to convert the data into a more acceptable or familiar form for tribunals.

(7) That more international efforts be made to pool national resources, capabilities and experience so that the development of new sensors and integrated sensor systems can be effected at minimum cost and take account of the diversity of requirements expressed by the community.

There is a wide spectrum of possibilities which may be acceptable from conducting joint experiments over test spills to the joint development of a complex integrated airborne system to carry out pilot projects for the participating nations.

(8) That countries explore the possibility of holding on the basis of their own individual sensors or integrated systems a <u>series of joint remote sensing trials</u> designed in particular to test, with controlled oil spills, the actual performances and operational limitations of their analytical techniques and facilities, and that to this end a <u>group of national experts</u> be set up so that countries can be kept regularly informed of oil spills that can lead themselves to remote sensing operations.

(9) That, as regards satellites (and inasmuch as these are provided with sensors that meet requirements) special attention be given to the possibilities:

* of carrying sensors optimized for oil pollution surveillance;
* of cover permitting frequent transits by day over limited parts of the shore and coastal areas;
* of real-time or near real-time transmission to marine environment surveillance authorities of information that is sufficiently processed for it to be used directly.

(10) That a special attention be given to the development and testing of promising sensors, keeping in mind that remote sensing techniques, similar to those found to be used for the detection and tracking of oil spills, may also be of use in spills of other hazardous substances.

(11) At a general point of view, it was generally agreed that a good remote sensing system which detects oil pollution violators should be well publicized since the meed fact that the system exists will deter violations. The more capabilities that the system has and the public is made aware of, the less chance a violation will occur.

Pooling resources; developing a fully equipped, integrated sensor system; obtaining baseline data; and publicizing sensor capabilities and penalty assessments are seen, at this time, to be the most cost-effective methods to prevent the intentional oil spill.

For the repeated, accidental oil spill, real-time, all weather, day/night synoptic data is needed to provide the on-scene clean-up personnel with spill location, movements, areal extent and thickness information.

Oil spills will continue to happen unless positive steps are taken to curtail them. Some positive steps are being taken in regards to the accidental spill, however, the detection and identification of intentional oil spill violators and their subsequent prosecution remains a problem. It was agreed that remote sensing can be a major element in reduction of such spills.

(12) That, in the light of the previous recommendation (8), and taking into account that the purpose of the International Standardized Oil Wake Experiments (ISOWAKE) is to provide the tribunals with reference material (recordings and imagery) so that the legal national authorities could reach a decision with full knowledge of the facts, the Marine Environment Protection Committee (MEPC) of IMO (International Maritime Organization) should be asked to act, in its enforcement forums, as the depository of remote sensing for oil spill data.

NATIONAL REQUIREMENTS

THE REMOTE SENSING REQUIREMENTS

FOR OIL SPILLS IN CANADA

J.D. Kingham
Director

Environment Canada
Ottawa, Canada

This paper describes some of the Canada's remote sensing
needs in the oil spill area and addresses some of our approaches
for solving these requirements. Remote sensing for our purposes
is considered to encompass a number of techniques and does not
only consist of airborne surveillance.

Surveillance, the process of looking for unreported
sources of pollution, is being performed at the present time
over certain areas of Canada. "Tracker" aircraft, flown by the
Department of National Defence on behalf of the Department of
Fisheries, patrol both the Atlantic and Pacific coasts on a
regular basis. These missions are flown for a variety of purposes, primarily fisheries management. Although these aircraft
do not contain sophisticated equipment for detecting oil spills,
they do contain photographic system useful for the documentation
of incidents. This effort has resulted in a number of prosecutions for oil spillage as well as the reporting of a number of
slicks. At the present time, such a surveillance system is not
available for the Great Lakes, Saint-Lawrence river and arctic
regions of Canada.

In Canada, we have concentrated our efforts on developing
the "monitoring" facets of remote sensing, that is the ability
to locate, track, and define oil spills whose presence is
already known. These efforts have been focused on two problem
areas: oil on water and oil in ice. A number of approaches are
presently available to detect and track oil in these situations.
The Canada Centre for Remote Sensing (CCRS) has undertaken the
development of an airborne sensor package for use both in ice-
infested and open waters. Details of this system appear in

other contributions presented in this volume.

Environment Canada, in co-operation with other agencies,
has also developed a spill-tracking buoy. This device has been
tested on several occasions and is held in stock by a number
of agencies. The buoy emits a radio-signal, which is then
received and tracked by personnel operating receivers from
either land, vessel or aircraft.

There is also a need to monitor the movement of
oil-contaminated ice, especially in the arctic regions of
Canada. To this end, Environment Canada has conducted studies
and in addition to the development of the airborne remote
sensing system noted above, are working on two systems.

One system, dubbed the "Macro system", consists of RAMS
buoys which are satellite interrogated for positional information. This system provides location of contaminated ice
with an accuracy of plus or minus 2 kilometers. The other
system, the "Micro system" is just being developed and will
consist of a small air-deployable buoy which will contain a
transmitter similar to the open-water spill tracking buoy.
This system will allow positional determination on the order
of a few meters.

The use of satellite data to monitor oil spills is
currently being investigated. The large oil spill from the
KURDISTAN in March 1979 enabled researchers to compare
satellite imagery with known spill locations. Results are
still forthcoming on this project. Satellite sensing is an
extremely interesting technique for Canadians, especially
for the vast and remote regions such as the Canadian Arctic.

In summary, Canada with her vast regions of sea and with
special problems such as ice, is dependent of a number of
remote sensing techniques to detect and monitor oil. A number
of problems are already well on the way to a satisfactory
solution; however, the implementation of operational system
is not complete at this time.

NATIONAL REQUIREMENTS

FOR AIRBORNE MARITIME SURVEILLANCE

J.M. Massin

Ministère de l'Environnement et du Cadre de Vie
Direction de la Prévention des Pollutions
Neuilly sur Seine, France

INTRODUCTION

The establishment of a program for aerial ocean
surveillance facilities and, even more so, its implementation
require certain prior decisions before a clear definition of
what is required can be obtained.

Two different types of action are concerned, relating
either to the general concept of surveillance or the general
concept of intervention, both of which are part of the responsibilities of all coastal states by virtue of their sovereign
rights recognized by the International Law of the Sea Conference
in the new economic zones and in furtherance of the international agreements regarding protection of the marine environment, shipping regulations and the safety of persons and
property.

Surveillance

This is concerned primarly with obtaining a picture as
complete as possible of a changing situation or a series of
changing situations in a given area. The actions to which it
gives rise are expressed in terms of surveillance, search and
reconnaissance.

Intervention

This may take different forms

- police duties: monitoring, prevention, reporting of

27

offenders and even rerouting or interception;
- assistance: assistance to shipping, navigation aids;
- relief: which covers both assistance and rescue, anti-
  pollution measures, pollution control, etc.

Some of these actions would seem to require priority.
This is the case for:

* The protection of natural resources with special reference to defence of the fish population.

Although it was decided by the members of the European
Economic Community that after 1st of January, 1977, there
should be joint conservation, management and exploitation of
fishery resources (except for French overseas territories and
Mayotte, which remain under French jurisdiction), the surveillance and control of fishing are still the responsibility of
each coastal state and will probably remain so for many years
to come.

Protection takes two forms:

- Monitoring: evaluation and conservation of fish communities;
- Control of fishing, enforcement of fishery regulations:
  authorized seasons and zones, methods and techniques
  employed.

* The control and surveillance of shipping enforcement of
existing national and international legislation on shipping
routes (especially that sterring from the provisions for the
separation of traffic), manoeuvring, speed, navigating lights
and signals, etc.

This type of responsibility may have two different but
complementary aspects, namely either round-the-clock surveillance of obligatory shipping routes or the identification of
a vessel which has committed an infraction or is off route and
has been detected by aircraft or shore facilities (e.g. surveillance radar centre).

* The detection and prosecution of deliberate acts of
pollution which contravene existing national and international
regulations. This concerns mainly the detection of illegal
discharges of oil at sea and the collection of evidence for
prosecutions.

* Search, rescue and assistance to shipping.

There are other actions which may be carried out in

addition to or in conjunction with these priority tasks, such as scientific surveys concerned either with basic research or prospecting for natural resources, the observation of special situations arising from the discharge of domestic or industrial wastes, with or without wastepipes, or the discharge of cooling water from high-capacity power stations situated along the coast, etc.

In addition to prosecutions for operational discharges from tankers should be added all actions designed to counter the effects of wide-scale pollution resulting from an accident at sea, for example.

Consequently, the consideration of the facilities necessary to carry out these tasks requires that a certain number of different factors should be taken into account:

* Environmental factors: geographical characteristics of the zone surveyed (area, cleanliness), area meteorological and oceanographic conditions, existing infrastructures (airfields, logistic facilities), resources, i.e. the economic interest of the zone, and possibly local political conditions (inter-state relations, possibility of joint actions with other similarly motivated countries, etc.).

* Surveillance factors: the approach may vary depending on the missions considered to have priority: scientific or industrial surveys, depollution, routine missions, to detect illegal discharges at sea or tracking of thermal discharges from nuclear power stations.

* Factors related to the use of facilities:

- operating modes: should surveillance have a deterrent impact through numerous local and selective operations of limited duration or should it be on a continuous basis?
- context: should the system be independent or part of a more comprehensive whole using both aerial and land means? Should the aim be an all-round capability or should the system be designed for a single role?
- economic aspects (cost/effectiveness ratio): the way to maximum cost/effectiveness for investments is through capitalizing on experience gained in France and abroad, integration into existing logistics and circuits and stiffer penalties for offenders.

EXISTING MEANS FOR OCEAN SURVEILLANCE AND POLICING

Surveillance, policing and pollution control at sea are

mainly carried out by the Navy, customs, the merchant marine,
the Gendarmerie, each of which having the aerial surveillance
facilities according to its missions (defence or police).
These bodies have concluded agreements among themselves, such
as the customs/merchant marine agreement on the co-ordination
of surveillance and co-operation which governs relations between these two bodies since 1st of January, 1978.

The facilities available to the Ministry of the Interior
(Civil Defence department) which are used for certain surveillance operations, are not taken into account because of their
heighly technical nature and because they are entirely limited
to the sea front.

Nevertheless, it is the Navy which bears the brunt of
maritime surveillance assignments, particularly where antipol-
lution and depollution measures, control of shipping (particularly since March 1978) and the surveillance of ocean fishing
are concerned.

It should be noted at the same time that the facilities
available to the Navy for civilian surveillance of the economic
zones are not always such as to enable it to carry out satisfactorily the assignments for which it is alerted. Either these
assignments are carried out at the expense of training or, on
the other hand, availability of the facilities cannot be guar-
anteed at all times and all places because of national defence
requirements. Furthermore, the sophistication of the observation equipment on aircraft (high-resolution cameras and
tracking radar, for instance) are ill-suited to civilian surveillance and greatly increase the cost of intervention.

The Navy moreover exercises its duties throughout the
length and breadth of the maritime approaches and the open sea,
while the facilities available to the other bodies referred to
above are suitable only for more purely in-shore operations,
either for technical reasons or because such is their mission.

The role of the Police and Gendarmerie is confined to in-
shore areas within the 12-nautical mile (22 km) limit and
that of the customs, which covers territorial waters and beyond
for the control of continental shelf installations, generally
extends up to 30 or 50 nautical miles from the coast.

The merchant navy has only a single aircraft, which although it is equipped with an infrared radiometry surveillance
system (SAT Supercyclope IR line scanner) and is used in connection with the "Centres Régionaux Opérationnels de Sauvetage
et de Surveillance" (C.R.O.S.S.), can only be flown within the
50-mile zone because of its operational characteristics.

CURRENT TRENDS IN SURVEILLANCE OF THE MARINE ENVIRONMENT

Generally speaking, it is obvious that the present organization results in several different systems for intervening
at sea (aircraft, helicopters, electronic equipment associated
with sensors, etc.) which use similar but different equipment
and procedures. Also, these systems each have independent
logistics back-ups which are organized and supplied differently
and have different specifications. And each of the administrations concerned must provide the reserves of personnel and
equipment of essential if the facilities are to be available
at all times.

Over the last few years, a number of factors have tended
to increase the difficulties of maritime surveillance. From a
juridical standpoint, changes in the Law of the Sea have made
for an increase in the sea-front responsibilities of coastal
states. Where the use of the sea is concerned, the increase in
the number and scale of activities at sea or connected with the
sea has led to ever stricter legislation which requires greater
supervision to inforce it.

Examples of this are:

* community fishing regulations which become stricter as
  each year goes by;
* the need for increasingly close control of shipping
  under the provisions for the separation of traffic which
  came into force on 20th July, 1977 (application of the
  international collision regulations of the "COLREG"
  convention).

Likewise, the Convention on sea-search and rescue will
lead to increased search and co-ordination activities.

Lastly, the increase in certain coastal activities (offshore prospecting and drilling, transport of petroleum or
chemical products which are highly toxic for marine biological
life) has led to far greater risks of accidents and pollution
which have to be faced as effectively and rapidly as possible.

THE MEANS REQUIRED

As a rule, and excluding certain very specific missions
which must still be carried out separately, most of the
responsibilities of the various administrations could be
combined into all-round surveillance operations.

Under these circumstances, in order to meet the increased
need for surveillance, a partial solution can be provided by

stepping up the use of existing facilities, bearing in mind,
however, the reservations expressed earlier regarding the drawbacks inherent in the deployment of different means.

This, however, is not an overall solution. The extension
to the high sea of maritime responsibilities of countries
entails the need for surveillance and monitoring missions for
which no specialized facilities exist at present.

It has therefore become necessary to define the aerial
facilities and equipment which best respond to the new requirements, bearing in mind that:

* the most effective means for rapid and comprehensive
  surveillance of extensive maritime areas is the fast-
  medium range aircraft (such as the MYSTERE 20/FALCON 20);
* certain inshore missions of short duration can be
  carried out by more modest aerial means (such as helicopters and light aircraft) which also provide for a
  minimum level of surveillance if the aircraft referred
  to before are not available.

With this in mind, an outline programme was recently
drawn up providing for the establishment of an integrated maritime and coastal surveillance system to come under a single
authority, with each administration concerned continuing to
determine the missions for which it has specific responsibility
and the equipment required.

Aerial Facilities in the Coastal Area

Helicopters: It is clear from experience that intervention
at sea in favour of ships in difficulty whose crews are in
danger or whose cargo constitutes a potential or known pollution risk can be entrusted to heavy helicopters.

Since such actions are exceptional, however, it has not
been considered necessary to mark specific facilities for this
kind of means.

The chosen solution therefore consists organizing an alert
system using existing facilities belonging to the Navy, army
and airforce (according to the maritime regions concerned).

Coastal surveillance aircraft: The positive results obtained using the CESSNA FTB 337 aircraft especially fitted for
the detection of oil spills and at present deployed by the
merchant navy around the French coast argue in favour of continuing to procure light coastal surveillance aircraft.

It is accepted that the following factors must be taken into account:

* The aircraft must be able to take-off and land from any coastal airfield, even rudimentary (grass runway) and if possible in a short reaction time (principle of craws on stand-by).

* Apart from anti-pollution and depollution missions and the surveillance of shipping, it must be also able to carry out search and rescue missions (80% of incidents at sea take place near the coast).

* It must have an all-weather capability.

* Lastly, its equipment must be able to integrate with the system presently used on the merchant navy's aircraft. This system is based on the ground processing, in real time, at Regional Operational Rescue and Surveillance Centres (CROSS) of data collected in flight.

Maritime surveillance aircraft: The maritime surveillance aircraft must have an endurance and range enabling it to carry out missions designed to obtain a picture complete as possible of the ships present and the activities being carried out within a sea area, detect and identify shipping by day and night, detect pollution and when necessary collect evidence of the law violators and, at least, carry out rescue operations in the event of accidents at sea.

The aircraft must be equipped to carry out all these missions and, in particular, must be able to inter-operate with the merchant navy's facilities.

The required performance is a minimum endurance of six hours, at a 200-nautical mile range from its base, or the ability to patrol, at the same 200-nautical mile distance, 900 nautical miles at low altitude. The FALCON 20 Jet and the NORD 262 aircraft both meet these requirements.

ASSESSMENT OF REQUIREMENTS

The assessment of the theoretical aircraft requirements has been made by zones (Channel, Atlantic and Mediterranean Sea) using parameters based on the local and regional characteristics referred to previously, i.e. extent of areas under surveillance, existing infrastructure, volume of shipping, natural resources and economic interests.

As a rule, the minimum hours which are necessary for

Table 1. Minimum hours to insure proper surveillance
          along the French coast

	Maritime surveillance aircraft	Coastal surveillance aircraft
Mediterranean Sea	600 h	600 h
Atlantic ocean	1,200 h	600 h
Channel } Straits of Dover }		600 h

proper surveillance are as shown in Table 1. It should be noted
that this would provide only for a coastal patrol every 72 hours
over the most vulnerable parts of the Mediterranean Sea and the
English Channel, and more particularly on the especially vulnerable Ushant region off the Brittany's coast.

Channel/Straits of Dover

The Channel and the Straits of Dover are one of the most

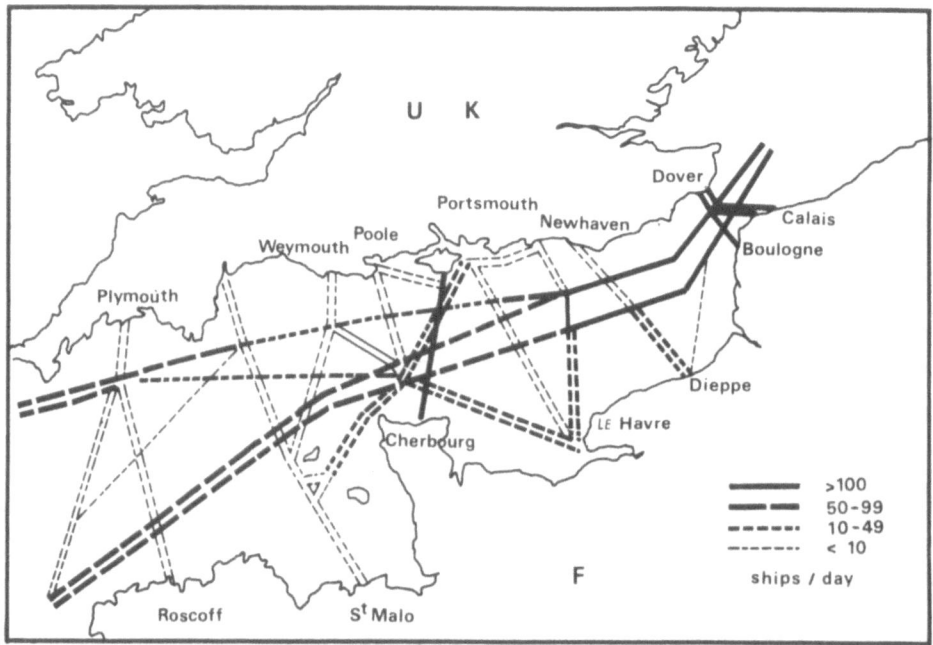

Figure 1. Major shipping routes in the English Channel

vulnerable areas because:

* of the density of shipping. The sea line which runs past Ushant and the Casquets before entering the Straits of Dover has a daily traffic of more than 200 vessels of different sizes, including 10 tankers (Figure 1);
* the often inhospitable meteorological and oceanographical conditions (mist, wind, high seas, currents);
* the navigating conditions which make it necessary for ships to pass at some points very close to the coastline (between 7 and 15 nautical miles).

The awareness of these risks have led to recent measures for the compulsory separation of traffic off Ushant and the Casquets (Figure 2).

A good statement of the difficulties which occur in the English Channel and of the risks assumed by shipping in this area is given by the number of accidents which have been occurring off the Brittany's coast during the last decade and which lead to serious pollution: OLYMPIC BRAVERY, BOEHLEN, AMOCO CADIZ, GINO and TANIO (Figure 3).

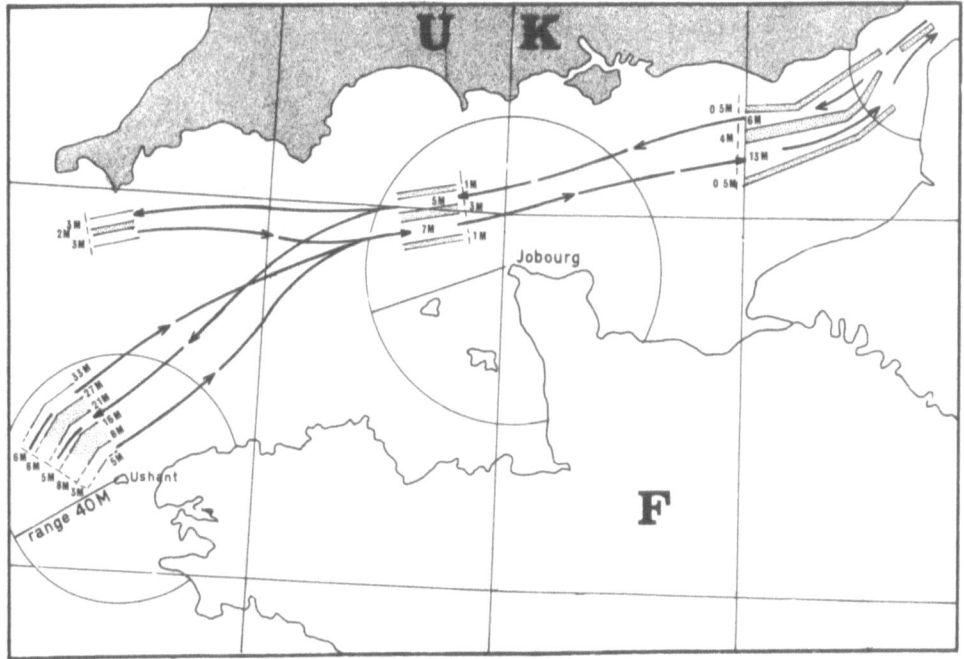

Figure 2.   Implementation of traffic separation schemes off Ushant and Casquets (English Channel)

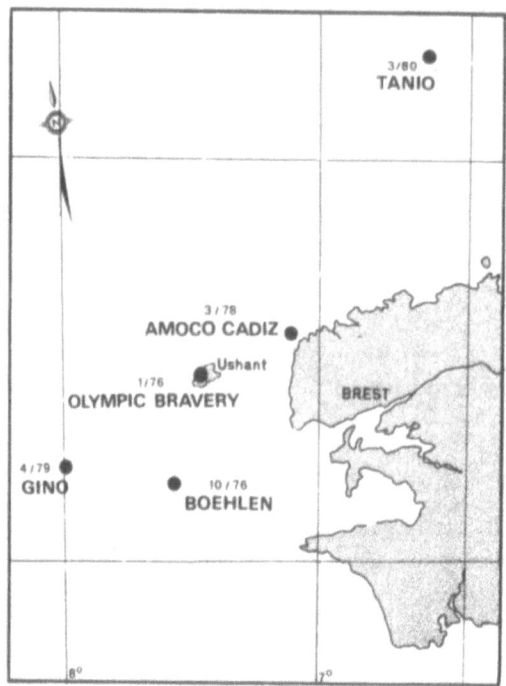

Figure 3. Major accidents off the Brittany's shore

        Lastly, the Channel is an extremely important fishing
ground, famous for shellfish raising (scallops in the Baie de
Seine and Baie de Saint Brieuc and clams in the Baie du Mont
Saint Michel).

        Surveillance of the Channel calls for the effective
presence of surveillance aircraft over the statutory shipping
areas and an ability to intervene rapidly in the event of an
accident or pollution resulting from incident at sea.

        It has been estimated that to cover this twin require-
ments one coastal surveillance aircraft operating for at least
600 hours per year (cumulative flight hours) would be considered as a minimum.

Atlantic Area

        The main tasks are:

    * surveillance of shipping, particularly in the northern
      section of the maritime area;

   * surveillance and control of fishing along the southern
     coast of the mouth of the Gironde, on the Plateau de
     Rochebrune and the La Chapelle bank up to the northern
     limit of the economic zone.

   It has been estimated that these tasks could be carried
out on the basis of 1,800 flight hours with two maritime surveillance aircraft and one single coastal surveillance aircraft.

   Consequently for the whole area including the Atlantic
Ocean and the Channel area, 2,400 cumulative flight hours
would provide essential surveillance.

Mediterranean Sea

   While fishing is at present subject to surveillance with
existing means, pollution control and depollution activities
constitute a very specific problem which could be tackled by
means of maritime and coastal surveillance aircraft.

   A total of 1,200 flight hours for surveillance and police
missions has been considered as a working hypothesis.

COST CONSIDERATIONS

   A preliminary estimate of the likely cost of setting up
such a surveillance system (maritime surveillance plus coastal
surveillance) can be made on the basis of the following costs
for new equipment:

Aircraft

   Maritime surveillance aircraft (i.e. FALCON JET,
COASTGUARDER HS-748 or FOKKER FRIENDSHIP F.27) ....50 mF

   Coastal surveillance aircraft:

   CESSNA FTB 337 ....3.5 mF
   CESSNA 404 or NAVAJO .... 5.0 mF

Equipment

   Infrared line scanner "Supercyclope" and airborne
peripherals (per unit) .... 1.5 mF

   Side-looking airborne radar (Ericsson type) ...... 1.5 mF

   Equipment for receiving IR information in the
control centres (CROSS) (per centre) .... 1.5 mF

   Taking in account the needs referred to above, the total

Table 2. Total estimated cost of a surveillance system
along the French coast

Equipment	Channel	Atlantic	Mediterranean
Maritime surveillance			
Number of aircraft		2	1
Total cumulative flight hours per year		1,200	600
Cost mF — Aircraft		100	50
Cost mF — Maintenance } Operating }		10	5
Coastal surveillance			
Number of aircraft	1	1	1
Total cumulative flight hours per year	600	600	600
Cost mF — Aircraft	5	5	5
Cost mF — Maintenance } Operating }	1.5	1.5	1.5
IR Radiometry equipment	1.5	1.5	1.5
SLAR + Equipment (periph)	1.5	1.5	1.5
CROSS equipment			
Gris-Nez/Jobourg Corsen Ouessant	3 1.5		
Etel/Soulac		3	
Agde/Toulon/Ile Rousse			4.5
Mobil Control Centre		1.5	1.5

estimated cost of the surveillance system network is estimated
as shown in Table 2. Note that for aircraft, unit costs include
30% spares; for the maritime surveillance aircraft, the cost
has been estimated on the basis of the average cost of the
three proposed aircraft.

METEOROLOGICAL AND OCEANOGRAPHIC CONDITIONS ALONG THE FRENCH COAST

To understand the difficulties which arise along

the French coast when we talk about surveillance programme,
it is necessary to consider all the meteorological and oceanographic conditions which exist along the whole French coast.

Channel and Atlantic Coasts

The prevailing conditions on French Channel and Atlantic
coasts are mainly conditioned by polar region turbulence generally moving from the west to the east across the British
Isles and bringing westerly winds and mild, damp weather.

The situation is completely different when the Eurasian
anticyclone envelops Western Europe in winter; then cold winds
may prevail for weeks on end.

The main annual temperatures vary between +10°C and +14°C.
The lowest temperatures are generally between -5°C and -12°C,
but have been known to fall to -18°C. The maximum temperatures
generally vary between +20°C and +36°C.

Surface temperatures: In winter, the surface temperature
of the North Sea is between +5°C and +6.5°C and that of the
English Channel between +7°C in the east and +10°C in the west.
In the Atlantic area, the minimum temperatures rise progressively from the north to the south (from +8.5°C on the southern coast of Brittany to 11.5°C at the Spanish frontier).

In summer, the surface temperatures vary depending on the
area from +15.5°C (Channel) to +20°C (Spanish frontier).

Humidity: In the western Channel, the relative humidity
is generally high (up to 80%) and seasonal variations fairly
slight. In the vicinity of the Straits of Dover (northern
part of the English Channel), humidity is particularly high:
average between 80% and 90%. However, even in winter, the air
can become extremely dry (20% humidity or less) when the wind
is blowing from the east.

In the Atlantic area, the relative humidity decreases
from the north to the south and more marked between the mouth
of the Loire and the Spanish frontier: 86% to 89% off the
Brittany coast all along the year, 75% in July and August in
the south.

Cloud formation: The Channel and southern Brittany coasts
are subject to heavy cloud cover. For more than half the year,
cloud cover is more than 75%. Recording of cover less than 25%
usually occurs in summer.

On the southern part of La Loire, average cloud formation

is less than one quarter and 28% of cover less than three
quarters.

   Rainfall: On the Channel coast and the Straits of Dover,
the mean annual rainfall varies from about 600 mm to 900 mm.
Rainfall in autumn and winter is approximately twice that
during spring. In wintertime, the monthly average for days of
moderate rainfall (1 to 10 mm) is between 10 and 14 and that
for days of heavy rainfall (10 mm and more) from 2 to 4. In
summer these figures are respectively between 6 and 9 and 1
and 2. Long periods of light or moderate rainfall are frequent
mainly in winter.

   From Brest to the River Loire, the mean annual rainfall
is between 600 and 800 mm. The period October/December is the
wettest period (75 to 110 mm by month). The dryest months are
May to September although there are important variations from
one year to another in the monthly averages (of the order of
1 to 3).

   Between the River Loire and the Spanish frontier, the
mean annual rainfall increases progressively from the north to
the south to a maximum of 1,500 mm at the frontier. This high
figure is explained by the intensity of precipitations rather
than its frequency. More than half the precipitations occur
between October and January.

   Visibility, Fog: Generally speaking, fog is fairly frequent
in the Channel area, mainly in the sector situated between the
Belgian frontier and the Seine estuary where radiation fog
often occurs, especially in the industrial zone, between
November and March. On the west part of the Straits of Dover,
the frequency of coastal radiation fog decreases gradually and
off Le Havre it occurs hardly more than two nights a month,
even in winter, and rarely extends out to the sea.

   Evaporation fog is also encountered in the cold season in
the eastern Channel area.

   During the period 1946/1955, the average number of foggy
days (visibility less than 1 km) was as follows: Dunkerque,
33 days (November to March mainly); Cap de La Hague, near
Cherbourg, 23 days (May to July); Ushant, 58 days (May to
September).

   Off the coasts of Gascogne, recordings of fog and mist are
frequent mainly in late spring and summer. On the coasts, the
figures for days of fog vary greatly from one region to another
one: Nantes, 40 days/year, maximum November to January;
Rochefort, 45 days/year, maximum November to January, Bordeaux,

82 days/year, maximum September to January.

Mediterranean Coast

The Mediterranean coast enjoys a desirable climate with hot and fairly dry summers and little cloud cover and rainfall (maximum in October and November).

The Mediterranean waters are remarkably clear along the coast and their colour varies from dark to light blue. Close to the coasts and mainly in the Golfe du Lion, there is a transitional zone resulting from particles in suspension where the water colour goes gradually from yellowish-green to greenish-yellow. Outflows from rivers can change the colour of the water often far from the coast.

The water is highly transparent and a white marker disc is visible between 25 and 35 meters; transparency improves from the west to the east.

NATIONAL REQUIREMENTS

IN THE FEDERAL REPUBLIC OF GERMANY

F. Günneberg

Bundesanstalt für Gewässerkunde
Koblenz, Federal Republic of Germany

The German coast consists of a large number of sand-formed islands, between which the tidal flats, the "Wadden Sea", extends. The coastal zone is very shallow, and about two thirds of the surface falls dry during ebb-tide.

The comparatively narrow navigable waterways from Elbe, Weser and Jade are bundled together south of Helgoland (Fig. 1). 100,000 ships per year pass through the German Bight, or 280 ships per day. About 30 million tons of oil per year are transported to Wilhelmshaven by tankers up to 250,000 tdw, and 22 million tons of oil to Brunsbüttel and Hamburg by tankers up to 100,000 tdw.

The islands and the coastal region are used as a recreational area to a great extend. The tourist trade here has a turnover of about three billion DM per year (Figure 2). The flounder- and the shrimp-fishing industry too is of great importance. A serious oil spill in this area would lead to severe economic losses. But even the small amounts of pollution originating from tanker cleaning reduce the attractiveness of the beach.

The responsability for fighting oil pollution is split between the Federal Government and the regional Länder-governments. These governments have formed an administrative board that organises precautionary measures ("Ölunfallauschuss"). An alarm centre was established in Cuxhaven. Several oil skimmers with a capacity of 400 tons per hour each are being bought. Among other things, half a million DM are provided for the acquisition of an airborne oil detection apparatus.

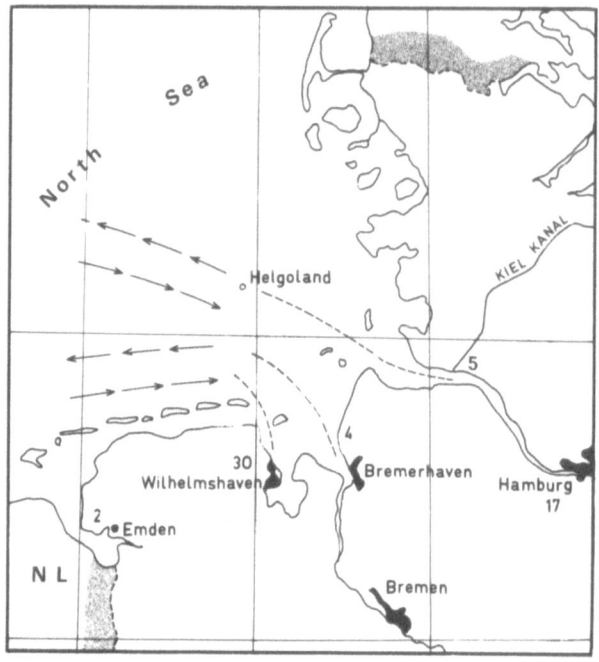

Figure 1.   Traffic density in the German Bight
            Numbers indicate amounts of oil un-
            loaded in harbours in Mt/year.

        It has not yet been decided what type of instrument this
is going to be.

        Naturally, there is a great interest in catching a delin-
quent ship before it can leave the territorial waters. This
is one of the reasons why the German problem require a real-
time information transfer from the remote sensing facility to
the prosecution authorities. Other reasons are the tidal movement of the waterbody, which exceeds 20 km during one tidal
cycle in the outer estuaries, and a large scale counter clock-
wise circulation in the southern North Sea, which transports
the water eastwards and then northwards along the German coast.

        Considering the larger number of ships frequenting the
German Bight, the identification of the type of oil spilled
(e.g. laser-induced fluorescence) would greatly reduce the
number of tankers under suspicion and make prosecution more
promising.

        Bad weather conditions prevail in the North Sea for more

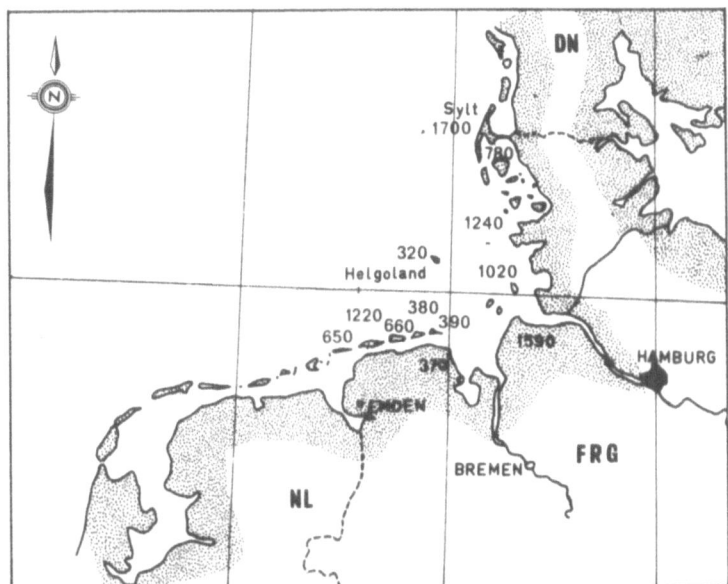

Figure 2.   Tourist trade along German coast.
            The numbers of overnight lodgings
            in thousands per year are reported
            per year for major towns.

than 70% of the year, so systems based on visible light imagery
appear to be less apt for German conditions.

     For the multiple use of a laser fluorosensor system, the
investigations of algae populations, eutrophication and of industrial pollution could be considered.

NATIONAL REQUIREMENTS IN THE NETHERLANDS

R. Spanhoff

Rijkswaterstaat
Directorate for Water Management
and Hydraulic Research
The Hague, The Netherlands

By the Agreement for Co-operation in Dealing with Pollution of the North Sea by Oil, concluded in 1969 between the North Sea countries, the Netherlands are responsible for 42,300 km^2 area of sea, the major portion lying within 50 miles along the 356 km long shoreline (Figure 1). Along the Dutch coast, an intensive shipping traffic exists; the density amounts up to 0.2 ship per nautical mile (Figure 2). The transport of oil by tankers is in order of 200 million tons per year (Figure 3). Thus, the risks for contamination of the sea by oil, either caused by deliberate oil spills or by accidents, are high.

The following arguments lead to the necessity of a system for oil detection and combat.

(1) Protection of the marine environment: notably the Dutch estuaries and the Wadden Sea reveal a copious biological wealth, which is, however, very vulnerable, especially in the seasons of biological reproduction.

(2) Protection of the fishery: oil pollution endangers the food chain in the sea (plankton). Furthermore, the birthplaces for the sea-fish might be affected. Deterioration of taste or even canceroid effects reduce the edibility of the products of fishery. Apart from fishery, the Netherlands exploit extensively oyster cultures and mussel-beds. Estimates of the financial consequences from pollution by oil are hard to give.

Figure 1.   Division into zones for the North Sea
            on behalf of the oil combat.

(3) Protection of the beaches on behalf of the recreation:
the entire Dutch shoreline is used for recreation. Assuming
that it takes 3 days to clean-up a beach contaminated by oil,
the short-term recreation losses are estimated to amount Hfl.
210,000 per 100 m^3 spilt oil. The costs for cleaning the beach
are estimated at Hfl. 300-600 per m^3 oil, while mechanical recovery from the sea only costs Hfl. 25-30.

The latter figures indicate that oil abatement at sea has
to be preferred. A system for oil detection and combat is based
upon this, taking into account the meteorological conditions.

Although 72% of the pollution by oil from shipping traffic
is caused by deliberate oil spills, and only 28% by accidents,
the direct consequences from the latter are the greatest by
fare. Taking as a rule of thumb for the chances of a tanker
accident within 50 miles from the shore:

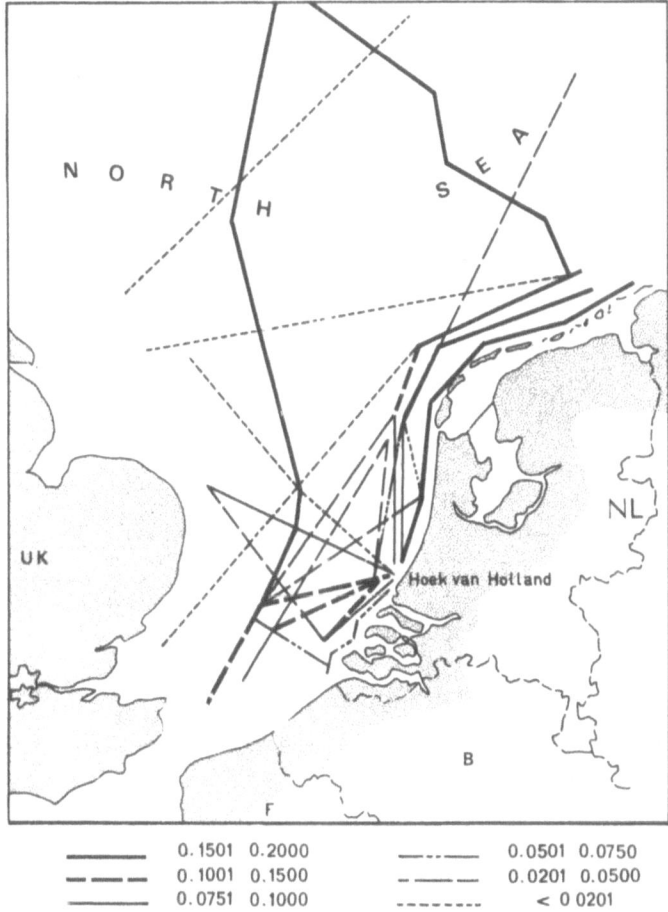

	0.1501 0.2000		0.0501 0.0750
	0.1001 0.1500		0.0201 0.0500
	0.0751 0.1000		< 0 0201

Figure 2. Ship density per nautical mile along route.

- 0.92 accident per 1,000 harbour visits, or
- 1 accident per 13.3 million tons transported oil,

then, 2-3 accidents per year in the Dutch zone can be expected.

The risk is highest at the crossing of the traffic over
the North Sea with the approach to Rotterdam, 40 km west of
Hoek ven Holland.

The most frequent (50%) wind force is Beaufort 4-5; the
most frequent (50%) wind directions lie within the range south-
west to north-west. In case of an accident at the above given
spot the average time available for dealing with the oil at sea

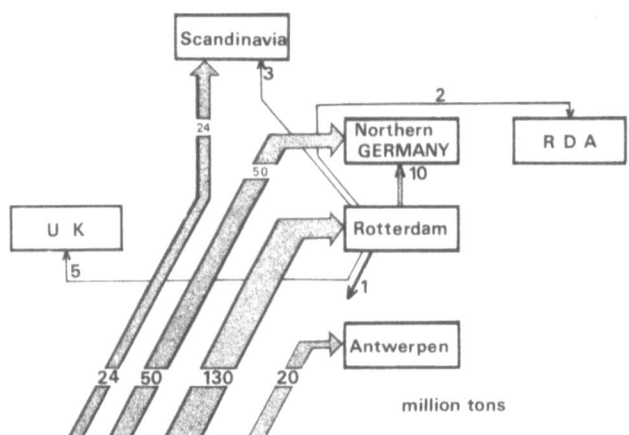

Figure 3.   Annual transport of crude oil
             over the North Sea (1970).

extends over 3 days, assuming a drift velocity for the oil of
2-4% of the wind-speed.

    To fight against the oil at sea, the Directorate "Noordzee"
has at disposal the ship SMAL AGT equipped with two sweeping
arms. The system has proved to be operative during 50% of the
weather conditions, namely up to Beaufort 4 and sea-state 3.
The experiences with this ship are very good. For those cases,
where mechanical recovery is not applicable as a consequence of
the weather conditions, or when the oil film is too thin or
rapidly approaching the shore, other methods are available,
like chemical dispersants, sinking and burning. When the wind
is offshore, another alternative is to let the oil degrade by
natural physical, chemical and micro-biological processes. The
drawback of the latter methods is that the harmful substances,
either oil or dispersants, are a thread to the environment for
a long time.

    Which of the alternatives has to be chosen depends criti-
cally upon the issues at stake and the actual parameters like
the weather conditions. It is evident that an airborne surveilling system is necessary in order to make the proper decisions
and to facilitate an effective oil-combat operation.

    As a consequence of the limited availability of remote
sensing surveillance systems, up to now only visual observations from an airplane are performed on an operational basis
during day-time.

The systems for surveillance and combat of oil spills are, of course, correlated, since they should operate simultaneously to achieve a common goal. Since there is no possibility to deal with the oil at sea states greater than approximately 5, the often quoted all-weather requirement for a monitoring system might be too severe.

Therefore, it is better to develop an around-the-clock (composite) remote sensing monitoring system, e.g. SLAR for the sea states up to 5 in combination with the IR and visible windows for the lowest sea states. The same remote sensing system will be operated to patrol over the Dutch zone of the North Sea. Thus far only visual observations are made. Civilian and military airplanes and ships inform a control centre of oil spills; in addition, the Directorate "Noordzee" carries out surveillance flights with an airplane since 1975, three times a week on the average. Due to the limited number of hours during which the plane can be used, the whole area of the Dutch zone cannot be covered. The results of statistical investigations into the occurrence of oil spills, performed in 1977 and 1978, each lasting two weeks, are used to determine the actual flight plans. In addition, the inspected area only extends to 30 km offshore, since it is estimated that it takes an oil spill 3 days or more to reach the beach from a larger distance. Many oil spills are missed in this surveillance. Thus, in order to be able to monitor the whole of the Dutch zone effectively, a remote sensing system is appropriate.

REQUIREMENTS FOR REMOTE SENSING OF OIL

ON THE SEA

H.D. Parker and D. Cormack

Warren Spring Laboratory
Stevenage, Hertfordshire, United Kingdom

INTRODUCTION

The two applications for remote sensing of oil on the sea
which have been considered in the United Kingdom are:
- to further enhance the current UK oil spill response
  capability;
- to detect illegal discharges of oil to the sea whether
  it is by ships, offshore platforms or industrial sites
  onshore.

At the present time, an operational system is being developed to fill particularly the first of these two roles.

About 500 million tons of crude oil pass through the
Dover Straits annually, which amounts to 5 or 6 VLCCs every
day. This is traffic both to and from Britain and Northern
Europe. In addition to this, there is a substantial traffic
along the coasts in refined products as well as exported North
Sea crude which now passes down both the east- and western
coasts of the British Isles. The coastal waters of the British
Isles and Northern Europe are therefore exposed to a high risk
of an oil pollution incident occurring.

This, taken together with the relatively short distances
between coastlines, calls for a rapid and sustained response
if any measure of clearance is to be achieved at sea. A sustained response implies 24-hour working or at least minimising
the down time due to poor visibility and darkness.

Thus, the United Kingdom requirement from an administrative

point of view would be for a system which would enable oil
clearance operations at sea to proceed through all weathers
and for 24 hours a day.

OIL SPILL MANAGEMENT: WIDE AREA SURVEILLANCE

In order to approach this requirement, clean-up vessels
must "see" oil at all times and must be positioned in the most
dense part of the slick. This calls for two levels or remote
sensing: wide area surveillance, to establish the appropriate
response and to locate the dense areas in a slick and local
area surveillance, to permit clean-up vessels to "see" the oil
to be cleared.

The information required from wide area surveillance is
as follows :

* Location of spill;
* Quantity of oil spilled;
* Distribution of this oil in the affected area, i.e.
  location of the areas of heaviest contamination (slick
  structure).

In general, the detection of an oil spill would not be a
function of a remote sensing system since this is usually reported by the casualty itself and the approximate location of
the oil would be known.

The use of satellites for oil spill surveillance has
largely been discounted for the following reasons:

* Excessive cloud cover for coastal waters: 4/8 cloud
  cover for about 60% of a year and considerably worse
  than this for the Northern North Sea;
* Inadequate resolution for oil spill management purposes;
* Interval between consecutive passes over the same
  position too long (18 days for LANDSAT and 4 days for
  NOAA satellites);
* Data retrieval time too long (real time required).

At present, this wide area surveillance is carried out by
light aircraft with experienced observers on board. The oil is
located, the areal extent of the spill mapped and an estimate
is made of the oil quantity, all visually. This information is
relayed to the incident control centre by radio link. The area
affected by the oil can be quite accurately mapped but the estimate of oil quantity is very subjective. After a short time
at sea, the oil forms "windrows": strips of oil interspersed
by clean water aligned along the direction of the wind.
Although it is very easy to identify these windrows from the

air, it is extremely difficult to accurately judge their
dimensions without some reference, e.g. a ship working in the
oil, and so estimates of oil quantity are not reliable.
Throughout the incident the control centre relies on verbal
communications of visual assessments of both the quantity of
oil to be dealt with and the degree of success achieved by
clearance operations.

The experience of both involvement at oil spills and sea
trials carried out by Warren Spring Laboratory (WSL) has shown
that even during daylight hours vessels at sea have a great
difficulty in distinguishing between sheen and heavy concentrations of oil. Under favourable conditions, the presence of
oil can be seen from a ship up to 2 miles away but it is not
until the vessel is within 1/4 - 1/2 mile of the oil that the
detailed structure of the slick can be appreciated. For this
reason, it is essential that oil spill clean-up operations are
controlled from an aerial vantage point.

The system under evaluation for aerial sensing utilises
two sensors, infrared line scanning (IRLS) and side-looking
airborne radar (SLAR) with the aim that the system will provide
an objective assessment of the spill. It is anticipated that by
combining these two sensors almost the same information on
slick thickness can be retrieved as observed visually from the
air. The SLAR image relates to thin films in much the same way
as the oil is observed visually from the bridge of a ship and
this should make it relatively easy for an observer on a ship
to relate to this type of imagery. The IRLS provides information on the detailed structure of the slick since it responds
only to the thicker regions. The resolution of the IRLS is
sufficiently high to pick up windrows in a slick since the
dimensions of these across the wind can be quite small. A
minimum ground resolution of 5 meters is required and the resolution of the IRLS at 2,000 feet is in fact about 2 meters.
A lower resolution can be tolerated for the SLAR.

The IRLS will not penetrate fog or cloud and although it
may be possible to fly the IRLS below the cloud base on most
occasions, the sensor will not penetrate in foggy conditions.
For 2-3 months of the year, during autumn, fog commonly occurs
for 2-3 days at a time, and indeed this may well be a contri-
butory cause to an accident resulting in an oil spill, e.g.
OLYMPIC ALLIANCE/HMS ACHILLES (November 1975). It is anticipated, however, that SLAR will penetrate sea mists, fog, and
most cloud structures to give sufficient information for the
movement of the oil to be monitored, but probably not
sufficient information to maintain clearance operations on
SLAR alone.

As mentioned earlier, the search function of the system is
not likely to be great and the system is only likely to be used
in this role at the initial stages of an incident. In the event
of an accidental oil spill, the approximate position of the
casualty at the time of the incident is known and movement of
the bulk of the oil can be predicted from a simple model developed by WSL:

Oil drift = Sea current + 0.03 wind (all vector quantities)

Use of this model can minimise the search area. In the
case of slicks generated by illegal discharges, these do not
generally result in large areas of the sea being affected by
oil since the quantities involved, in comparison to an accidental spill are, in general, low.

Furthermore, the system capability has to be directed at
the same scale of operations as the UK oil spill response
capability. The view is taken in the UK that it would be un-
reasonable to invest in a response capability to cope with a
major disaster on the scale of the TORREY CANYON or AMOCO CADIZ
incidents. For the purpose of assessing the appropriate
response, the UK has been divided into 9 regions and the risk
analysed for each region. The response capability in each of
these sectors varies from 1,000 to 3,000 tons of oil a day
according to the amount of traffic passing through that sector.
The risk was assessed from world-wide experience and has been
borne out by the scale of major incidents occurring over the
last few years, with the exception of course of the AMOCO
CADIZ; OLYMPIC ALLIANCE (1975), 2,000 tons; ELENI V (1978),
5,000 to 6,000 tons; CHRISTOS BITAS (1978), 4,000 tons. These
quantities are the same order as the loss of cargo from one
tank of a VLCC (10,000 tons).

The area of sea affected depends predominantly on the wind
direction during the incident and the proximity of the casualty
to a coastline. The heavy concentration of oil is generally
found at the down-wind edge of the slick and for the quantities
of oil under consideration here this area of heavy contamination is unlikely to exceed 100 square miles. From the clean-up
point of view, it is only this area which is interesting since
the remainder of the slick, upwind of this, consists only of
thin films which will dissipate rapidly through natural forces.
A dual range SLAR might be considered as best fitting the above
situation: a long range (25-mile swath width) for rapid assessment of the search area, and a shorter range (5-10-mile swath
width) to complement the IRLS over the working area.

Given that the IRLS and SLAR sensors respond to oil on
the sea (and it has been demonstrated in a later report that

they do) then at least three important advantages over the present method of surveillance can be identified:

(1) The ability to continue surveillance through the hours of darkness which, even neglecting any further improvement brought about by weather penetration, immediately doubles the present capability;

(2) A scaling factor is automatically introduced into the imagery, permitting an accurate measurement of the area of sea most heavily contaminated;

(3) It is possible to transmit this imagery so that the incident control centre is able to control operations remotely from an aerial vantage point.

The final system envisaged would include data links to both the incident control centre ashore and also to a command vessel at the scene of the spill. Ideally, a composite image would be produced by superimposing imagery of one sensor onto the other, generating an image close to that visually observed. However, it was felt that the problems of synchronising the two sensors for the same field of view would be insurmountable, and instead the imagery will be presented on adjacent monitors.

The prime requirement is for real-time presentation to allow the most effective manipulation of resources in a rapidly changing situation, and in addition the shore station will require facilities for video taping and for producing rapid hard-copy of the imagery. This later requirement is to facilitate the mapping of the spill on conventional charts and also to provide copies for the various interested parties involved in an oil spill. Careful synchronisation of the navigation systems aboard the aircraft and ship are necessary for this to be completely successful.

To enable the incident control centre to compile and plot the overall picture it is essential that the imagery includes the following annotation: date, time, aircraft heading and air-craft position. Ideally, it would have the ground scale super-imposed as a grid but in any case some mechanism for retrieving this information must be provided. One further important requirement is that the imagery should be relatively easy to interpret since personnel involved in oil spill clearance, although fully trained, will not be familiar with handling imagery on a day-to-day basis. It would be recommended, there-fore, that a skilled interpreter be present at the control centre to advise the incident control team.

Such a system would provide an incident control centre with a first hand picture of the scale of the problem and also provide feedback on the clearance rate being achieved. Onshore clearance teams would receive advance warning of the location

and reliable estimates of the quantity of oil likely to come
ashore.

OIL SPILL MANAGEMENT: LOCAL AREA SURVEILLANCE

It is not envisaged that clean-up activities would be
possible at night using this system alone. Such a system should
be able to place vessels within 1/4 to 1/2 mile of the working
area, however, and then a shipborne system would take over for
local detection.

A number of sensors have been evaluated for this purpose
with, as yet, little success. The most obvious approach is to
illuminate the area around the ship but, if the light source is
ship-mounted, the transmitted light is reflected away from the
ship so that the oil does not become visible except at very
short distances (25 meters).

A thermal band pyroelectric camera showed no response to
oil on the sea when mounted aboard ship, presumably because
insufficient radiation was received at such an oblique angle
of observation. Similarly, it was found that light levels were
too low to operate low light television (LLTV). However, LLTV
was successfully used to pick up oil fluorescence on irradiation
with UV on a pilot scale in a test tank. Work is now proceeding
in an effort to transmit a wide beam of UV light over the required distance.

In order that the UK requirement for this particular
application of remote sensing be fully realised, i.e. that
mechanical recovery or dispersant spraying operation may
proceed through the night, this final link between the clean-up
vessel and the oil has to be made. This calls for the provision
of an inexpensive short-range sensor for each ship involved in
clean-up.

DETECTION OF ILLEGAL DISCHARGES: CONTINUOUS SURVEILLANCE

This level of surveillance is more properly termed routine
surveillance, the purpose of which is to detect illegal
discharges of oil with a view to prosecuting the discharger.
The use of remote sensing for this application has not, as yet,
been actively pursued in the UK for the following reasons:

(1) Other measures are available for controlling and
monitoring discharges;
(2) The discharger has to be "caught in the act" of
discharging, and for this to be fully effective all the
shipping routes around UK would require routine surveillance,
a very expensive operation;

(3) There is some doubt as to whether or not the evidence of remote sensing imagery alone would secure a conviction.

Under UK law, it is necessary to show essentially that:

* It is oil that has been discharged.
* The identity of the discharging vessel can be indisputably established.
* That relevant legislation has been violated.

The legislation in the UK is the 1971 Prevention of Oil Pollution Act which forbids discharges of any oil within territorial waters (3-mile limit) and implements the provisions of the 1954 International Convention as amended in 1969, i.e. 100 ppm outside 12 miles and 60 litres per mile outside 50 miles.

At the present time, routine surveillance missions are carried out for oil rig and fisheries protection purposes and sightings of oil during these missions are reported. Indeed, there is a standing instruction to all civil aircraft and ships that any sightings of oil on the sea should be reported to the Department of Trade. Recently, the Shetland Islands Council has instigated routine airborne patrols at the approaches to the North Sea oil terminal at Sullom Voe, but this is the only surveillance dedicated to the detection of oil pollution apart from that required in the event of an oil spill.

A number of successful prosecutions have been brought primarily on photographic evidence, through the use of an expert witness. Such experts, experienced in the aerial observation of oil, rely on data published initially in the API Manual on Disposal of Refinery Wastes (1969), which has more recently been corroborated by Hornstein (Table 1).

Using this data RMS ANDES was prosecuted in 1969 on photographic evidence showing a discharge of No. 6 fuel oil, provided by a French Navy aircraft. This evidence taken together with irregularities in tank residues, was sufficient to secure a conviction. Photographs provided by the Canadian Armed Forces were the main evidence used to secure convictions against the HUNTINGDON, a dry-cargo vessel, in 1971, and a small tanker M/V HALCYON DAYS, 12,600 tons, in 1973.

Other prosecutions have followed, some on photographic evidence alone, but more often with some supporting evidence. Normally this takes the form of a chemical comparison of oil on the sea with oil carried aboard the vessel suspected of the discharge.

Table 1.   Thickness-appearance relationship.

APPEARANCE	THICKNESS RANGE (nm)	DESCRIPTION
Colorless Films	Up to 150	Films reflect more light than does water and look brighter. May need adjacent bare water for comparison. Apparent brightness increases with thickness. At about 75 nm and thicker, a pearly or metallic luster is usually apparent.
Onset of color	Approx. 150	First color seen in a warm tone, more bronze than yellow. As film deep violet or purple appears - these colors begin the first set of rainbow bands.
Pure rainbow colors	150 to 900	The set of bands around 300 nm are in the sequence : bronze, purple, blue, green, in order of increasing thickness. These colors are pure and intense.
		The set of bands around 600 nm are slightly less intense than at 300 nm and have a modified color sequence : yellow, magenta (reddish violet), blue, green. They are quite pure.
Dull impure colors	900 to 1500	Main characteristic is reduction in number and purity of colors. Colors at 900 nm are a rich terra-cotta (brick-red) and turquoise (rather bright blue-green). At 1200 nm and 1500 nm these colors are progressively duller or less pure looking. These sets of bands may also contain a trace of white or pale yellow.
Light + dark bands with little color	1500 to 3000	Any color present is merely a tint in the light and dark alternating bands.
		At 1800 nm, the contrast between light and dark bands is strong, but weakens as thickness increases.
		At 3000 nm, it is apparent that interference effects are weak, and they will quickly disappear as thickness increases.

(From B. HORNSTEIN, The visibility of oil-water discharges" in Proceedings of Joint Conference on Prevention and Control of Oil Spills, API 1973, pp. 91-99, Washington DC Reproduced by permission of the U.S. Environmental Protection Agency)

For a successful prosecution on photographic evidence, the photographs must be of high quality, and normally include at least three photographs:

* An overall shot showing the slick amenating from the vessel.
* A shot from which the vessel can be identified, i.e. a close-up of the vessel.
* A shot of the slick in detail.

The third photograph is of particular importance since it is used to determine that oil has been discharged and the thickness.

Obviously colour film is preferable but, in fact, generally black-and-white photographs have been presented. For best results, this third shot is taken from about 500-1,000 feet at a position close to the vertical with the sun behind the observer. These photographs, together with the pilot's and co-pilot's statements describing the appearance of the oil and the dimensions of the resulting slick, have in the past met the three main requirements for a successful prosecution.

Reference to Table 1 shows that the film thickness used to calculate the quantity of oil discharged is extremely low. These are typical thicknesses resulting from the discharge of oily water and appear as a sheen on the water surface. The appearance of the sheen itself does not amount to a violation of international legislation i.e. outside 12 miles, but by reference to the descriptions given in Table 1 an assessment can be made as to whether legislation has been violated or not.

The extension of this technique to remote sensing imagery is unlikely to provide reliable evidence for a prosecution:
* because of anomalies caused by the ship's wake;
* because these film thicknesses are below the threshold thickness for infrared line scanner (as will be shown in an other report).

The ship's wake will cause anomalies for both IRLS and SLAR sensors. The wake remains stable over several miles under some sea conditions and appears quite clearly in SLAR imagery, so that oil would remain undetected until the wake subsided. In the case of IRLS the wake appears as a cold feature, as does oil, so that some ambiguity is likely here too. It may be possible by careful interpretation of imagery to distinguish between clean and oily wakes, however, particularly if observations can be made over a period of time.

IRLS will detect a gross discharge of oil, and provided some method is available to identify the vessel a prosecution should be possible, particularly inside the 3-mile limit.

Where it is merely required to show that oil has been discharged, i.e. inside the 3-mile limit, a UV fluorescence technique might also be used with success.

The adoption of such remote sensing techniques for routine use would enhance the present reliance on visual observations by extending a capability into the hours of darkness, albeit a reduced capability. It is possible that greater attention will be paid to the use of such techniques once the provisions of the MARPOL 73/78 Convention come into force.

CONCLUSIONS

The UK requirements for a remote sensing system which could enhance the management of oil-spill clearance operations and ultimately permit clean-up operations to continue through the hours of darkness appear clearly.

Such a system would be provided by two levels of remote sensing: wide area and local area.

The wide area system has to:

* operate in an environment at 50% cloud cover, 60% of the year and have a capability day and night and in fog;
* have a ground resolution better than 5 meters;
* permit transmission of suitably annotated data to control bases as afloat and ashore, and,
* provide imagery that is easily interpreted.

A system combining infrared line scanner (IRLS) and side-looking airborne radar (SLAR) will fill this requirement.

For local area detection, an inexpensive short range sensor is required, to be mounted aboard ship involved in clean-up.

The system selected for oil spill management is unlikely to be suitable for the detection of illegal discharges except in the case of gross quantities of oil.

REFERENCES

Hornstein, B., 1973, The visibility of oil-water discharges in: "Proc. of Joint Conf. on Prevention and Control of Oil Spills", API, Washington, D.C.

Parker, H. D., Cormack, D., 1979, "Evaluation of Infrared Line scanning (IRLS) and Side-looking Airborne Radar (SLAR) over Controlled Oil Spills in the North Sea", Warren Spring Lab. Report LR 315 (OP).

"Accidental Oil Pollution of the Sea", 1976, Pollution Paper No. 8, Dept. of Environment CUEP, London HMSO.

"Earth Resources Imagery Annotation", 1977, Note No. 9, Dept. of Industry, SARED, London.

"International Convention for the Prevention of Pollution from Ships", 1979, International Maritime Organisation, London.

THE UNITED STATES COAST GUARD'S

REMOTE AERIAL SENSING PROGRAM

R.L. McFadden, G.S. Voyik and W.E. Plage

United States Coast Guard Headquarters
Washington, D.C., United States

ABSTRACT

The passage of the Federal Water Pollution Control Act
necessitated the United States Coast Guard to develop an airborne, real-time, all-weather, day/night remote sensing system
that would detect oil pollution and identify violators. This
paper deals with the development of an aerial system for its
inception to the present time.

THE U.S. COAST GUARD'S REMOTE SENSING PROGRAM

The Coast Guard's surveillance authority with respect to
the aerial remote sensing program is mandated by the Federal
Water Pollution Control Act. This Act prescribed the development of a National Contingency Plan for the removal of oil and
included, among other things, "a system of surveillance and
notice designed to ensure earliest possible notice of discharge
of oil and hazardous substances". The scope of our aerial oil
surveillance program is divided into two areas, harbor and
coastal.

Harbor surveillance for oil pollution is conducted by
aircraft, boats, vehicles and in the near future, in situ
sensors. Our present surface requirements include one patrol
during daylight and night hours of the main harbors, and a
monthly patrol of remote harbors. In some instances vessels
and vehicles are used. Aircraft harbor patrol frequencies are
based on the quantity of petroleum handled in the port area.
Ports handling over 10 million tons of petroleum yearly are
assigned surveillance at frequencies dependent upon the actual

quantity. Basically, ports with large cargo handling will
receive more overflights than the smaller ones. In addition,
those ports with air stations near by receive a minimum of two
overflights per week. Ports just over the 10 Mt get one patrol
per week, while the largest, New York, is required to be pa-
trolled 6 times a week.

Offshore surveillance requirements are based on pollution
potential, derived from analysis of shipping densities and
location of offshore oil facilities, and on the environmental
sensitivity of the area. With the exception of the Great Lakes
primarily emphasis is placed on the coastal area inside the
contiguous zone's outer boundary (12 miles). Occasional random
surveillance patrols are flown over the prohibited zone (out
of 50 miles).

The purpose of coastal surveillance is multi-faced: detection of discharge, identification of source, determination of
quantity, aerial mapping, documentation for response and enforcement, and to provide a deterrent. In regards to these
purposes, near all-weather capability, with real-time output,
hard-copy documentation and annotation, and long range detection capabilities are needed. The present surveillance does
not provide adequate coverage during night-time or during
adverse weather. The necessity of the all-weather, day/night
capability prompted the development of remote sensors, and the
Coast Guard began long range planning to provide these capabilities. In the interim, the Airborne Remote Sensing System
(ARSS) was developed with off-the-shelf equipment. Six systems
were procured, consisting of IR/UV scanners, and placed at five
various air stations. It served as an interim measure until
1976.

The present aerial sensor package, Airborne Oil surveillance System (AOSS), was originally installed in an HU-16E
aircraft until that airframe was retired. It is now installed
in a C-130 long range search aircraft based at Elisabeth City,
NC. System integration of this second generation sensor
package (referred to as AOSS II) was completed in October 1976
and accepted by the Coast Guard in February 1977.

The AOSS consists of a sensor package, inertial navigation
system, and data annotation system which are integrated into a
complete package by a computer controlled system which processes
all data and constructs imagery as commanded by an operator.
Real-time data displays are provided, as well as hard-copy
documentation.

The sensors include a passive microwave imager (PMI),
multispectral line scanner (IR/UV), side-looking radar (SLAR),

and a high resolution aerial camera. Together, this sensor
package provides long range, day/night, all-weather detection
capability, mapping and measuring, and source identification.
A brief description of each sensor is included.

The Side-looking Airborne Radar (SLAR)

The side-looking radar, which produces excellent detection
results has an oil detection range of approximately 12 NM to
either side of the aircraft; provided antennas are installed on
both sides of the aircraft. SLAR can detect oil slicks in all
types of weather, day or night, and operates on suppressing
action disturbances. This action produces a contrast in backscatter levels which is observed by the radar operator.

During prototype testing, several violations were detected
in adverse weather. In addition, one SAR case was closed successfully when SLAR located the small, wooden vessel. SLAR does
not offer coverage directly below the aircraft (of a swath equal
to twice the altitude) and is ineffective for detections of oil
in calm water. These discrepancies are overcome by other sensors
in the package.

The IR/UV Line Scanner

The IR/UV line scanner (three channels; two infrared and
one ultraviolet) provides high resolution spill detection directly below the aircraft. The IR detects thermal differences
between oil and water while UV detects reflective differences
between water and oil and will detect very thin slicks which
may escape IR detection. The UV sensor also provides for discrimination of false alarms occurring in the IR channels due
to the surface thermal gradients (ship wakes). UV is not usable
at night and neither UV nor IR operated in fog, clouds or haze.

The Passive Microwave Imager

The passive microwave imager does provide adverse weather
capabilities. It also has a limited quantifying capability. The
PMI senses radiation in the microwave portion of the spectrum.
The energy radiates from objects as a function of surface
roughness and dielectric constant or emissivity. Thin oil films
will appear radiometrically cool (low "brightness temperature")
while larger, thicker spills (although still smoothing the
surface) exhibit a larger dielectric constant and therefore are
radiometrically warmer than surrounding waters. Film thickness
can be inferred from relative brightness levels.

The final sensor, a high resolution aerial camera, provides
hard-copy data and source identification evidence. It replaced

the low light level television of the AOSS I system which was
not successful due to problems of "blooming" from vessel lights,
wake or other sources.

AOSS II became operational in April of 1977 and has been
successfully utilized in a variety of programs since that time.
The aircraft has participated in International Ice Patrols
(IIP) during the last two springs, utilizing the SLAR sensor
as a method of detecting and mapping the ice as it moves south.
In August of 1977, while on a random patrol over the Gulf of
Mexico, the a/c detected a severed feeder line from an offshore
facility to the mainland. This past fall the aircraft has been
used extensively in random patrols throughout the East Coast.
These random patrols are conducted at various times of the day
or night and have been very successful, especially the night
patrols. There are presently two cases pending from patrols in
the Florida area, one of which will be heard using the AOSS
imagery as the only evidence.

The future uses of AOSS are to continue the aircraft as
the Coast Guard primary remote aerial sensor package. In conjunction with these patrol missions plans include further
testing and evaluation of the AOSS data imagery, tests to determine the usefulness of the SLAR for search and rescue, and
of course International Ice Patrol activities.

Although there is only one system at the present time,
the concept has proven to be successful and has a great potential to marine environmental protection and the savings of tax
payer dollars. Realizing this potential, the Coast Guard is
proceeding to the third generation of remote aerial sensing
with its new AIREYE program. This project will incorporate
successful AOSS subsystems while employing current state of
the art electronics technology to integrate them into a more
compact and efficient package.

The Aireye aircraft will contain side-looking radar (SLAR),
infrared/ultraviolet line scanner (IR/UV-LS), a KS-87B aerial
reconnaissance camera, and automated data annotation system
(ADAS). These sensors will be improved versions of thoses already in operation on AOSS. Undergoing concurrent development
for incorporation into the system is an active gated television
(AGTV). This sensor will be able to identify and document suspected violators with high resolution at night. Aireye will
have an improved surveillance capability over that of AOSS.

Initially, the system will be outfitted on six of the
Coast Guard's new MRS aircraft; however, all 41 aircraft will
be configured to accept the system, thus providing a "quick
change" feature. The Aireye equipped aircraft will be stationed

at various Coast Guard air stations around the country. From these bases, they will fly regular surveillance patrols off the nation's coasts. Designed primarily for marine environmental protection, Aireye, like AOSS, will interface with and aid a number of other Coast Guard mission areas.

Though Aireye is still a couple years away from operation, remote sensing is fast becoming a reliable and accepted part of the Coast Guard's surveillance programs. Law enforcement is one of the service's major missions and the day has arrived when one cannot be reasonably assured he will not be seen violating U.S. law even on the darkest night. A sensor technology advances improve, more sophisticated systems will be adopted thus increasing the risks for those who frequently abuse the environment.

NATIONAL PROGRAMS

A DANISH AIRBORNE OIL POLLUTION MONITORING
AND COASTAL SURVEILLANCE SYSTEM:
STATE OF ART AND BEYOND

B.M. Sørensen

Intradan Environmental Consultants A/S
Copenhagen, Denmark

INTRODUCTION

All countries with direct access to the sea have an
interest in obtaining continuous information about the activities within their territorial waters. In addition, they wish
to monitor the water quality and pollution level in the marine
region that has a potential impact on the economical and ecological interests of their coastal zone.

To desire to have a synoptic view of a large coastal water
region cannot be satisfied economically and efficiently by
using land-based observation stations or patrol boats.

Certain water quality parameters can be monitored from
space, but satellites do not offer the flexibility and high
resolution required for operational coastal surveillance.

Many countries have therefore turned to aircraft as the
best platform for the operational monitoring of ship traffic,
fishery inspection, oil pollution, sea ice hazards, and for
custom patrols.

Initially, pilots flew only "eyeball" surveillance, trying
to observe features of interest by looking out of the cockpit
window with no other aid than binoculars. It is evident that
this procedure is inadequate at night and in poor weather conditions. Therefore a number of countries has carried out
research to develop systems which satisfy the following basic
requirements:

- that they can be operated from a relatively low altitude
  to avoid frequent climb and descent;
- that they can survey a large area in a short time
  (thousands of square kilometers per hour);
- that they are weather-independent, i.e. capable of
  penetrating clouds;
- that they can be operated both in daylight and at night.

In Denmark, such a system is developed through a joint
effort between the Electromagnetic Institute of the Technical
University of Denmark, Terma Elektronik A/S and Intradan Environmental Consultants A/S, which represent hardware technology
research, technology development, and user orientated research
development respectively.

The system may be composed of some or all of the following
sensors:

* Side-looking airborne radar (SLAR)
* Infrared/ultraviolet scanning radiometer (IR/UV scanner)
  or infrared scanner;
* Forward-looking airborne radar (FLAR);
* Dedicated airborne handheld photographic camera with
  position and date/time annotation;
* Forward-looking low light level television camera (LLLTV)
  with mini-monitor for pilot or manoeuverable LLLTV with
  display on operator's monitor;
* Microwave radiometer (MWR). This sensor when fully developed in 1984 will make the IR/UV scanner redundant for
  most applications;
* Laser fluorosensor (LFS). This sensor is not yet developed for operational use, but the technology exists to
  build it.

APPLICATIONS

Oil Spill Detection and Tracking

Ideally, an airborne oil pollution monitoring system
should be able to detect, quantify and classify oil in the sea.
Since no single instrument can meet these requirements, it is
necessary to compose a package of sensors.

Detection: The SLAR is quite unique for detection of oil
on the sea surface. It is a so-called active sensor which
transmits a series of pulses across the track to one or both
sides of the aircraft giving one co-ordinate. The other co-
ordinate is established by the motion of the aircraft parallel
to the track. The pulse travels with the speed of light, and

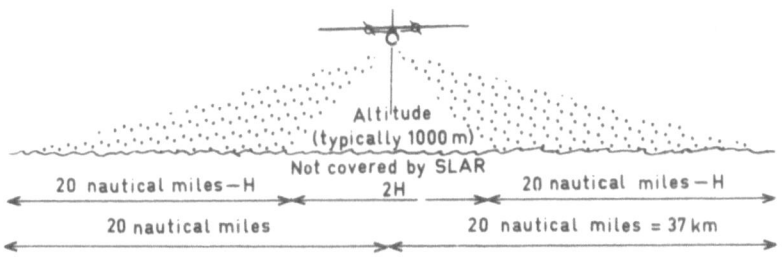

Figure 1. SLAR coverage for oil pollution monitoring.

the time for it to reach an object and be reflected back to the aircraft is proportional to the range (distance) of the object at right angles to the flight track.

SLARs used for oil pollution monitoring typically cover a range of 18-20 nautical miles (33-37 km) to each side of the aircraft with an uncovered area underneath the aircraft equal to approximately twice the flight altitude (Figure 1). For other purposes such as ship traffic surveillance, the SLAR may cover 40 nautical miles or more to both sides.

The intensity of a returned pulse is, among other things, highly dependent on the roughness of the ground sea surface from where it is returned. With regard to the sea, wind-generated waves will return a pulse with an intensity related to the wind speed and direction, while in the case of a completely smooth surface with no waves the transmitted pulse will bounce off from the sea surface and disappear away from the aircraft (Figure 2).

The gravity waves (sea waves) are commonly known to be related to wind speed. These waves are covered with small ripples called capillary waves, which cause a strong radar signal return. However, when the sea surface is covered with

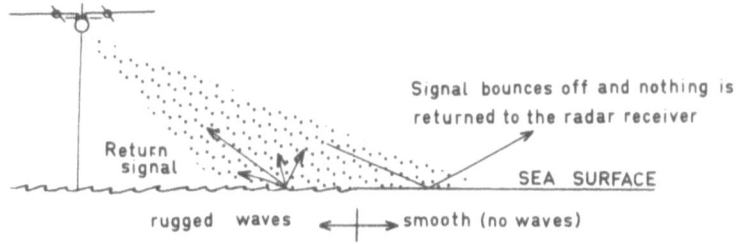

Figure 2. The meaning of surface roughness for signal return.

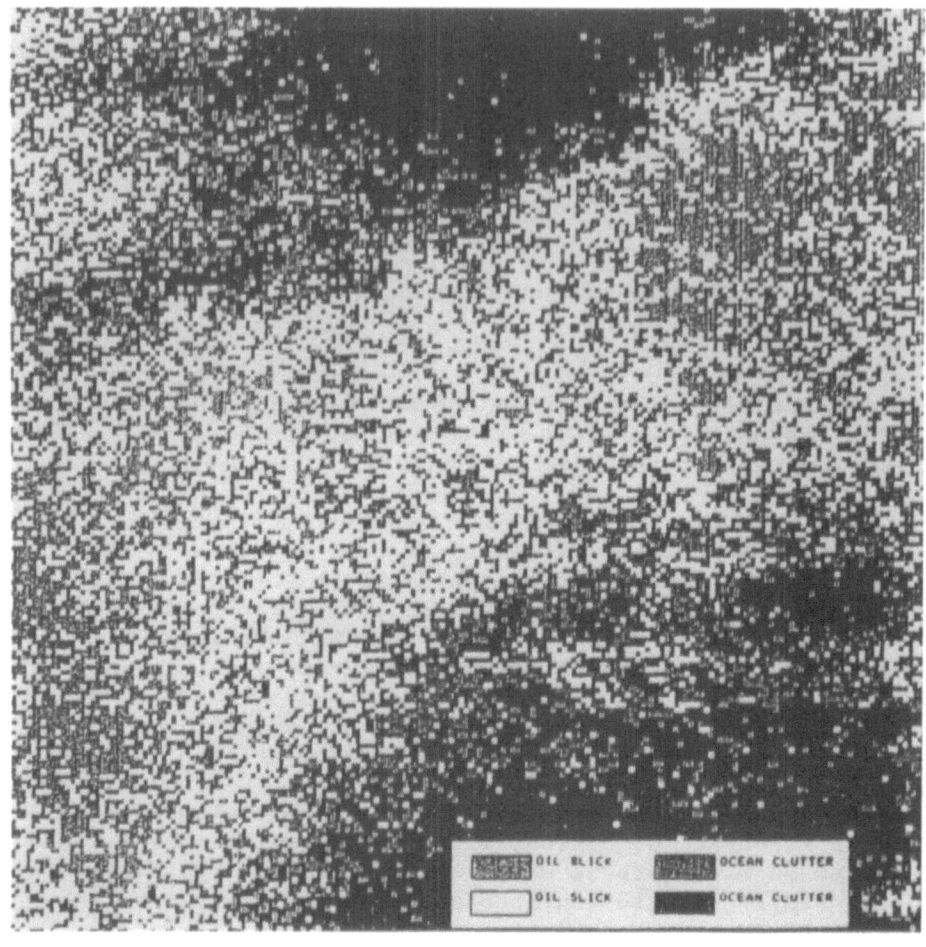

Figure 3.   Image produced by Terma SLAR on February 22, 1983
             off Burin Peninsula, New Foundland. Aircraft altitude
             was 5,000 ft, image size is 7 km horizontal by 15 km
             vertical. Data acquisition and processing by Intera
             Environmental Consultants Ltd., Ottawa, Canada.

oil (even if the thickness is only a few micrometers) these
ripples are suppressed or eliminated to the point that little
or no signal return is received by the radar system from such
an area. It is this difference in signal return between an oil
covered and non-covered sea surface that makes it possible to
detect oil on the sea surface. A SLAR image showing oil is
given in Figure 3.

The SLAR's ability to cover large areas within a short time is unmatched by any other sensor. As an example, an aircraft moving at ground speed of 400 km/hour will cover an area of more than 25,000 sq km/hour for oil pollution monitoring and considerably more for some other applications.

The only major limitations of the SLAR as an oil pollution detection device are its inability to "see" oil below the sea surface and to detect oil in completely calm weather.

No all-weather sensor can detect oil below the sea surface.

A detailed description of the Danish Terma SLAR is given later.

Quantification: After having detected and determined the position of an oil spill, the next step is to establish the amount of oil by measuring its areal extent and thickness.

Again, it is desirable to have a sensor which operates under (almost) all-weather conditions and during both day and night.

The passive microwave radiometer (MWR) satisfies these requirements as far as oil on the sea surface is concerned.

A MWR makes use of the fact that the apparent microwave brightness temperature is greater in the region of an oil slick

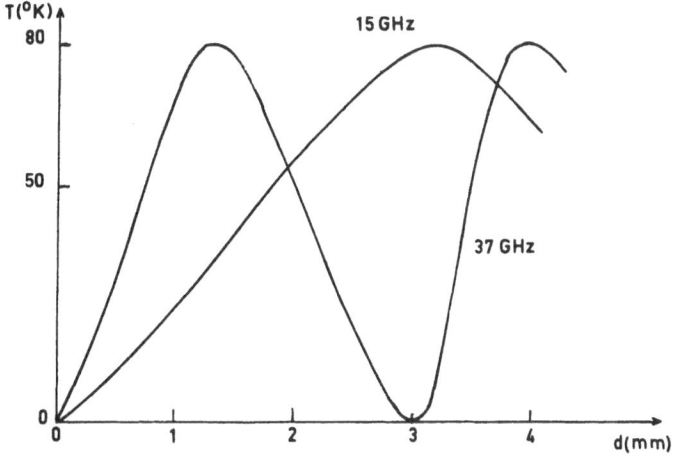

Figure 4.   Changes in brightness temperature at 15 and 37 GHz for variable oil thicknesses. Vertical view.

than in the adjacent unpolluted sea by an amount depending upon
the slick thickness. As the thickness of the oil film increases,
the apparent microwave brightness temperature at first increases and then passes through alternating maxima and minima
due to the standing wave pattern of the sea surface. By using
two or more microwave frequencies, thickness ambiguities introduced by the oscillations may be removed and the film thickness determined. Figure 4 shows the change of the original
brightness temperature as a function of the oil thickness. By
using two frequencies, such as the 15 GHz and 37 GHz shown
here, it is possible unambiguously to determine oil thickness
from 01.-4.0 mm.

A MWR system with a scanning antenna can be configured to
produce a brightness temperature contour map from which the
volume can be determined.

It is quite common in cases of larger oil spills that more
than 90 percent of the oil is confined in a compact region comprising less than 10 percent of the area of the visible slick.
The MWR can therefore be a useful tool in assisting oil combatting vessels in the clean-up operation by pointing out the
biggest oil concentration within the spill area.

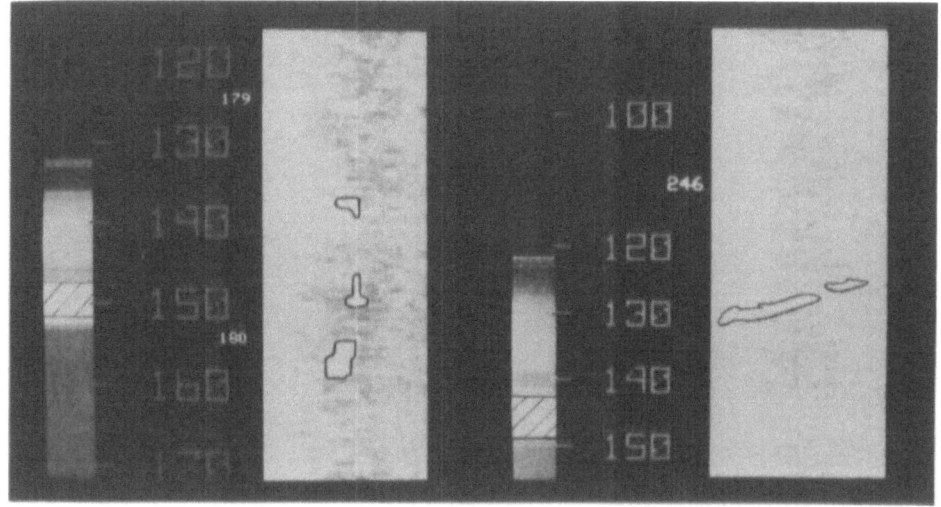

Figure 5a. Microwave radiometer images showing "raw" 34 GHz
            data. The grey scales correspond to brightness
            temperatures in °K. Two perpendicular flight
            passes over a 6 ton controlled oil spill. 179 and
            180 are reference numbers, corresponding to every
            24 seconds of flying. Pixel size ≃ 25 x 35 m.

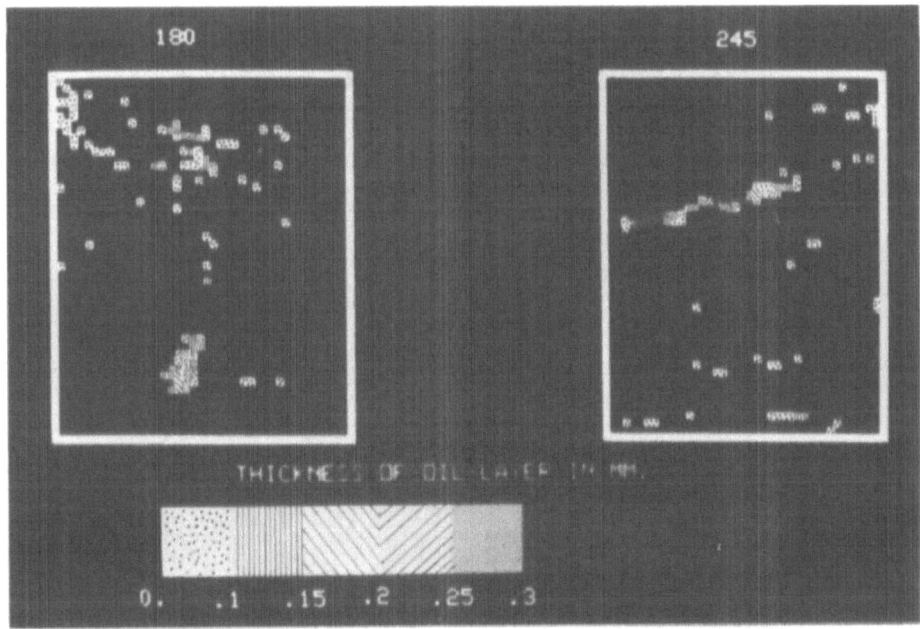

Figure 5b.   Oil thickness calculated from the brightness tem-
             peratures data using only the 34 GHz frequency.
             By adding an X-band frequency, thickness of up to
             4 mm (or even 7 mm) can be mapped.
             (Courtesy of the Technical University of Denmark).

Because of the MWR's unique ability to quantify oil under
all-weather conditions during both day and night, it may be
used to determine whether an oil slick detected by the SLAR is
sufficiently significant to justify an oil combatting vessel
being sent to the scene for a clean-up operation. The radiometer may therefore be used to save costly shiptime and prevent
false alarms.

The three groups involved in the development of the SLAR
are now developing an imaging microwave radiometer which will
go into production in 1984. At this stage of development, the
technical specifications cannot be published, but imagery produced from the "base sensor" (not the prototype) are shown in
Figure 5.

Other sensors or combination of sensors have been used in
an attempt to quantify oil on the sea surface. Scanning radiometers operating together in the ultraviolet (UV) and thermal
infrared (THIR) spectra can be used to delineate the areal

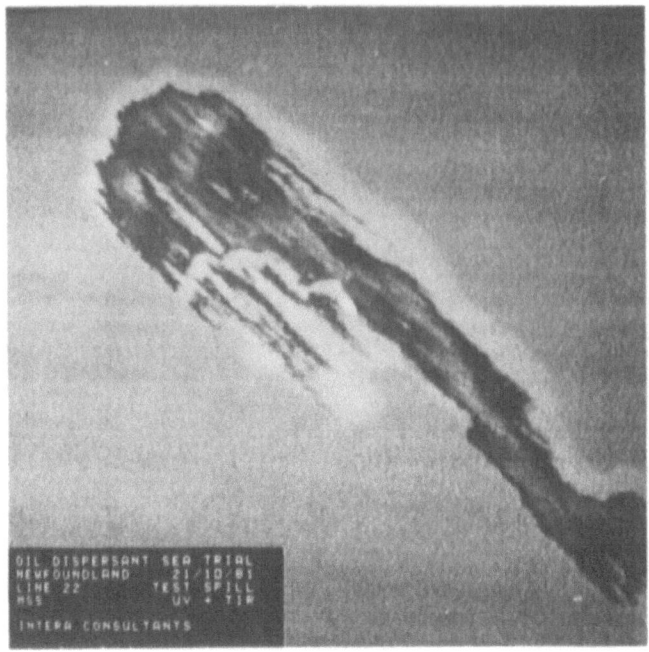

Figure 6a. Overlay of UV and THIR images of a controlled
            marine oil slick showing thin and thick layers.

extent of an oil spill. Since oil absorbs very strongly relative
to water in the UV spectrum oil on the sea surface will appear
in strong contrast to the surrounding non-polluted water, even
at a thickness of a few micrometers. Only the thicker layers
will be detected by the THIR detector. It is therefore possible
to point out the position of the biggest concentration within
a spill area. In fact, a skilled interpreter may be able after
the flight to distinguish a few thickness ranges using various
computer enhancement techniques (Figure 6).

The UV/THIR scanner will be replaced by the microwave radiometer in 1984, either alone or in combination with a UV
video system to maintain a moderate daylight water penetration
capability.

Classification: Identification of oil has several meanings.
It is often used in the context of distinguishing oil in sea
water from other substances which may appear like oil to a
particular sensor. It is also used with regard to the discrimination between various categories of types of oil.

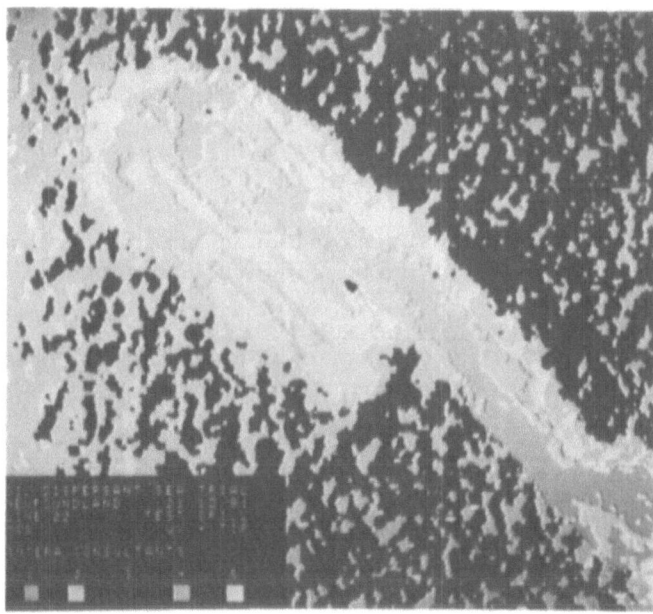

Figure 6b. Computer enhanced image of Figure 6a, originally
displayed in colour (Courtesy of Intera, Canada).

Once a sea surface "slick" has been detected by the SLAR,
it is fairly easy to determine if it is caused by oil by flying
over it with a microwave radiometer, a UV/THIR scanner, a laser
fluorosensor, a UV video camera or a combination of two or more
of these sensors. Needless to say, this identification is made
before quantifying the oil.

The identification dealt with here concerns discrimination
between oil categories (light fuel vs heavy fuel vs crude,
etc.) and the determination of oil types (e.g. Kuwait crude,
Arabian light, etc.). To that end, several instruments are
being tried to research level, such as the correlation spectrometer, the Fraunhofer line discriminator and the laser fluorosensor. The laser fluorosensor stands out among the potentially
capable oil type/category discriminators as the most promising
system. Several laser fluorosensor experiments have been carried out in U.S. and Canadian waters during the last few years.

The laser fluorosensor (LFS) is an active sensor which
means that it provides its own source of illumination. For oil
pollution classification a narrow beam of ultraviolet light
is emitted co-axial with the receiver optics, which collect

the fluorescent light from the target (Figure 7).

    The fluorescent return signal, at wavelengths longer than
the laser wavelength, is detected by a range-gated spectrum
analyser. Time-gating the detector in synchronisation with the
back-scattered radiation pulse permits the system to be operated
in full day-light.

    A Canadian laser fluorosensor based on a nitrogen laser
which excites at the 337.1 µm wavelength has shown promising
results.

    Development of a laser fluorosensor which works in the
deeper violet spectrum is planned to start in 1984.

    Other capabilities: Users of coastal surveillance systems
may wish to incorporate other sensors in the instrument package.

    A dedicated airborne, handheld photographic camera linked
to the aircraft navigation system for time and position annotation may be useful to produce evidence of an oil spilling
vessel's unlawful activity.

    Pilots who have difficulties in flying right over an oil
slick of limited areal extent for quantification or classification may find it useful to have a forward looking video camera
(low light level/UV) with a monitor on the aircraft instrument
panel.

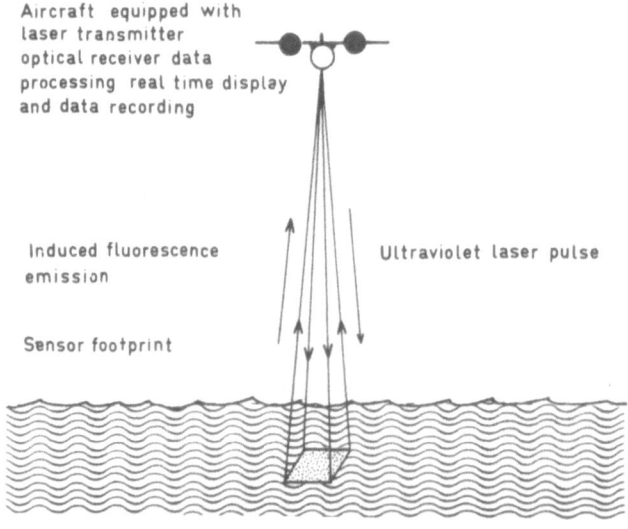

Figure 7. Laser fluorosensor operation principle.

The data from all the sensors which produce an electronic output (SLAR, MWR, LFS, Video) are displayed on a real-time television monitor and recorded on video tape or digital tape (CCT). All these data can be transmitted in real time to a shore-based pollution control centre or an oil combatting vessel, using a powerful VHF downlink system developed by Intera

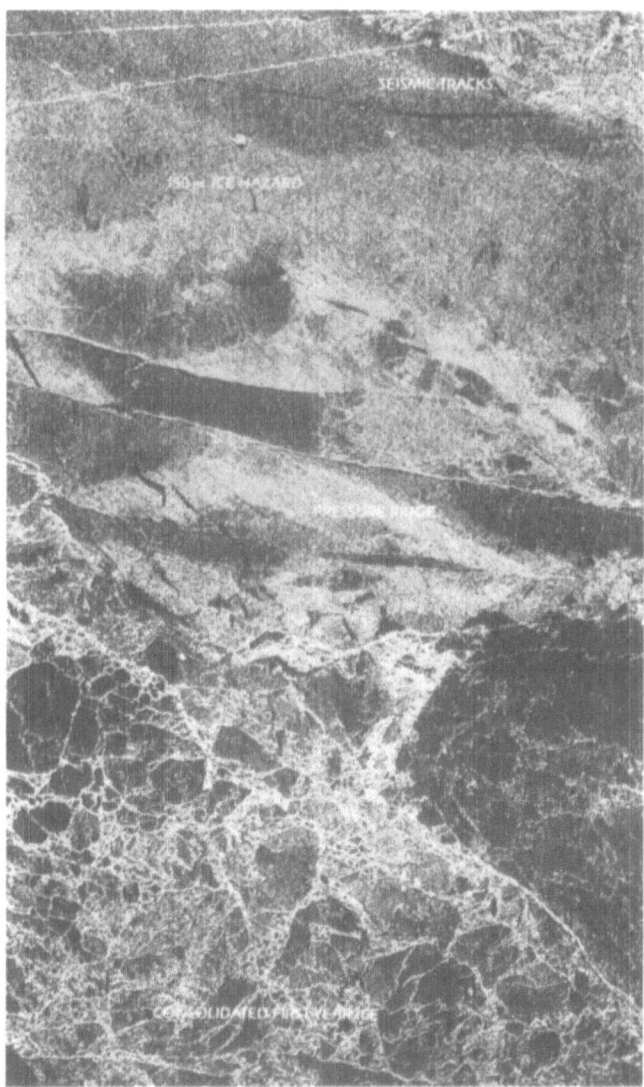

Figure 8.   SAR image showing a sea ice area of approximately 5.5 x 9 km at 6 m resolution. It shows small ice hazards (10 x 150 m in size), large and small pressure features, refrozen leads and consolidated first-year pack ice.

Environmental Consultants Ltd., Canada.

Sea Ice Reconnaissance

Sea ice can be an obstacle to ship traffic in arctic regions and such semi-enclosed waters as the Gulf of Bothnia, Baltic Sea, Norwegian fjords, etc. Drifting icebergs and big sheets of ice also frequently pose a threat to drilling rigs and their supply boats in arctic waters.

The most efficient sea-ice mapping system in the world to-day is the STAR-1 SAR (Synthetic aperture radar) developed and owned by Intera, Canada. The system provides data from a swath of 25 or 50 km with 6 m resolution (Figure 8).

The Danish Terma SLAR, however, is often adequate for mapping of major icebergs and determination of sea-ice boundaries. It has already flown numerous such missions in Alaska and Arctic Canada.

Although discrimination of sea ice types based on SAR data is possible in many cases, a microwave radiometer will dramatically reduce classification ambiguities. An example of sea-ice classification with a Danish MWR is shown in Figure 8.

Other Applications

In addition to pollution monitoring and sea-ice reconnaissance, the Danish surveillance system can be configured for many other applications, such as fishery and custom inspection, search and rescue, sea traffic surveillance, land use mapping, mineral exploitation, deforestation mapping and remote sensing research.

Dedicated systems for ship identification at night can be included.

TERMA SLAR

All configurations of airborne surveillance and monitoring systems described in the previous chapters are built around the Terma side-looking airborne radar which is the most powerful and versatile SLAR available in its lightweight category to-day.

Installed in an executive type twin engine aircraft, it can cover up to 70,000 square kilometers per hour.

Table 1 lists the pricipal operating parameters specified for the Terma SLAR system.

Table 1.   Terma SLAR main specifications.

Antenna	
Type	12 ft, dual-sided antenna
Gain	>34 dB
Horizontal beamwidth	<0.6°
Vertical beamwidth	19 +/- 3°
Polarization	Horizontal (vertical optional)
Transmitter	
Frequency	9375 +/- 30 MHz
PRF	0-2000 Hz dependent on the speed of the aircraft
Pulse width	0.25 µs (125 ns optional)
Peak power	20 kW
Receiver	
Mixer	Two balanced mixers with Gunn diode local oscillators for signal and AFC circuits
IF	Linear
Bandwidth	5 +/- 1 MHz
IF gain	>102 dB
Video output	1.5 - 5 V
Noise figure	−8 dB
Weight and Power	
Total weight	97 kg
Total power	28 V, 15 A

Data and Display System

The data processor interfaces with the navigational system of the aircraft (INS, ONS, DECCA a.s.o.). Position, heading, speed and altitude are automatically recorded and displayed on the TV monitor. The radar image on the TV monitor will normally be displayed in a rolling fashion corresponding to the progress of the aircraft. The image can be frozen and enlarged, and several transfer functions can be selected for optimum display of the information from the 8 bit (256 grey levels) refresh memory (Figure 9). The image can be stored by a TV video recorder for later review and by a digital tape recorder for further computer treatment.

The main characteristics of the imagery are listed below:

* Picture elements   512 x 512

```
* Pixel size  (10 nautical miles)     50 x 40 m
* Pixel size  (20 nautical miles)     100 x 80 m
* Pixel size  (40 nautical miles)     200 x 160 m
* Grey levels                         8 bits  256 levels
```

System Configuration

The Terma SLAR basic (Figure 10) which requires only
40" x 19" x 16" of rack space comprises:

* Rack mounted drawers with all electronic circuitry
* Operator keyboard
* Dedicated airborne videocassette tape recorder
* Television monitor for real-time display of radar
  image
* Table and shelves for maps and log sheets.

While the basic SLAR is suitable for oil spill monitoring,
search and rescue, and ship tracking, most other applications
will normally require a motion compensation system to correct
for aircraft motion to fully benefit from the high resolution
of the system. The Terma SLAR is available with an aircraft
motion compensation system which uses a digital microprocessor
and a large memory to correct for drift angle and high frequency yaw motions.

The motion compensation system makes the Terme SLAR the
only lightweight side-looking airborne radar system suitable
for sea-ice surveillance, terrain mapping and remote sensing
research.

Other options are CCT output, real-time data down-link

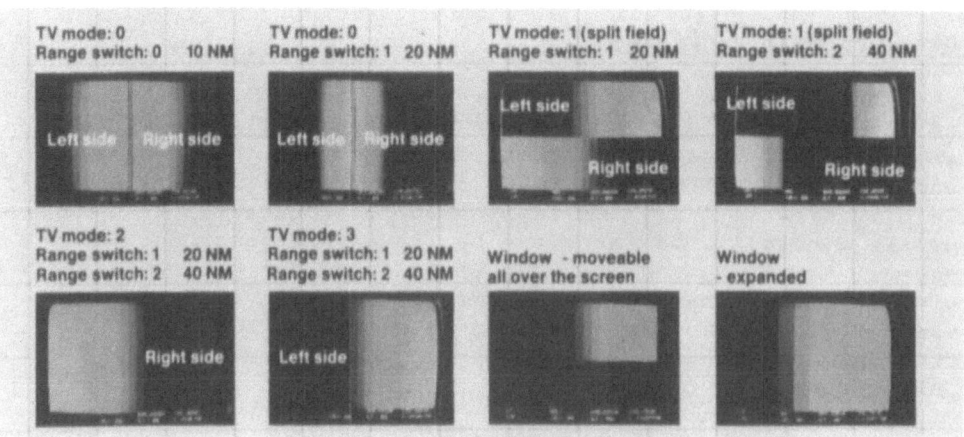

Figure 9. Terma SLAR monitor modes display.

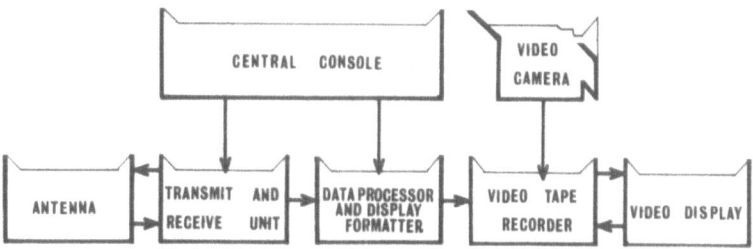

Figure 10. TERMA SLAR basic.

and hard-copy output (Figure 10). The computer compatible tape output allows for various image processing and enhancement techniques.

The data down-link is particularly useful where oil pollution combatting vessels and ship navigating in ice-infested waters need the information obtained by the aircraft with little or no delay. The SLAR can be connected to a powerful data down-link system which transmits all the data acquired by the aircraft to ships or ground in real time.

For land applications and remote sensing research, a real-time hard-copy will enable the mission manager to check

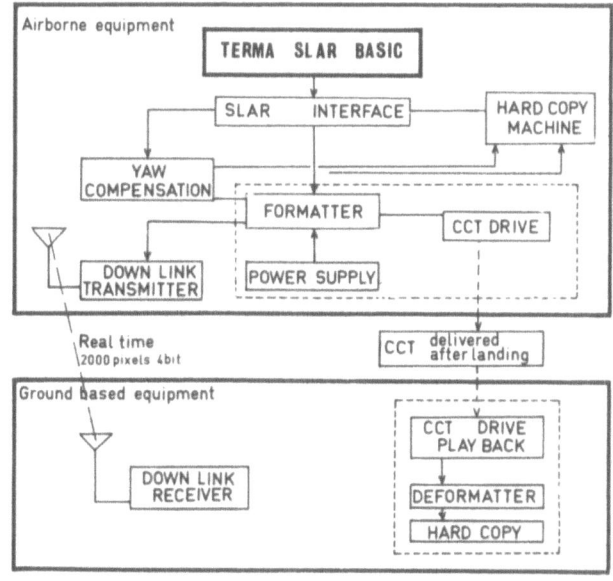

Figure 11. TERMA SLAR basic with all options.

the general quality of the data and to check that the target
area is properly covered.

    A real-time continuous strip hard-copy unit developed for
aircraft operation can be integrated into the SLAR system offering various scales on dry silver paper with 200 mm width.

EXPERIMENTAL RESULTS AND OPERATIONAL EXPERIENCE

    The Terma SLAR has been operated on numerous sorties both
on experimental and operational missions. Table 2 summarizes
the missions in terms of SLAR's state of development (P: prototype, 7' antenna; D7: Fully developed, 7' antenna; D12:
Fully developed, 12' antenna; CCT: Computer compatible tape
output), time of mission and purpose. It also includes some
experiments planned for 1983.

Table 2.   Terma SLAR missions.

Place	Time	Purpose
Greenland	1978 1983	P Ice reconnaissance; assistance to ships in ice-infested waters
North Sea (N1)	1981	P Isowake
North Sea Ekofisk oil-field	1982	D7 Experiment
North Sea Nth. of Skagen	1982	D7 Isowake
North Sea (FRG) German Bight	1982	D7 Experiment
North Sea (N1)	1982	D7 Tanker accident "KATINA"
Mediterranean Sea	1982	D7 Protecmar exercise
North Sea (DN)	1982	D7 Operational pollution monitoring
Alaska	1982	D12 Operational ice reconnaissance, ice edge mapping; iceberg drift
New Foundland (Canada)	1983	D12 CCT Experiment
Atlantic Ocean	1983	D7 Oceanographic research
Greenland	1983	D12 Marginal ice zone experiment MIZEX 1983
Nova Scottia	1983	D12 CCT Oil experiment
Atlantic Ocean (Brittany coast)	1983	D12 CCT Protecmar Experiment
North Sea off the coast of Holland	1983	D12 CCT Archimedes Experiment

THE OPERATIONAL OIL POLLUTION SURVEILLANCE SYSTEM

BEING USED IN FRANCE

R. Burkhalter and C. Meyer

Laboratoire National d'Essais
Paris, France

INTRODUCTION

The long and varied coastline of France bordering three sea areas of very different character gives rise to a considerable diversity in the nature and relative importance of the resources. So, a great deal of importance is attached by the public and the French Government to the protection of the marine environment and of the coasts.

To protect the marine environment, two different types of action are concerned relating either to the general concept of surveillance or the general concept of intervention, both of which are part of the responsibilities of all coastal states by virtue of their sovereign rights recognized by the International Law of the Sea Conference in the new economic zones and in furtherance of the international agreements regarding protection of the marine environment, shipping regulations and the safety of persons and property.

Over the last few years, a number of factors have tended to increase the necessity and the difficulties of marine surveillance and intervention:

(1) from the juridical standpoint, changes in the Law of the Sea made for an increase in the sea front responsibilities of coastal states;

(2) where the use of the sea is concerned, the increase of the number and scale of activities at sea or connected with the sea has led to ever stricter legislation which requires

87

greater supervision to enforce it.

In fact, the intervention missions may take different
forms:

* police duties: monitoring, prevention, reporting of
  offenders and even rerouting or interception of vessels;
* assistance: assistance to shipping, navigation aids;
* relief which covers both assistance and rescue, anti-
  pollution measures and pollution control.

Some of these actions would seem to require priority. This
is the case for control and surveillance of shipping and for
the detection and prosecution of deliberate acts of pollution
which contravene existing national and international
regulations. This concerns mainly the detection of illegal
discharges of oil at sea and the collection of evidence for
prosecution.

In fact, elsewhere along the French coast, the highest
risk of tankers accidents and deliberate oil spills exist
mainly at the approaches to the major oil terminals and in the
Channel between France and the southern coast of England. This
area is particularly one of the most congested waterways of
the world and a number of serious accidents have confirmed
the fact that the Channel is one of the highest risk zone for
collisions and groundings and one of the most polluted area,
along the French coast, by deliberate discharges.

These discharges are generally attributed to oil tankers
discharging ballast water but all the other ships, including
small fishing and pleasure craft are concerned by bunkering,
pumping of bilges, cleaning of fuel tanks, etc. Over 300
vessels pass through the Channel each day. Excluding yachts,
there are normally 500 vessels in the Channel at any time.
Of these, approximately 50 are tankers carrying crude oil or
refined petroleum products. It has been estimated that two
fully laden VLCC's pass through the Channel each day, going
to or coming from the North Sea ports.

In this context, the French Administration gave special
attention to the setting up of technical means to fight
against oil pollution.

In France, surveillance policing and pollution control at
sea are mainly carried out by the Navy, Customs, the merchant
navy and the Gendarmerie, each of which having the aerial
surveillance facilities appropriate to its role (defence or
police). But it is the task of the Ministry of the Environment
to take responsibility for the general control of marine

environment quality. So, the Ministry of the Environment has been appointed to set up technical means to fight against deliberate or accidental oil pollution, in connection with the "Centre National pour l'Exploitation des Océans" (CNEXO) and the "Institut Français du Pétrole" (IFP). For these purposes, remote sensing was chosen by the Ministry of the Environment, and CNEXO and IFP were appointed to test existing remote sensing techniques and to design an airborne oil pollution surveillance prototype for coastal zone, keeping in mind that this system had to provide:

* detection of deliberate or accidental oil spills;
* identification of polluting vessels;
* establishing of data necessary for legal action.

From 1974 to 1976, several experiments have been carried out by CNEXO and IFP, and in 1976 an airborne oil pollution surveillance prototype has been operational. Then, the first long term cruise was made to monitor illegal disposal of oil off the English Channel and the Atlantic coast. These operations were conducted in conjunction with the merchant navy and the French military Navy.

In 1978, an Interministerial Instruction gave the "Service des Affaires Maritimes" (external department of the Merchant Navy) the duty to perform the tasks of:

(1) gathering all information related to sea pollution from ships, in order to assess the risks to the environment;
(2) transmitting to the competent authorities the above information to enable them to choose appropriate measures;
(3) co-ordinating the intervention which aims establishing and curbing breaches of regulations on the disposal of oil at sea;
(4) and more generally, centralising and circulating all information on sea pollution by oil.

Therefore, the merchant navy has been entrusted with purchasing a remote sensing system corresponding to the results of research carried out in this field by CNEXO and IPF and with using it as an airborne surveillance system. This airborne scheme was integrated within a land-based surveillance system, within the "Regional Operational Surveillance Centres" (Centres Régionaux Opérationnels de Surveillance et de Sauvetage) (C.R.O.S.S.) stationed along the whole of the French coast controlled by an head office of the Merchant Navy Department. These different centres have been specifically entrusted oil pollution duty, in particular due to the fact that they have important radio communication facilities which could be adapted to the task of keeping watch, especially as regards transmission and exploitation of the data gathered.

DESCRIPTION OF THE SURVEILLANCE SYSTEM

The system is composed of two sub-systems:

* An airborne detection system including an infrared line
  scanner, with video display and recording devices;
* An on-ground system for data reception and real-time
  exploitation.

The Airborne Detection System

The aircraft: The surveillance system is installed on
board a twin-engined aircraft (CESSNA FTB 337) built in France
by Reims Aviation (Figure 1).

The choice has been partly dictated by the operational
qualities of this aircraft (maximum speed: 200 knots; work
speed: 140 knots; maximum range: 1,000 nautical miles; flight
duration: 4 hours) as well as its low working cost.

Moreover, the aircraft was adapted by Reims Aviation as
already in 1975 to the monitoring of oil discharges at sea,
especially with regard to navigational equipment and sensing.

At the present time, the aircraft is equipped with a
navigational system which operates at the world network ULF/
OMEGA. This equipment enables to know at every time the aircraft's position and thus the exact position of the vessels
below within the range of one nautical mile. Furthermore, it
is equipped with the following special equipment:

Figure 1.   The twin-engined aircraft Cessna FTB 337

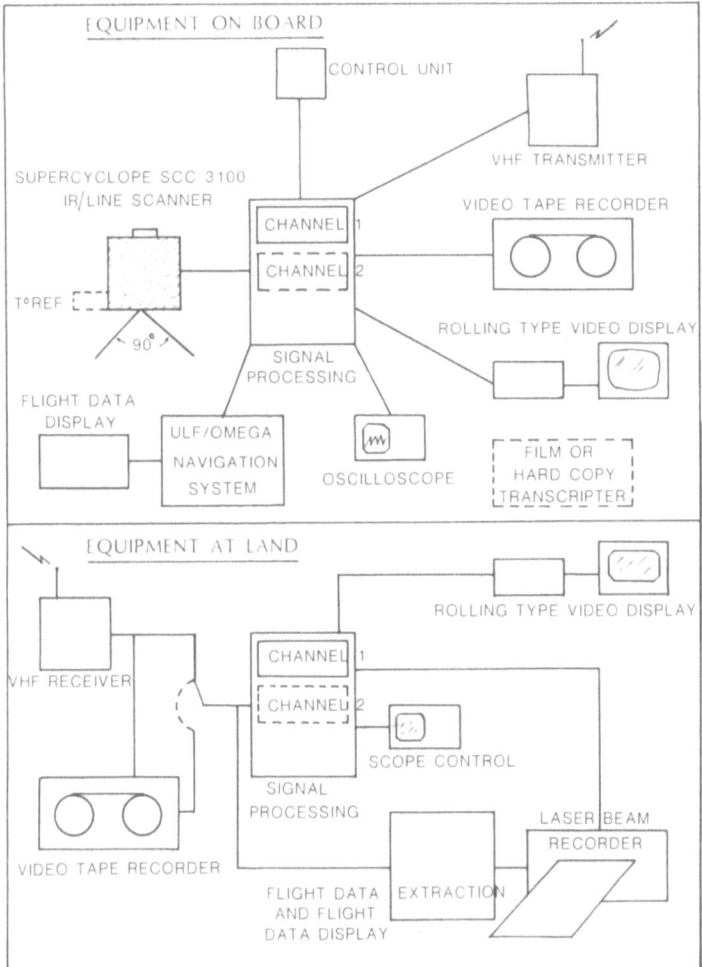

Figure 2.   Supercyclope block diagram.

A VHF and HFBLU marine band receiver/transmitter maintains permanent contact with the surveillance vessels or the CROSS reception facilities and a special transponder enables the same CROSS to follow the aircraft on its marine surveillance radar screen. A weather/navigation radar can also be installed on board the aircraft.

Equipment on board: A Supercyclope SCC 3100 infrared line scanner manufactured by Société Anonyme de Télécommunications (SAT) is set on the outside part of the aircraft, in a special container mounted on the lower part of the fuselage to provide a broad field of view.

This infrared line scanner includes (Figure 2):

* An analysis head whose main characteristics are:

Spectral band   8-12 micron
Detector   HgCdTe cell cooled with liquid
                     nitrogen
Scanning angle   90°
Scanning speed   240 scans/second
Spatial resolution   1.5 milliradian
Roll correction   ± 15°

* An electronic rack which receives the electric signal coming from the detection cell and which generates a video/synchro signal.

* A control box.

* An oscilloscope to control the video signal.

* A video tape recorder which performs magnetic recording. It records simultaneously the video channel, the multiplexed navigation parameters (longitude, latitude, course, speed, altitude, time, date) as well as an audio track.

* A rolling-type video display which reproduces in real time on a monitor a thermal picture of the area under analysis.

* A radio transmitter which ensures the immediate transmission to the ground of the information directly gathered in flight (video signal and flight parameters). When some difficulties arise with the direct transmission, such as too long distance of the aircraft from the reception base, electromagnetic disturbance of the atmosphere, the same system can be used for delayed transmission of the information recorded previously on the video tape recorder.

Different modifications can be added to the infrared line scanner detection system such as addition of a temperature calibration, a second detector working for instance in the UV channel and a film display.

To complete the system, two Hasselblad 500 EL cameras (70 mm) can be installed under the wings of the aircraft and ensure vertical views with different combinations of filters and films. In addition, 24 x 36 camera equipped with an automatic system which superimposes identification data (time and date) onto all exposures enable the crew to photograph the polluting vessels. To identify vessels at night, a television camera can be linked to the lighting aircraft system.

Fig. 3a & 3b: The mobil reception unit (out- and inside views).

The ground reception unit

    Ground stations are the essential complement of the air-
borne capabilities. They permit immediate exploitation of the
information gathered during flights.

    Two similar systems are now being built: a fixed one, to
be used by CROSS personnel and a mobil system installed in a
small truck (Figure 3).

    The system is composed in the following way:

    * A radio-receiver receives the information from the air-
craft. The antenna is remotely controlled to ensure a perfect
reception of the information.

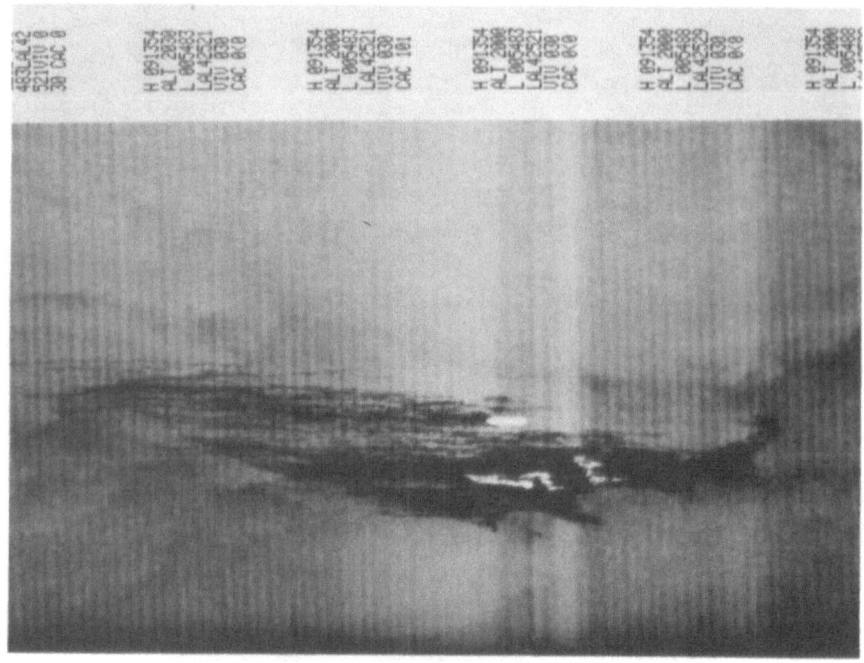

Figure 4.   IR imagery of a controlled oil spill
(Protecmar exercise).

    * An interfacting rack which processes the signal so as
to separate video signal, synchronisation signal and flight
data, respectively.

    * A rolling video type display.

    * A laser beam recorder designed and marketed by Soro
Electro Optics, which restores in near real time (less than
10 seconds) the transmitted imagery on a dry processed recording medium which can be developed in 8 seconds and then
immediately produces recordings with both high contrast and
resolution (76 gray levels, 2,000 spots per line of scan)
(Figure 4).

    This device uses a low power, industrial type and relia-
ble He Ne 6328 Å laser. The laser beam is modulated according
to the signal received by an accousto-optic modulator controlling the exposure with respect to the intensity modulation
response and gamme curve of the medium.

    Flight data are recorded on the edge of the imagery. Thus

the operator can obtain in near real time high definition documents without any particular constraints such as those due
to films processing in damp atmosphere. The documents can
therefore be directly put in the file describing the offence.

* A video tape recorder similar to the video tape on board
the aircraft.

SURVEILLANCE OPERATIONS

During operations, the crew is composed of two people: a
pilot and an operator.

The aircraft patrols generally at an altitude between 100
and 1,000 meters below the cloud cover and flies over the different ships found in the previously selected area. Their
wakes are systematically analysed. The data are displayed on
board the aircraft, recorded and transmitted permanently to
the nearest CROSS station or to the mobil unit. If an abnormal
image of wake appears on the screen, the decision to identify
the suspected vessel is immediately taken jointly with the
ground centre, after examination of the recorded imagery or
the hard-copy image. Then the aircraft descends to a lower altitude so as to identify and photograph the vessel.

The observations are then transmitted to the control
centre (CROSS). The control centre may then decide to confirm
the pollution, using sampling buoys, for example, or sending a
surveillance vessel to the suspected polluted area. A decision
to take legal action may then be made: immediately if the suspected ship is reaching a French harbour or by transmitting
a file concerning the offending vessel to the foreign authorities if the suspected ship is not reaching a French harbour.
Such a file is composed of photographs, IR imagery (flight
data annotated) and sample analysis, if possible.

SYSTEM PERFORMANCES AND OPERATIONAL LIMITS

The present surveillance system has, of course, operational limits due to at once the weather conditions. The limits of
infrared sensors are largely depending of meteorological conditions: fog, cloudiness, sea conditions. In fact, good results
have been obtained with the Supercyclope infrared line scanner
in light mist or rain. The image quality was adequate for detecting oil slicks on the surface of the sea in these conditions.

Some other limitations exist, in relation with the IR
band-pass. The principle of recording radiometric temperature
variations or thermal emissivity over the observed area can

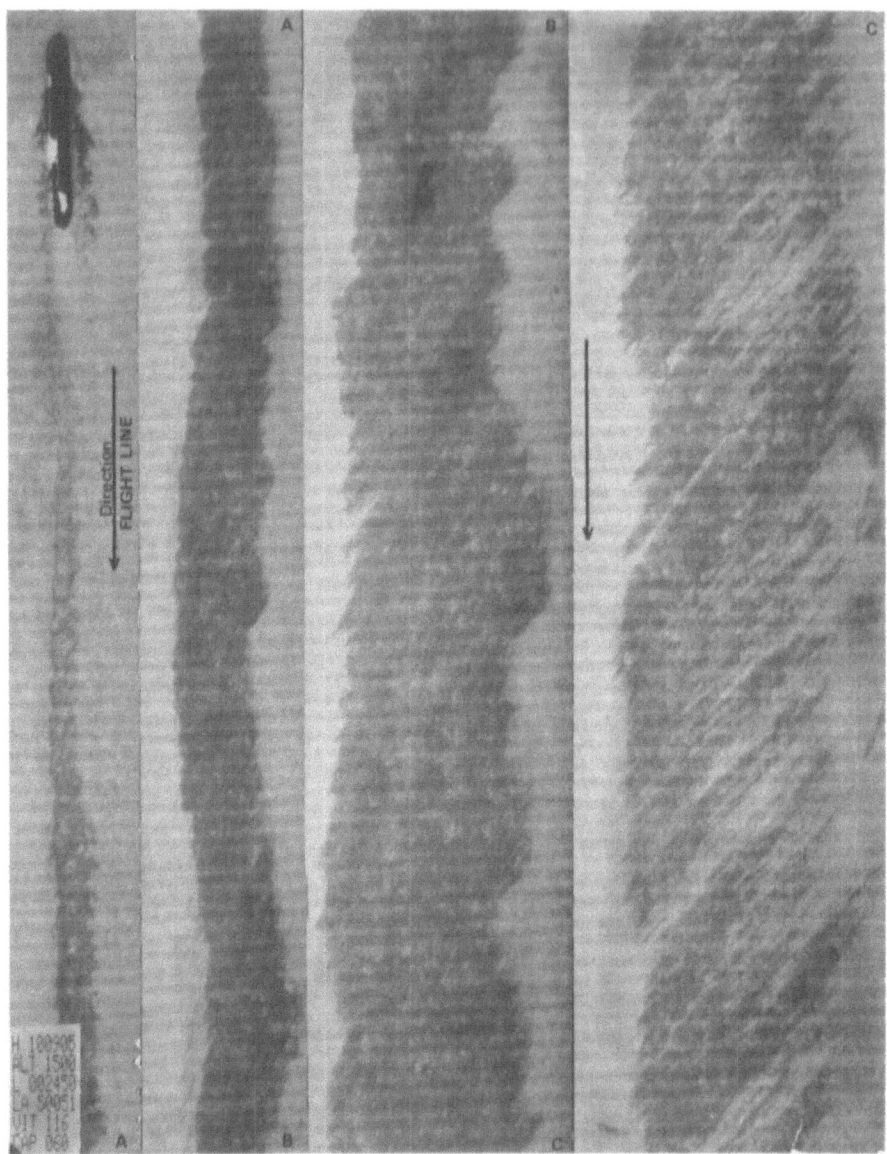

Figure 5. Windrow formation in an oil polluted ship's wake
         (Infrared imagery).

induce interpretation errors, especially when other phenomena
interfere and modify the characteristics of the environment:
rising of cold water in the ship's wake in summer, disposal of
warm oil, cloudy screen of irregular shape between the aircraft

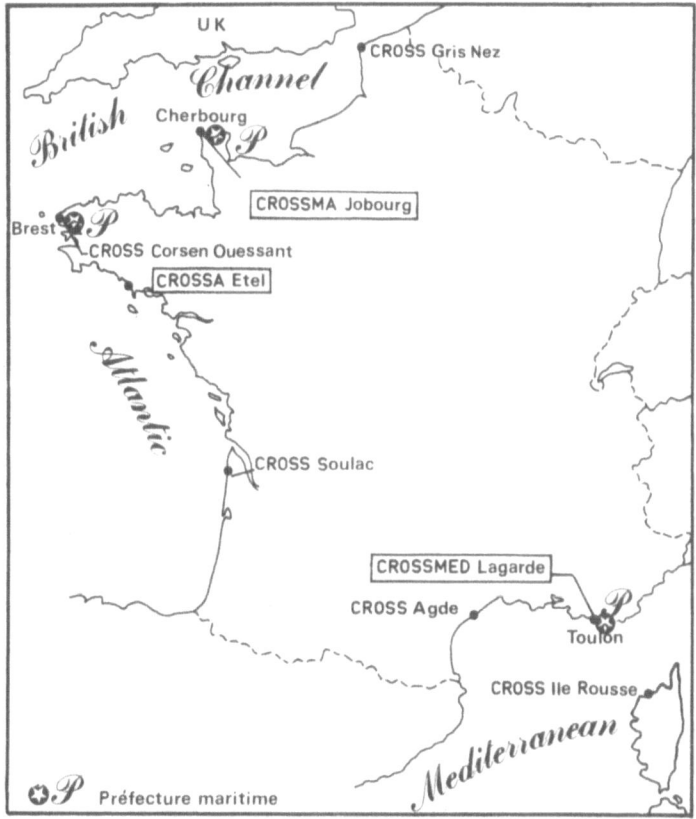

Figure 6. CROSS distribution along the French coast.

and the ships. But shape characteristics of the thermal imagery
can permit to correctly interpret the data (Figure 5).

Experience has shown that there is no complete correlation
between what is visible for the naked eye and infrared sensing.
Traces of oil may be observed by the operator without providing
IR detection. This happens when the thickness of the oil slicks
is somewhat below the wavelength at which the radiometer works.
But the comparison of visible and infrared imageries can provide some informations about the average thickness of the oil.

At least, the transmission range (about 80 nautical miles
at an altitude of 600 meters) not only depends on the flight
altitude but also on the position and the orientation of the
CROSS antenna towards it.

Considering the task accorded to this surveillance system

(detection and prevention of deliberate oil spills), these limitations are less important than they appear at an operational
efficiency point of view. In fact, on the one hand, the identification of a polluting vessel is practically impossible at the
present time in bad atmospheric conditions; on the other hand,
a sufficient number of CROSS stations have been installed along
the French coast to cover reception of the whole surveillance
area. The majority of interventions seems to take place in an
area within the 50-mile limits (Figure 6).

CONCLUSIONS AND FUTURE DEVELOPMENT

The conclusions drawn from operations carried out up to
now are as follows:

In 1975, a test operation in the English Channel enabled
the checking of 207 ships in 34 hours and the identification
of 8 confirmed pollutors, one of which at night.

In 1976, 153 hours of surveillance were undertaken in the
same area and along the Atlantic coast. Checks were made on
1,148 ships and fourteen were identified as pollutors.

In 1977, the whole of the French coast underwent surveillance. During 231 hours, 1,200 vessels were checked on and 17
pollutors detected.

In 1978, 300 flying hours were devoted to the marine surveillance. More than 1,500 vessels were checked on, not only in
the Channel area but also in the Atlantic and the Mediterranean
areas. Only 7 violators were detected.

In 1979, after this period of apparently decreasing number
of violators and during the period following the AMOCO CADIZ
disaster (March 1978), 400 flight hours were undertaken to
control more than 2,000 vessels. 22 pollutors were detected and
identified.

This being, the aim of an operational oil pollution surveillance system is:

* To detect and discourage illegal oil discharges along
  the coastal areas so as to provide necessary information
  for legal action following an infringement (proof of
  the infringement and identification of the vessel) and
  for pollution treatment actions if the spill is too important.
* To intervene in the case of an accident (collision or
  grounding) when large quantities of oil are spilled.
* To map the oil slicks and to know their movements so as

to assist with pollution control system on land and at sea.

The present system seems to satisfy the first aim by using a very simple and cheap equipment, but is not completely adapted to the second aim. As a matter of fact, for this purpose, it is absolutely essential to see clearly the oil slick in all weather conditions (accidents often occur in bad weather conditions). At the opposite, identification of ships is not essential.

In fact, in order to detect oil spills under optimum conditions and with maximum efficiency, and considering the scientific aspects alone, it would be useful to get remote sensing systems able to operate under all weather conditions and able to transmit in real time pertinent information which could permit without having to check in-situ (by sampling for instance) the identification of the hydrocarbons spilled on the sea surface, either to provide evidence or, when dealing with crude oil, to identify the geographic origin of the product.

Some possibilities seem to emerge from research carried out in France and other countries. In particular, it seems that the possibilities of remote sensing could be considerably increased, with regard to oil pollution, by using both passive and active sensing systems.

According to these conclusions, a possible evolution of the present system might be as follows:

* The installation of light weight systems (similar to the system described above) along the French coast to reinforce the number of surveillance hours, and thus to act as a deterrent.
* The design and installation on board the surveillance aircraft of a single more sophisticated system equipped with an all-weather detection sensor for occasional operations in the case of accidents or large oil spills.

In this more complete system, another device like the multispectral line scanning radiometer manufactured by MATRA could be installed.

The MATRA Thematic Scanner

The Thematic Scanner manufactured and marketed by MATRA is a multispectral line scanner. In a standard version, it is equipped with 6 channels: four visible and two infrared.

Its modular concept assures flexibility. The different filters used in the visible channels can be switched over and

commutated in flight. The system allows digital or analog data
recording. The in-flight calibration of the infrared signal is
made by two blackbodies. Real-time display on two screens may
be provided as well as an airborne restitution on photographic
films.

The main Thematic Scanner characteristics are as follows:

Infrared response band:	8-14 μm
Detector:	Hg.Cd.Te photoconductor
	$7.5 \cdot 10^{-2} \times 7.5 \cdot 10^{-2}$ mm
IR temperature resolution:	0.1°C
Visible response band:	0.4-0.9 μm
Detector:	Photo diode PIN Si
	0.2 x 0.2 mm
Optical aperture:	2.6 f/71.4 mm
Angular resolution:	$1.7 \cdot 10^{-3}$ rad
Instantaneous field of view:	better than 3 mrad
Line ground coverage:	90°
Number of points per line:	900
IR calibration:	blackbodies, cold and hot regulated
Scanning speed:(scans/s):	4,55 - 9,1 - 18.2 - 36.4
Data transmission speed:	37.2 - 74.5 - 149 and 298 and 298 Kbits/s on each of the six channels.

In the same way, the future airborne surveillance system
could be completed by the multipurpose synthetic aperture
"Varan-Anaconda" radar designed and marketed by Thomson CSF for
the marine surveillance purposes.

The sensors and flight data transmission to mobile and
fixed ground stations will be in any case maintain to get
speedy information near the polluted zone or near the operational control centres.

ADDENDUM

Since the Washington workshop where this paper was presented, the decision has been taken to design a new radar as
the complementary tool for the Supercyclope line scanner. This
radar will equip the aircraft which will be acquired by the
customs, one of the French administration which has in charge
to assume the marine surveillance. This radar does not compete
with the Varan-Anaconda radar system whose performances are
beyond those strictly required for oil spill and shipping surveillance.

RESEARCH AND DEVELOPMENT PROGRAMS

RELEVANT TO THE REMOTE SENSING OF OIL SPILLS

F. Günneberg

Bundesanstalt für Gewässerkunde
Koblenz, Federal Republic of Germany

In 1978, the Deutsche Forschungsgemeinschaft established a new concentration of efforts program concerning the remote sensing of the hydrosphere. 24 projects are being financed in this context, but only one is concerned with the remote sensing of oil. Professor Luther at the Ossietzki University in Oldenburg is developing a laser fluorosensor that can be carried by a helicopter. He is aiming for the identification of oils, of algae populations and for the tracking of rhodamin-labelled water bodies. This work began in summer 1978.

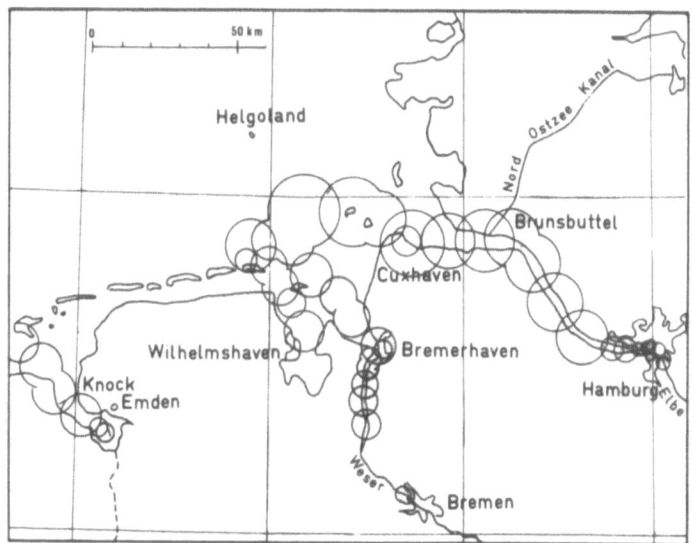

Figure 1. Radar control networks along coastal waterways.

Figure 2.   Tidal flats in the Jade Weser area.

        Günneberg in the Bundesanstalt für Gewässerkunde is just
beginning to work with a flash-lamp pumped dye laser, which is
less apt to excite the fluorescence of oil. He plans to survey
the fluorescence-background of the larger rivers and the coastal
waters and its variations through the course of the year from
board a ship. This will serve as the basis for assessing the
prospects of advanced investigations.

        Earlier, Dr. Hellmann in the Bundesanstalt investigated
the aging of oil exposed to the atmosphere, employing gas-chro-
matography.

        There is a multinational effort in progress to test the
feasibility of a Swedish idea to label oils with heavy metal
compounds.

        The navigable waterways in the German estuaries are con-
trolled by a network of radar stations (Figure 1). Their
frequency is 8.8 to 9.2 GHz (X-band), their polarisation is
vertical or switchable. The antennas are mounted on towers at
a height of about 40 meters above sea-level. Therefore sea-
clutter and the edges of waterbodies of different salinities
can be visualised to a distance of 3 kilometers only (Fig. 2).

The circles on Figure 2 indicate 2-mile zones around radar
stations up to which sea-clutter may be observed with antenna
mounted at 30 m height. The detection of an oil spill has not
yet been tested, because we fortunately have had none. Auspices
are not altogether good. The low water-depth modifies the wave
forms and sea states depicting the ground profile, including
wakes. So we cannot expect an even background of sea clutter,
on which an oil spill would be clearly discernible. This also
applies to an airborne radar system, as far as the zone of
water-depth less than 10 m is concerned.

Thermal scanners and multispectral scanners are also at
hand in the Federal Republic of Germany if a tanker accident
should occur on a clear day.

REMOTE SENSING PROGRAM FOR OIL DETECTION

IN THE NETHERLANDS

H.W. Brunsveld Van Hulten

Rijkswaterstaat
Directorate for Water Management and
Hydraulic Research
The Hague, The Netherlands

INTRODUCTION

To aid in the task of surveilling the Dutch Continental
Shelf, Rijkswaterstaat (Department of Water Control, Directorate
for Water Management and Hydraulic Research) investigates
remote sensing techniques as a possible surveillance system.
To build up techniques for adequately handling the large amount
of data inevitably produced by such activities, a research
program was started in 1971, supported by the Physics Laboratory
TNO and the National Aerospace Laboratory (NLR). Emphasis was
laid on sensor-object interaction studies. The initial sensor
was an infrared line scanner (IRLS), followed in 1974 by an
X-band side-looking radar (SLAR) of the real aperture type with
horizontal polarization and a large footprint (30 m) carried on
a QUEEN AIR BEECHCRAFT from NLR. Because of the limited size
and payload of the aircraft, both sensors cannot be flown
simultaneously. Only conventional cameras and an inertial
navigation system can be flown with the SLAR and the IRLS. Of
primary importance was the use of SLAR to measure contrast
produced by waves and oil spills.

Since the program and facilities were limited, the research
program concentrated on selected objects, such as waves, oil
pollution and traffic density in shipping lanes. From the SLAR
imagery it soon became clear that the sensor is eminently
suited for the detection of very slight modulations (to 1 dB)
of waves, generally existing under slight to moderate wind
conditions. In rough weather, the imagery becomes too speckled
and no structured wave pattern can be detected. Oil patches

105

on the contrary then become more distinct, because of the
suppression of the speckled structure.

The SLAR imagery collected over the range of wind speeds
from 0 to 13 ms^{-1}, provides information on the detectability
of gravity waves. In addition, sand waves and internal waves
have been detected through their effect on the sea surface.
The latter two will be referred to as wave-like patterns
because of their appearance in the imagery, although they
cannot be described by a wave motion.

Sea-truth measurements of waves have been carried out by
wave rider buoys providing wave power spectra; spectra derived
from radar imagery are compared with such power spectra.

Characteristics of the relationship between radar measurements in a wave tank and the input spectrum for the wave generator are considered. In order to develop a better understanding
of the detectability of oil spills by remote sensing, studies
related to wave detection are mentioned.

It is realized that results obtained by remote sensing
techniques largely depend on the characteristics of the sensors
applied which is in particular true for radar systems.

Reference is given to De Loor and Brunsveld (1978), in
which most of the results are comprised and related references
are found.

EXPERIMENTS

Objectives of the experiments carried out in the research
program are to assess the following features:

* the capability of the available SLAR system (X-band,
  HH polarization and 30 m footprint, dynamic resolution
  of 1 dB);
* the contrast of oil slicks compared with that of other
  sea surface phenomena showing up in the imagery, in view
  of discrimination and false alarm rates;
* the set of requirements which a surveillance system
  should meet for effective control of oil spills.

Radar experiments have been carried out in the North Sea
(1975) and in collaboration with the French Petroleum Institute
(IFP) in the Atlantic (1975). The SLAR system has also been
applied in a French experiment in the Mediterranean Sea (1976).
In a wind-wave tank at the Hydraulic Laboratories, experiments
with an FM/CW system and a typical North Sea wave spectrum
have been carried out. In all these experiments, light crude
oils have been applied.

Some results have been gathered during a limited experiment around the scientific platform Noordwijk (1977) with FM/CW scatterometers. In this case, oleic alcohol have been applied, which is supposed to be quite a perfect substitute for oil without environmental impact. One experiment in which a thermal infrared line scanner has been applied has been carried out in the North Sea (1975).

The sea state conditions under which the experiments have been carried out ranged from lower than 1 to about 4. The upper bound in sea state for the detection of oil spills is drawn from studies by the author on mechanical oil recovery systems. In these studies, the labor conditions for the crew and the effectiveness of the cleanup systems and the floatability of the oil with increasing sea surface roughness have led to a realistic figure of sea state 4 to 5. If for particular reasons it is necessary to obtain data at higher sea states, more sophisticated systems have to be developed and have to be adopted at the cost of real-time information.

The application of IRLS is not sufficiently investigated, because of the abundancy of results in the literature, and because of the poor condition of the scanner and the inadequacy of the inertial navigation system. The positive and negative deviations of the radiation temperature of the oil slick from that of the sea surface provide only limited information such as presence of oil and the area covered.

RESULTS

The results obtained during the research program can be summarized as follows:

Relative to radar, the independant samples within the defined illuminated cell tend to be Rayleigh distributed, the heigher their number is. This indicates that a large footprint in combination with a low pulse repetition frequency (PRF) provides statistically reliable information. In particular, the low-frequency components, i.e. gravity waves, are dominant. A small footprint provides non-Rayleigh distributed information and hence the signal will be spiky. Imagery obtained by such systems show up speckled.

In the case of light to moderate wind conditions, a SLAR with HH polarization and a high dynamic resolution can detect very slight modulations of the capillary waves. Besides by gravity waves, such modulations can be produced by other phenomena, such as sand dunes at the bottom in combination with

a strong tidal current, and internal waves in the ocean. Such
patterns show up in the imagery and may obscure the oil slicks,
what necessitates well-trained interpreters.

In the case of moderate to severe wind conditions, the
SLAR imagery shows up very spiky. In the oil-polluted area, the
spiky character is strongly reduced and the contrast of the oil
slick with the background sea clutter exists. A contrast of
5-7 dB has been established at sea and in the wave tank.

The total dynamic range for sea clutter is estimated to
be 20-25 dB. However, in a particular case, the dynamic range
drops down to 10 dB, as indicated by our X-band SLAR system.
The contrast between oil and the background and the relatively
small dynamic range necessitate a high dynamic resolution in
the order of 0.5 dB.

The trade off between dynamic and spatial resolution is
easened by lowering the all-weather capability requirement to
an "around the clock" requirement combined with an upper sea
state limit of approximately 5 to 6. For very low wind conditions without capillary waves present, IRLS and MSS systems
can be applied, although the latter only for daytime conditions.

CONCLUSIONS

On basis of the results obtained and some reasoning of the
constraints concerning the situation in which a limited oil
spill surveillance system has to operate, the following can be
concluded:

* The requirements of "all weather tool" is over done in
  many cases;
* An "around the clock" detection capability suffices, for
  which radar is considered to be most useful; a contrast
  of 5 to 7 dB of the oil spill and the surroundings have
  been found (De Loor and Brunsveld, 1978);
* Above sea state 4 to 5, the oil will be easily emulgated
  into the water, in which case no surface phenomenon will
  be detected. The combatment of oil spills than also is
  quite impossible because of the sea surface roughness.
  If for particular reasons, detection of oil spills under
  higher sea states has to be successful, more sophisticated remote sensors have to be applied at the cost of
  real-time information performance;
* Under light wind conditions (sea state less than 1),
  radar fails because of the absence of capillary waves,
  in which case IR ("around the clock" capability) or MSS
  (day time conditions) can be applied;
* The dynamic resolution of a radar system has to be in

the order of 0.5 dB due to a total dynamic range of
10 dB in any particular situation.

* Interpretators of the remote sensing data must be thoroughly trained in recognizing oceanographic phenomena
  showing up in the imagery, in order to bring down the
  false alarm rate.
* A remote sensing surveillance system has also to be
  applied as a data information system for effectively
  conducting oil combatment operations at sea by ships.
* HH polarization for a SLAR system in combination with
  a large footprint (30 m) provides adequate information;
  these systems can be relatively simple because of the
  real-aperture mode.
* The trade off between dynamic and spatial resolution is
  optimized by adopting an upper bound for sea state conditions in which detection has to be carried out. We
  indicate a limit of sea state 5 to 6.
* The presence and the area covered by oil can be detected
  by SLAR, IRLS and MSS systems. For rough classification
  of the oil type and the estimation of an average thickness, other types of sensors have to be added to the
  sensor package as will be found in other abstracts presented during the Washington, D.C. and Paris workshops.

(This contribution was presented at the Washington, D.C. workshop and revised by the author in 1983).

AERIAL PHOTOGRAPHY INTERPRETED FOR CONTINGENCY PLANNING,

SPILL PREVENTION, COMPLIANCE MONITORING AND SPILL SURVEILLANCE

R.W. Landers

Environmental Protection Agency
Las Vegas, United States

ABSTRACT

The EPA's Environmental Monitoring and Support Laboratory in Las Vegas is producing photo interpretation keys which are aerial photographic examples of hazardous substances spills and potential spill conditions within typical chemical processing and storage facilities. Color aerial photography, acquired over a variety of chemical processing facilities along the Lower Delaware River estuary and the Baltimore Harbor area, provides the primary source of data for the keys. Typical in-plant facilities covered by the keys include :

* product storage facilities, including storage and holding tanks and dry product stock piles;
* product transfer facilities, such as marine and river terminals, tank car and truck racks, in-plant pipelines and conveyors;
* plant drainage and wastewater treatement facilities.

The keys are designed to aid both experienced remote sensing specialists and EPA regional oil and hazardous substances spill prevention and control personnel in monitoring chemical facilities for compliance with anticipated spill prevention regulations to be issued under authority of the Federal Water Pollution Control Act (FWPCA) as amended in 1977.

Manuscript originally printed in the Proceedings of the National Conference on control of Hazardous material spills, April 1978. Available from Information Transfer Inc., 9300 Columbia Blvd., Silver Spring, MD 20910

INTRODUCTION

     At present exists no comprehensive set of Federal
regulations designed to prevent spillage of hazardous substances
(chemicals) into the Nation's navigable waters. However, it is
anticipated that in the near future the Environmental Protection
Agency (EPA) will issue regulations patterned, at least in
part, after the Oil Spill Prevention, Control and Countermeasures (SPCC) regulations established in January, 1974.

     Implementation of the SPCC regulations in 1974 presented
the 10 EPA Regional Offices with the task of monitoring
thousands of facilities for regulation compliance. Limited
manpower resources to cover the tasks of the Regions greatly
restricted effective compliance monitoring. At the request of
the EPA's Oil and Special Materials Control Division (OSMCD) in
Washington D.C., the Environmental Monitoring and Support
Laboratory (EMSL-LV) in Las Vegas, Nevada, implemented an
aerial photographic reconnaissance program to augment the SPCC
compliance monitoring efforts of each of the EPA regions. In
the first years and a half of operations, EMSL-LV conducted
aerial surveys in 9 of the 10 EPA regions. These surveys
covered 56 large refineries, 235 bulk oil storage terminals,
100 chemical-related facilities, 29 power utilities, 23 rail-
road yards, and nearly 150 other industrial sites having some
type of petroleum product storage.

     This program of applying remote sensing technology to
assist SPCC compliance monitoring efforts has been
enthusiastically welcomed by most of the EPA Regions and has
led to requests for additional assistance. As an example,
aerial photography obtained in the San Francisco Bay area,
EPA Region IX, indicated some 90 instances of potential SPCC
violations. Follow-up ground inspections found 31 violations
and appropriate citations were issued. Probably the greatest
value of the remote sensing program is the potential savings
in ground inspection time. Utilizing the remotely sensed data,
an inspector can concentrate on those facilities or those
locations within a facility that appear to be out of compliance
and thereby avoid those facilities that appear to be in
compliance.

     In 1972, EPA produced an interpretation key to assist
in monitoring oil spill conditions in petroleum refineries
and in 1974 it was recognized that interpretation keys would
be essential in order to adequately exploit aerial photography
over chemical production and storage facilities. With the
sucess of these oil SPCC monitoring support efforts, it is
apparent the same remote sensing techniques could be applied
to monitoring the same or similar spill and spill threat

conditions within chemical storage and production facilities.
When viewed through the lens of an aerial camera, storage and
holding tanks, secondary containment structures, pipelines,
and product transfer structures are essentially the same for
both oil and chemical facilities. In light of these similarities
the Las Vegas Laboratory, under the auspices of the OSMCD, has
initiated a project to produce aerial photographic examples of
chemical spills and spill threat conditions, in particular,
engineering deficiencies in storage and secondary containment
structures that can be photographed from an airplane.

The purpose of these photographic keys is to assist both
experienced photo interpreters and on-site inspectors in
recognizing structural deficiencies and other spill threat
conditions when viewing and interpreting aerial photographs of
a chemical facility. When hazardous substance spill prevention
regulations are implemented, the Las Vegas Laboratory
anticipates an increase in reconnaissance support requirements
from the Regional EPA Offices.

AERIAL RECONNAISSANCE OF CHEMICAL FACILITIES

The typical aerial reconnaissance operation over a
chemical facility is patterned after the technique used for
oil SPCC reconnaissance. The reconnaissance aircraft is
equipped with a standard aerial mapping camera with a 6-inch
color corrected lens and a 9-inch film format. A high speed,
color reversal film (KODAK SO-397) is normally used over oil
and chemical facilities. Experience has shown that color film
is essential to differenciate many spilled products and
product stains from water and/or wet ground. The aircraft
operates at altitudes ranging from 2,500 feet to 4,200 feet
above the ground producing image scales from 1 inch to 417
feet to 1 inch to 700 feet. Color photographs taken at higher
altitudes are incapable of resolving the spatial parameters
of the engineering deficiencies under surveillance.

Photo-interpretation is performed on the film transparen-
cies with the assistance of zoom-stereomicroscopes. The micro-
scopes not only provide the necessary magnification but also
allow viewing each facility in three dimensions. With this
analysis capability, the adequacy and integrity of containment
structures and the slope and direction of surface drainage can
be properly assessed.

SPILL AND SPILL THREAT CONDITIONS

The primary facilities within a chemical storage and
production complex that are amenable to aerial surveillance
include:

(1) Product storage facilities such as storage and holding tanks, and dry product stock piles;

(2) Product transfer facilities such as marine and river terminals, tank car and truck racks, in-plant pipe-lines and dry product conveyors;

(3) Plant surface drainage and waste treatment facilities.

It is recognized that materials produced, stored, processed and transported within a chemical complex are not all "designated hazardous substances". However, the contents of tanks, pipes, etc., cannot be determined directly from photography, thus all anomalies detectable from the film are considered as potential spill condition.

Specific features that can be readily detected in color aerial photography can be divided or grouped into two broad categories. The first group of features or conditions are those that relate directly to a spill or spill threat. The list of these features includes:

* the total absence of secondary containment;
* a breach in secondary containment;
* inadequate secondary containment;
* product spillage and/or stains outside a containment structure;
* leaking or deteriorating tanks;
* point of entry of product spill into the natural drainage system;
* damaged vegetation.

In the group are features which, while not directly related to spill conditions, they are often the site of numerous spills and are considered high spill risk areas. These areas warrant careful scruting when the aerial film is interpreted. Within this group are such features as:

* tank car and truck loading racks;
* large drum storage areas;
* wastewater treatment facilities and waste dumps;
* wastewater outfalls.

Storage tanks and the condition of their associated containment structures are the principal targets for a spill reconnaissance. Most secondary containments are earthen dikes that are readily discernable in the aerial photography. The structural and containment integrity of most of these dikes can be assessed with little difficulty. In a more confined situation such as a processing area, containment may consist of concrete or steel walls. These are more difficult to assess because of their smaller size and they are often obscured by shadows from nearby tanks or other structures.

The total absence of containment structure can be determined easily. Where no containment structure is visible, the
image analyst determines if a spill of any significance from
the tank would be confined to the facility property or enter
into the natural drainage system. Breached dikes are not un-
common, many are deliberately breached to facilitate entrance
of vehicles and equipment for tank repair or construction.
However, when evidence of construction or repair activities
are absent, a breached dike is considered evidence of a spill
threat.

Adequacy of a containment dike is based on the condition
of the dike, i.e. badly eroded or in poor condition, and
judgement as to the capacity of the dike to hold at least
100% of the volume of the largest tank within the dike. In
addition, large amounts of standing water within a containment
dike reduces its capacity to contain a product spill, thereby
posing a spill threat.

Many dry (non-liquid or gaseous) products and raw
materials found within a chemical production facility are
often stored in the open with little more than a temporary
wall erected as a containment structure. Spills can occur from
overloading the structure, collapse of the structure, or
simply leaching and wash-out during heavy rains. Evidence of
spills of this nature are often discernable with aerial photography.

Spills stains outside secondary containments are not
always evidence of a threat to the environment. Spills stains
are common throughout chemical production facilities. Most are
small and confined to areas with adequate protection to the
drainage system. However, where spill stains are numerous and
heavy, close scruting of the film is warranted to determine
the source of the spills and whether or not any spillage has
entered the natural drainage system.

Leaking or deteriorating tanks are easily detected on the
color film; however, no spill threat is posed if the tanks are
adequately contained and there is no evidence that leakage is
escaping the containment by way of the drain valve or gate.

All drainage features in the vicinity of production and
storage facilities warrant close surveillance. Evidence of
product leakage or spillage from the facility may include a
sheen, slick or discoloration on the water surface. Once the
point of entry of the discharge is identified, nearby
facilities are carefully scrutinized in order to detect the
potential sources.

Figure 1.   Spill conditions associated with dry product storage.

        Damaged or dead vegetation is often evidence of a product
spill or leakage from nearby facilities. Damaged vegetation can
be readily seen in color photography and is a good indicator of
a present or past spill condition.

        Numerous product spills and resultant stains are quite
common at points of transfer, in particular at tank car and
truck loading racks, and river and marine terminals. The
loading racks require careful inspection to verify that large
spills would not pose a threat to nearby drainage. River and
marine terminals pose the most direct threat to the drainage
system.

        The film is analyzed carefully to determine the presence
of spill prevention features. Containment booms (for floatable
substances) should be in evidence at river and marine terminals
either deployed or located nearby for quick deployment should
a spill occur. Boom deployment often requires a motor launch
and one should be located nearby in case of an emergency. Wood
plank piers or docks offer little containment should a hose or
pipe rupture on them. Concrete piers with an elevated apron
provide better containment for potential spills. Drip pans are
also essential, but because of their small size, they generally cannot be seen in the photographs.

Drum storage areas are usually the site of many spill
stains, yet more often than not, these areas have no secondary
containment structures. Therefore, once a drum storage area is
detected in the photography, the surrounding areas are carefully searched for evidence of spillage or leakage reaching the
nearby drainage. Wastewater treatment facilities and waste
dumps are carefully inspected to ensure that waste material has
not leaked, leached, or otherwise spilled into the nearby drainage system. Damaged vegetation in the vicinity of these facilities often indicates the presence of product spillage. All
waste-water outfalls are also inspected to see if they emanate
from the waste treatment facility, or possibly an untreated
source.

PHOTOGRAPHIC KEYS

The figures are examples of typical aerial photographic
keys of product spills and spill threat conditions in chemical
production facilities along the lower Delaware River. The facilities were overflown in August 1976.

The large facility shown in Figure 1 produces plastics,
resins and general compounded products. This overview of the

Figure 2. Waste treatment and drainage facilities at a large
          chemical processing plant.

plant shows it is bounded by water on two sides, and any major
spill here would likely enter the river. The two largest tanks
at this plant have no secondary containment. A closer look of
at a small part of the facility (Figure 2) reveals additional
product spills and spill threat conditions. Product is being
transferred between the river barge and terminal. Note that
containment booms are not deployed around the barge and none
are in evidence on the pier. Spill stains are clearly evident
on the pier. The four large tank barges are serving as heated
product storage containers (venting steam is visible) without
any secondary containment. Note the spill stains on the barge
surface clearly indicating that product spillage has reached
the river. Spill stains at the tank car rack are clearly
evident. There is no containment evident at this rack and a
large spill could be easily reach the river.

Heavy product spillage and associated vegetation damage
are evident throughout the industrial chemicals production
facility photographed in Figure 3. Note the uncontained tanks
and the dry product storage area. A retaining wall on two
sides contains much of the dry bulk product, but two breaches
in the wall can be seen. There is extensive dead vegetation
near the wall and it is likely that some of this product has
leached into the nearby river.

Extensive waste treatment facilities can be seen at the

Figure 3. Overview of a large chemical manufacturing facility.

Figure 4.   An enlarged or closeup view of a small section of
            the chemical plant.

large plastics and resins manufacturing facility in Figure 4.
However, much of the plant surface drainage bypasses the treatment facilities and is discharged into the river. Note the dis-
colored water at the point of discharge. Spill stains can be
seen near the drainage ditch and several tanks are without
benefit of secondary containment. Tanks with adequate containment are also shown.

CONCLUSIONS

     A significantly large proportion of any hazardous material
production and storage structures can be photographed from the
air, and many facilities can be photographed in a single day.
Color film is well suited for recording the presence and structural integrity of secondary containment structures, for identifying product spill stains and vegetation damage resulting
from product spills, and for recording evidence of spillage
and leakage entering nearby water bodies.

     The use of color aerial photography as a reconnaissance

tool also has great potential for improving the efficiency of
federal, state and local inspection efforts. After carefully
reviewing the photography, an inspection itinerary can be set
up to reduce the time needed to conduct an inspection by con-
centrating on those facilities and areas within facilities
with suspected spill problems.

It seems apparent that industry or individual companies
could utilize these same techniques to conduct their own reconnaissance inspections to assess the adequacy of spill prevention measures for a facility as a whole as well as for individual components within the plant.

AIREYE: A NEW GENERATION OIL POLLUTION SENSING

FOR THE 80'S*

J.R. White and R.E. Schmidt

U.S. Coast Guard Headquarters
Washington, D.C., United States

ABSTRACT

The U.S. Coast Guard is developing an airborne, real-time, all-weather, day/night remote sensing system that will detect oil pollutants and identify violating vessels. The system has been designated AIREYE and will be installed on six of the 41 new FALCON 20G jet aircraft (military designation HU-25A) which the Coast Guard has purchased to replace the aging HU-16E Grumann ALBATROSS as its medium range surveillance aircraft. The sensor system will include a side-looking airborne radar, two channel infrared/ultraviolet line scanner, aerial reconnaissance camera, airborne data annotation system, and a control, display and record console. To identify polluting vessels at night, an active gated television (AGTV) is also being developed for inclusion in the Aireye system. The AGTV will use a one watt, pulsed, lead vapor laser illuminator and will be capable of recording vessel names at night from a slant range of 500 meters. In addition to an active and passive mode, the AGTV will be capable of both computer and manual target acquisition and tracking. Each of the sensors will produce annotated, hard-copy imagery suitable for prosecution of polluting vessels.

THE AIREYE SYSTEM

Early research in the infrared, ultraviolet, visible and

---

* The opinions or assertions expressed herein are those of the authors and do not necessarely represent the views of the U.S. Government.

Table 1.   HU-25A characteristics.

Length	17 m	Max. take off gross weight	14.5 t
Wind span	16 m	Internal cockpit/cabin vol.	21 m^3
Height	5.4 m	Endurance at 320 km/h	4 h
Wing area	40.5 Sq m	Dash speed	670 km/h
Engines	2 Garrett ATF		

microwave areas of the electromagnetic spectrum demonstrated
that the detection, identification and quantification of oil
spills was feasible with remote sensing techniques but that a
multisensor system would be required to achieve all-weather
and lighting conditions performance (White and Breslau, 1978).
These initial investigations led to the development of the
Airborne Oil Surveillance System (Edgerton et al., 1975) and
its subsequent modification and installation in a Coast Guard
HC-13 OB aircraft (Meeks et al., 1977). This experimental
system demonstrated that a multi-sensor airborne system could
reliably detect and map oil spills in most weather and lighting
conditions.

Because of the success of the experimental system, the
Coast Guard decided to develop an operational remote sensing
system for installation on its new medium range aircraft.
Current plans are to acquire six Aireye sensing systems, which
will have the capability to detect, and map oil spills, and to
identify illegal discharging vessels. The aircraft are
modified FALCON 20 G's (military designation HU-25A), which
are currently being purchased by the Coast Guard. Table 1 gives
a summary of the HU-25A characteristics.

Each of the 41 aircraft will come equipped with complete
provisions to carry the Aireye system. These provisions include
such things as mounting points for the sensor pods, an optical
window for the camera and a console to hold sensor equipment.
The Aireye system is being designed to be easily transferred
from one aircraft to an other in less than 4 hours. Because of
the relatively small size of the aircraft, Aireye will be
limited to a weight of 720 kilograms, and to an internal volume
of 2.13 cubic meters.

A competitive sollicitation was released in November 1978
to obtain a contractor to assemble, and install the first
system. This contract was awarded in August 1980, and flight
testing should begin in mid-1983.

Aireye will include (Figures 1 and 2)
          * Side looking airborne radar (SLAR)

Figure 1. The Aireye system: short range sensors.

* Infrared and ultraviolet line scanner (IR/UV)
* Aerial reconnaissance camera
* Airborne data annotation system (ADAS)
* Active gated television (AGTV)
* Surveillance system operator console (SSO)

Surveillance System Operator Console (SSO Console)

The SSO console (Figure 3) is the control point for the integrated sensor system. The "heart" of the SSO console is the sensor computer. The sensor computer will perform 5 primary functions. They are:
* Automatic pointing of the active gated television;
* Generation of maps displaying sensor and "target" position;

Figure 2. The Aireye system: long range sensor.

* Implementation of display and control functions;
* Digital interface of the data annotation system to the aircraft data buss;
* Monitoring of sensor and system performance and failure alerting.

The front-panel functions to be located on the SSO console are:
* Trackball (used to make small "corrections" to television pointing while in the video-track mode, and to position computer generated cursors/vectors on the multipurpose display);
* Joystick (manual television camera pointing);
* Active gated television control panel (high-use controls);
* Side-looking airborne radar control panel;
* Active gated television power control panel (low-use controls);
* System velocity/altitude (V/H) signal control panel;
* Reconnaissance camera control panel;
* Infrared/ultraviolet line scanner control panel;
* Multipurpose display (MPD) control panel;
* Control display unit (CDU).

This control and display unit (CDU) is used by the operator

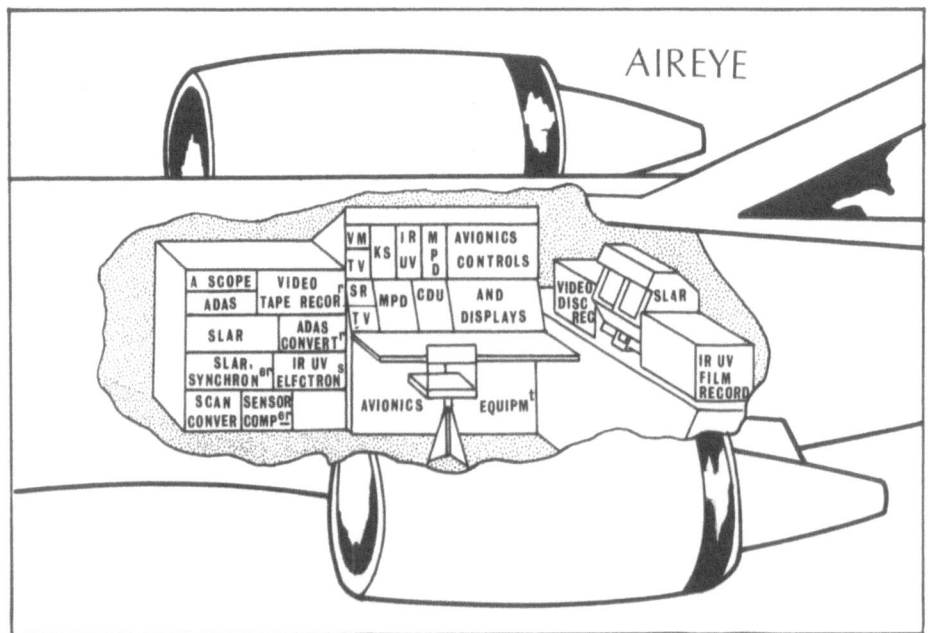

Figure 3. The surveillance system operator (SSO) console.

to enter alphanumeric data into the computer. The CDU can also
display abbreviated maps as well as alphanumeric data and is
physically identical to the aircraft avionics system CDU
located in the cockpit.

The display functions which are available to the sensor
operator at the SSO console are:
* Real-time imagery from the side-looking radar, infrared
  line scanner, or ultraviolet line scanner, displayed in
  the form of a moving map in the multipurpose display;
* Real-time television imagery will also be available on
  the multipurpose display; Figure 4 shows the wide angle
  acquisition mode of the camera during which the target
  is selected. Figure 5 depicts the shift to a narrow
  field of view which allows identification of the target.
* A computer generated map can be called by the control
  and display unit and presented on the multipurpose
  display. It will depict relative positions of up to 10
  search radar targets and 10 aircraft waypoints.
* A cursor, displayed as a unique symbol, such a triangle,
  will be used to designate important targets. The sensor
  system operator will be able to move the cursor anywhere
  on the multipurpose display using the trackball (Fig. 6).
  The data block in the upper left corner indicates that
  the target is 15 kilometers from the sensor aircraft.
  The data block also indicates the relative and magnetic
  bearing from the aircraft, and the longitude/latitude
  position of the tanker. The data block in the lower
  right corner is used to provide the heading and speed
  of a television target;
* A vector can be displayed as a line with one end fixed,

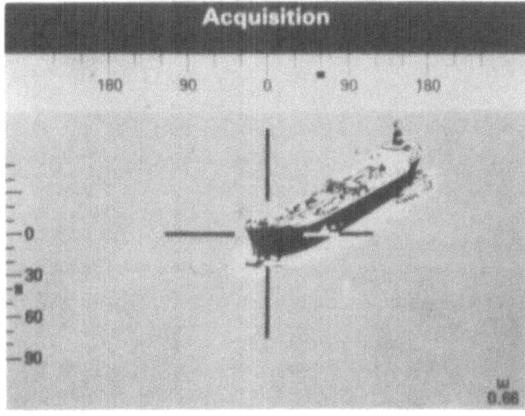

Figure 4.   The wide angle acquisition mode of the TV camera.

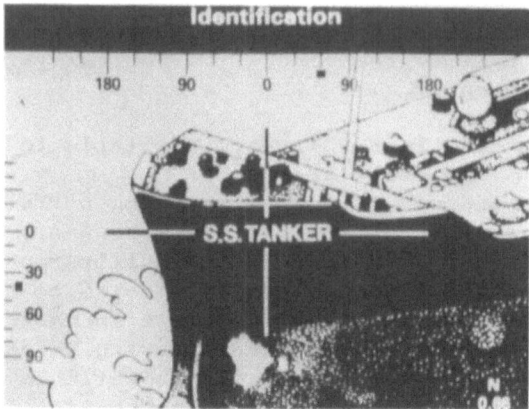

Figure 5.   Identification of a vessel by means of the TV camera.

normally at the sensor aircraft position, and the other
end varying with the cursor;
* Full resolution stop action/frame freeze recording and
playback of any video displayed on the multipurpose
display will permit frame-by-frame examination of the
imagery.

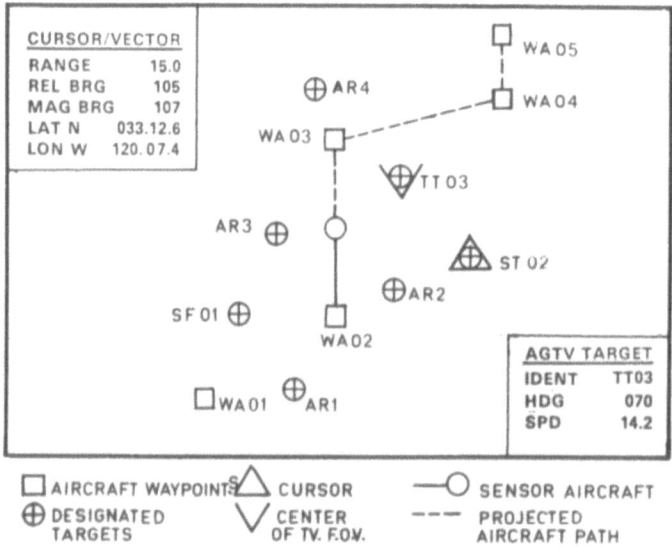

Figure 6.   Computer generated map on the multipurpose display:
relative positions of radar targets and aircraft
waypoints.

Table 2. Principal SLAR parameters.

Peak transmitter power	200 kW
Frequency	9250 MHz
	(center of 9210 to 9290 MHz tuning range)
Pulse width	0.2 µs
Pulse repetition frequency	750 pulses s^{-1}
Antenna characteristics	
Azimuthal half-power beamwidth	0.9° one-way
Elevation shaped pattern coverage	-1.5 to -45°
Peak gain	35 dB
Depression angle of beam peak	1.5°
Receiver noise figure (referred to GaAs FET preamplifier input)	4.16 dB
Receiver bandwidth	6 MHz
Transmitting microwave losses (from magnetron to antenna)	2.2 dB
Receiving microwave losses (from antenna to GaAs FET preamplifier input)	1.95 dB
Ranges available	24, 32, 64 and 128 kilometers

Side-looking Airborne Radar (SLAR)

The SLAR to be integrated into the Aireye system is a new generation of the AN/APS-94 radar which has been in the Department of Defence inventory for 15 years in various versions. It is currently being manufactured by Motorola Inc. The principal improvements in the Aireye SLAR over that used in previous Coast Guard sensor systems are:

* The peak power of the side-looking airborne radar has been increased from 40 kilowatts to 200 kilowatts;
* A microprocessor has been incorporated to the control radar system timing circuits, built-in-test equipment circuits, and longitude/latitude calculations;
* A dry-silver film processor will replace the wet processor of previous systems;
* Longitude/latitude grid lines will be superimposed onto the film imagery (Figure 7);
* A new 2.4 m antenna and pod designed to be physically compatible with the HU-25A aircraft. The antenna will be a back-to-back dual array which will allow simultaneously mapping on both sides of the aircraft. It will be yaw stabilized to $\pm$ 3°.

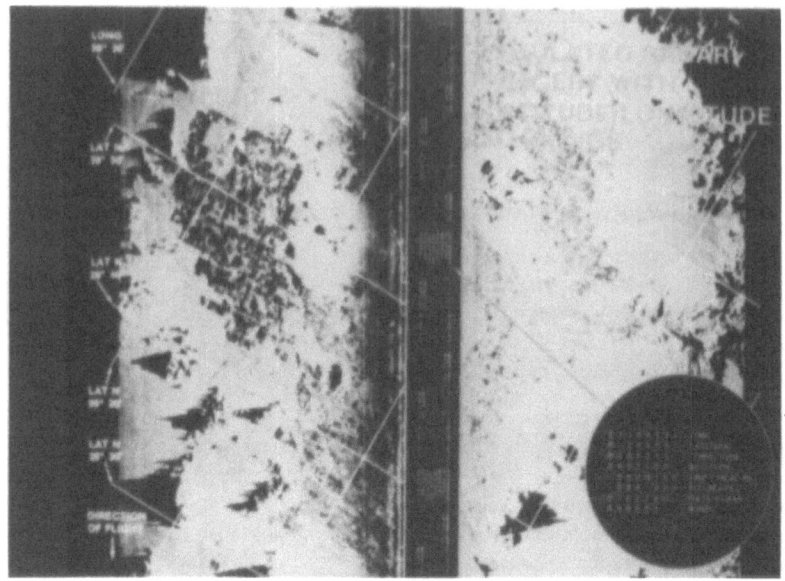

Figure 7.   Side-looking airborne radar (SLAR)
             Imagery annotation format.

The SLAR has been the backbone of past generation Coast
Guard integrated sensor systems, and is expected to continue as
the key sensor in the system. The operating characteristics
which distinguish the SLAR as the prime Aireye sensor are:

Long range: detection of oil slicks at distances up to
16 km has been demonstrated on past SLAR systems. Oil slick
detection range with the new generation Aireye SLAR is estimated
to be between 24 and 48 km depending on sea state and aircraft
altitude. This could provide an oil detection swath of up to
80 km.
    All-weather, day/night operation: the SLAR can produce
good imagery in most weather conditions and is unaffected by
clouds or darkness. Since the detection of oil slicks by radar
depends on the damping of capillary waves by the oil, slicks
normally cannot be detected on flat calm seas or on heavy seas.
    Wide area mapping: the output imagery of the SLAR is a
wide area map on 23 cm dry-silver film, and video for presentation on the multipurpose display.

Table 3.   IR/UV line scanner operating parameters.

Infrared response band	7.6 - 13.8 μm
Infrared detector angular resolution (DAR)	2.5 mrad
Infrared noise equivalent temperature	0.2°K
difference (NETD)	0.2°K
Ultraviolet response band	0.32 - 0.40 μm
Ultraviolet detector angular resolution	5.0 mrad
Ultraviolet channel sensitivity	
(measured at output of video amplifier)	17 dB SNR*
Altitude operating range	150/3,600 m
Velocity/altitude (V/H) ratio for	0.20 radian/s
contiguous line-to-line IR ground coverage	

* For scene irradiance of 0.2 W/m^2 and target reflectance of
  20%

Infrared/Ultraviolet (IR/UV) Line Scanner

The IR/UV line scanner to be integrated into the Aireye
system was specified to be operationally similar to the RS-18
line scanner successfully deployed on AOSS-II. The IR/UV line
scanner has a two-fold purpose on Aireye. It is used to scan
the area directly below the aircraft that is missed by the
side-looking airborne radar and is used to provide multispectral
information on any target of interest that is detected by the
SLAR. With data from three portions of the electromagnetic
spectrum, wake scars and kelp beds can be differentiated from
oil slicks. The output imagery from the IR/UV line scanner will
be recorded on film and also can be shown in real time on the
multipurpose display. The optical/mechanical portion of the
line scanner will be mounted in a pod carried on hardpoints
on the starboard aircraft wing. A roll stabilization system
will provide both IR and UV channels with compensation for aircraft roll manoeuvers of ± 15 degrees. A closed-cycle cryostat/
cryogenic-cooler will cool the mercury-cadmium telluride (Hg
Cd Te) infrared detector elements to their optimum operating
temperature.

Table 3 lists the principal operating parameters specified
for the IR/UV line scanner.

Airborne Data Annotation System (ADAS)

The ADAS is a key element in the production of imagery
that can be used to prosecute pollution law violators. The ADAS

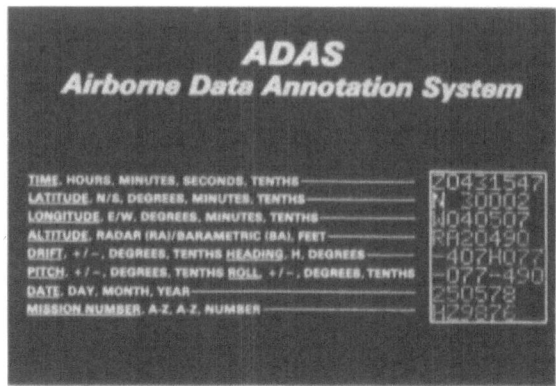

Figure 8.   Airborne data annotation system (ADAS).

generates an alphanumeric data block (Figure 8) that will
annotate IR/UV line scanner and reconnaissance camera film.
(The SLAR has integral provisions for data annotation and the
computer will generate the annotation for the multipurpose
display, and thereby the video tape recorded from it). The
AN/ASQ-154 airborne data annotation system, being procured as
part of a large Air Force contract, will be provided to the
integration contractor as Government-furnished equipment (GFE).

The following flight parameters will be annotated on the
IR/UV line scanner and reconnaissance camera film:
> Barometric altitude
> Drift
> Mission number
> Heading
> Longitude
> Latitude
> Pitch
> Radar altitude
> Roll
> Time
> Date

Self-test and built-in-test provisions are being incorporated into the ADAS to monitor principal voltage levels and
other critical equipment parameters. The system integrator is
being tasked to interface the ADAS with the aircraft navigational data buss and the annotated sensors.

Aerial Reconnaissance Camera

The aerial reconnaissance camera will be a KS-87B camera

Table 4. KS-87B camera characteristics.

Lens			
Cones	15 cm	30 cm	45 cm
Focal length	15 cm	30 cm	45 cm
Resolution	50 lines/mm	45 lines/mm	65 lines/mm
View angle	41° 6'	21° 14'	14° 14'
Number of exposures	maximum of 6 per second		
Film format	11.4 cm by 11.4 cm		
Effective shutter speeds	variable 1/60 to 1/3,000 s		

which is presently in Department of Defence inventory. It is
produced by Chicago Aerial Industries and is a pulse operated
frame camera containing interchangeable lens cones, high speed
focal plane shutter, integral automatic exposure control, data
recording and forward motion compensation. The film handling
system will be a combination cassette-magazine drive capable
of operation with full 150 m standard film spools. An optical
window in the HU-25A will permit operation of the camera in two
positions, either vertical or 60° oblique from the vertical.
Table 4 lists detailed camera characteristics.

Active Gated Television (AGTV)

The AGTV sensor is required to provide night-time identification of polluting vessels and their activities. It is being
developed by General Electric's Aerospace Electronic Systems
Dept. in Utica, New York.

The AGTV is a small, light weight, low power system that
will be mounted in a pod on the left wing of the aircraft. It
will have a 39 cm turret with a stabilized pointing system.
In the passive mode, it will function as a low light level
television and have a maximum field of view of 18°. It will be
capable of detecting deck lights of 320 candella at a range of
21 kilometers. As the aircraft approaches the ship at approximately 150 m altitude and a speed of 250 km/h, the field of
view is reduced to 6.5° and finally to 2.3°. At a slant range
of 4 km, the AGTV is switched to the active mode, and at this
range, small ships (6 x 18 m) can be observed.

The intensifier assembly in the camera head module consists of a second generation proximity focus intensifier,
fiber-optically mated to a first generation intensifier. The
Coast Guard is evaluating the possibility of incorporating the
newly developed third generation intensifier tube into the AGTV.

Table 5.   AGTV characteristics.

Gimbal coverage	+15° - 84° elevation
	± 220° azimuth
Angular velocity	60°/second (slow 100°/second)
Drift	less than 0.1°/s
Diagonal field of view	18.2° / 6.5° / 2.3°
Output power at 855 nm	0.8 watt at 15.75 KHz PRF
Resolution	28.5 lines/mm at 1 x 10^{-4} ft-candle
Photocathode response at illuminator wavelength	30 MA/W
Weight (with pod)	173 kg
Power	214 W, 28 VDC
	104 W, 110 VAC, 400 Hz

The increased sensitivity of a third generation intensifier would result in an increase in range performance of 15 to 20% .

In the active mode, illumination is provided by a 0.8 watt, pulsed gallium arsenide laser operating at 855 nanometers (nm). The laser pulse width is 300 nanoseconds and the laser beam pattern is 10 x 10 milliradians. The camera operated as a gated receiver. After the laser is pulsed, the camera waits a period of time proportional to the range to the target, and then the camera intensifier is gated "on" to receive the reflected light from the target. The range information can be provided either manually or by the Aireye computer. At a slant range of 500 m, 30 cm high letters can be resolved and 15 cm high letters are readable at 300 m. The imagery will be presented on the Aireye multipurpose display and recorded on a video tape recorder and a video disc recorder. This will enable the operator to play back the passes on the ship and "freeze" a frame to identify it.

The AGTV can be operated in either a computer track, manual track or automatic track mode. In the computer track mode, the gimbal angle is controlled by the Aireye computer after adjusting for the aircraft motion. In the manual mode, the sensor operator has direct control of pointing the AGTV through either a track ball or joy-stick. In the automatic track mode, a correlation tracker in effect memorizes a scene and stores its location in a digital memory. As the aircraft moves, the location of the scene is moved to remain aligned with the new information. Error signals are generated which move the gimbals to keep the tracked scene in view. Targets can be tracked at rates up to one radian per second. A corre-lation tracker is particularly necessary for the Aireye system

because it must be capable of "locking on" to the name of the
vessel which may not provide a sharp edge or boundary required
for a more simple edge or centroid tracker. The correlation
tracker can maintain track through mode transitions and on
targets with only 15% contrast

Detailed characteristics of the AGTV are listed in Table 5.

(This contribution was presented at the Washington, D.C. workshop and revised by the authors in 1983, to be included in this
volume)

AIRBORNE OIL SPILL SURVEILLANCE SYSTEMS

IN SWEDEN

L. Backlund

Swedish Space Corporation
Sölna, Sweden

ABSTRACT

Two different oil spill surveillance systems have been
developed by the Swedish Space Corporation for the Swedish
Coast Guard and the Swedish Board for Space Activities. These
systems are installed in two Coast Guard Cessna 337 aircraft.
One of the systems is based on a side-looking airborne radar,
developed by L.M. Ericsson Telephone Company under contract
to the Swedish Space Corporation, and the other on a Daedalus
infrared/ultraviolet line scanner. In addition to the remote
sensing equipment, both aircraft have a Decca tactical air
navigational system and a camera system on board. The technical
evaluation and field testing of the two surveillance systems
was carried out during 1978. The systems gradually will be put
into operational use by the Swedish Coast Guard for coastal
surveillance.

BACKGROUND

The Swedish Space Corporation (SSC) is a state owned
company. SSC is, among other things, responsible for technical
execution of the national space science and remote sensing
programmes. This task includes both in-house research and
development as well as award of contracts in these fields to
Swedish industry and other agencies. To promote the use of
remote sensing techniques in Sweden, priority has been given
to the development of operational remote sensing systems for a
few defined applications. Ocean and coastal surveillance is
one of those, and at an early stage oil spill monitoring became
the prime area of interest.

Many field experiments both in Sweden and in other countries have been carried out to test different sensors for oil
spill detection under varying conditions and to verify calculated performance. As a result of studies, field experiments
and operational requirements, two different prototype airborne oil spill surveillance systems have been developed by
the Swedish Space Corporation for the Swedish Coast Guard and
the Swedish Board for Space Activities. One of the systems
uses a SLAR as sensor while the other uses an IR/UV line
scanner. The systems are installed on two Coast Guard Cessna
337 aircraft (Figure 1). The SLAR system was developed by the
L M Ericsson Telephone Company under contract to the Swedish
Space Corporation. The IR/UV scanner system was developed in
house by the SSC.

SYSTEM DESCRIPTION

Aircraft

Since 1976, the Swedish Coast Guard has operated two
Cessna 337 Skymaster aircraft (Figure 1). Both have tactical
air navigation system (TANS) computer, and a camera system.
Additionally, one remote sensing system has been installed on
each aircraft. The Decca TANS provides accurate information
on aircraft position. Time of day, date, aircraft heading,

Figure 1. The Swedish Coast Guard two Cessna 337 surveillance
           aircraft. The box under the fuselage of the left
           aircraft contains the IR/UV scanner; the "cigar"
           under the right aircraft contains the 3 m antenna
           for the side-looking airborne radar (SLAR).

Figure 2. Camera system.

mission identification and other information are available from
other units in the aircraft. Thus, a complete set of navigational and house-keeping data (nav-data) is provided for unambiguous identification and co-ordination of all imagery and
recording from both cameras and remote sensing systems. The
camera system superimposes nav-data on each frame. The remote
sensing systems record and display all relevant nav-data together with the generated imagery.

     In general, the Coast Guard operates the aircraft with
only two pilots. Thus, the co-pilot also has the tasks of observer, operator for the remote sensing system, and photogra-
pher. The demand for easy-to-operate functions and easily
understood characteristics of all equipment is, therefore,
obvious. A typical flight surveillance mission lasts three to
five hours. Both aircraft operate five days a week.

Camera System

     The camera system has one hand-held camera for oblique
pictures and one vertical camera. Both have winders, automatic
exposure systems and a data box which superimposes all nav-data
onto the film (Figure 2). The hand-held camera uses 36-expo-
sures film; the vertical camera has a large magazine for 250

Figure 3.   A vessel illegally cleaning its tanks in the Baltic
            Sea. The photo is taken with the hand-held camera.
            The vessel is seen at a distance of about 4 nautical
            miles (about 7.5 km). The interpretation of the data
            block is shown below (Figure 4).

exposures. Both cameras use 24-by-36 mm miniature film.

    Operational experience proves that the cameras are indis-
pensable for routine surveillance, identification and documentation. Figure 3 illustrates use of the hand-held camera. The
nav-data block, found in the upper-right of the photograph is
interpreted as shown in Figure 4.

78 04 18	2	Year, Month, Day; Coast Guard Region	
15 06 44	2	Hour, Minute, Second; Camera ID: 1 vertical, 2 hand-held, 3 external	
59 174	088	Latitude, Degrees, Minutes and 1/10 Minutes $59^\circ 17.4'$; Mission ID	
18 524	064	Longitude, Degrees, Minutes and 1/10 Minutes $18^\circ 52.4'$; Exposure ID	

Figure 4.   Interpretation of the data block shown in Fig. 3.

Side-looking airborne radar

The basic design philosophy behind this prototype system
was to stress simplicity, economy and ease of handling rather
than technical perfection and maximum performances. The antenna pod, about three meters long and 15 centimeters in diameter,
is mounted under the fuselage looking to the right (Figure 1).
The rest of the system is installed in the cabin.

One of the design requirements was to display real-time
imagery from the SLAR with high grey level dynamics and at low
cost. The tradional methods of recording SLAR imagery on photographic film were, therefore, ruled out and the system was configured around a digital signal processor and a "rolling map"
black-and-white television display.

At the top of all SLAR imagery (Figure 5) is a calibration
grey scale which is a read-out of the 64 levels of radar video
quantization. To the left is the alphanumeric display of data:

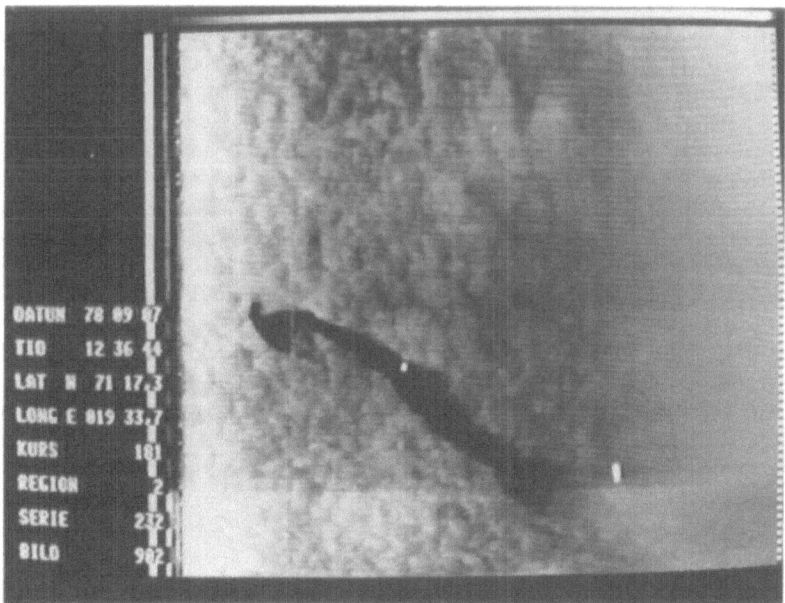

Figure 5.   Radar image of an oil slick from a controlled dis-
            charge in the North Atlantic. The slick was dumped
            24 hours ago and contains 25 tons of Ekofisk crude
            oil. Wind 4-7 m/s. The length of the slick is about
            10 km and the width 1-2 km. The white spot within
            the oil slick is a research ship while the white
            spot lower right is a trawler.

date, time, latitude, longitude, heading, Coast Guard region,
mission number and picture number. The aircraft position is at
the upper left of the picture. The area displayed is 20 by 20
kilometers.

Some technical data on the SLAR system:

* Type   Real aperture side-looking airborne radar
* Antenna   Slotted waveguide, 3 meters long, unstabi-
              lized, vertical polarization.
* Frequency   X-band, 9.4 GHz
* Peak power 10 kW
* Display   Real-time TV display 20 by 20 km.

Flight Test Program: The SLAR system was installed in
January 1978 and was successfully flight-tested during the
rest of the year. The equipment has remained in the aircraft
to gather further experience and gradually is being put into
regular use by the Swedish Coast Guard.

The flight test program has proven the system to be a
valuable tool for coastal surveillance. The major applications
so far tested have been oil spill detection, sea-ice reconnaissance, ship detection, monitoring of fishing activities,
and search and rescue operations.

A general conclusion is that the performance of a relatively low-cost real aperture SLAR is sufficient to fulfill
the requirements for coastal surveillance. The SLAR is able to
produce excellent radar maps and experience shows that these
can be achieved even with the radar installed in a small aircraft and without sophisticated antenna stabilization.

An oil spill makes the sea surface considerably smoother
than the surface of the surrounding unpolluted water. The
smooth surface gives less radar backscatter and oil slick will
appear as a dark area on the radar image. The SLAR has been
flown several times in field experiments with controlled oil
discharges. Figure 5 shows recording from one of them.

Infrared/ultraviolet Scanner System

Field experiments using an IR/UV scanner for oil spill
detection have been carried out several times since 1972. As
experience showed that the technique ought to be operationally
applicable, a decision was made in early 1977 to develop a
prototype system adapted as far as possible to Coast Guard
requirements. The system is intended for operation in clear
weather, the IR channel being available both day and night and
UV channel only during day-time (Figure 6).

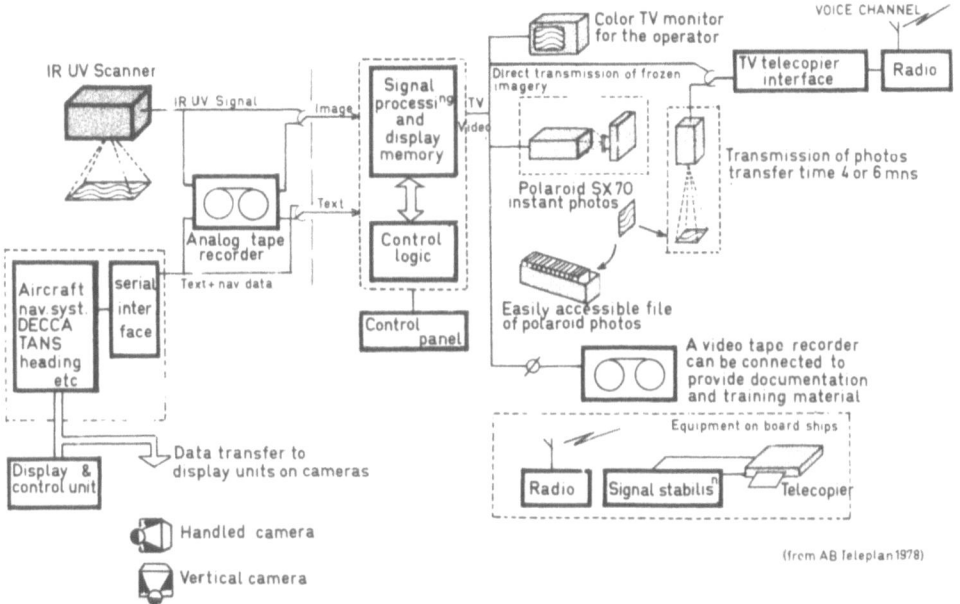

Figure 6.   IR/UV scanner system - Block diagram.

The scanner is a Daedalus DS-1200 series line scanner system. Outputs from the scanner system are analog IR and UV channel signals.

Imagery from the system is presented to the operator and co-pilot on a color TV monitor. The scanner and nav-data signals are processed either in real time or from the analog tape recorder through the signal processor and display memory.

In parallel with the operator's monitor, a small five-inch black-and-white monitor is built together with a Polaroid SX-70 instant camera. By this arrangement, the operator may easily take polaroid photos of registrations that need to be documented.

The picture shown on the operator's monitor can be transmitted from the aircraft to, for example, a Coast Guard vessel in a clean-up operation. Transmission is made via voice channel link. The transfer line is four or six minutes. A video tape recorder can be connected parallel with the operator's monitor to provide documentation and training material.

Some technical data on the IR/UV line scanner are listed below:

* IR channel   8 to 14 micrometer (μm)
* UV channel   0.3 to 0.4 μm
* Resolution   5 milliradians
* Field of view   80 degrees
* Scan rate   40 scans per second
* Sensitivity   better than 0.5°K

Flight test program: The IR/UV scanner system was installed in the aircraft in May 1978. Since then, an initial
flight test program has been carried out to verify basic system
characteristics. The results obtained so far are very encouraging. The test program will continue along with training of
Coast Guard pilots. After tests and training, the system will
be used operationally in the Coast Guard service. Imagery from
the test program is shown in Figures 7 and 8.

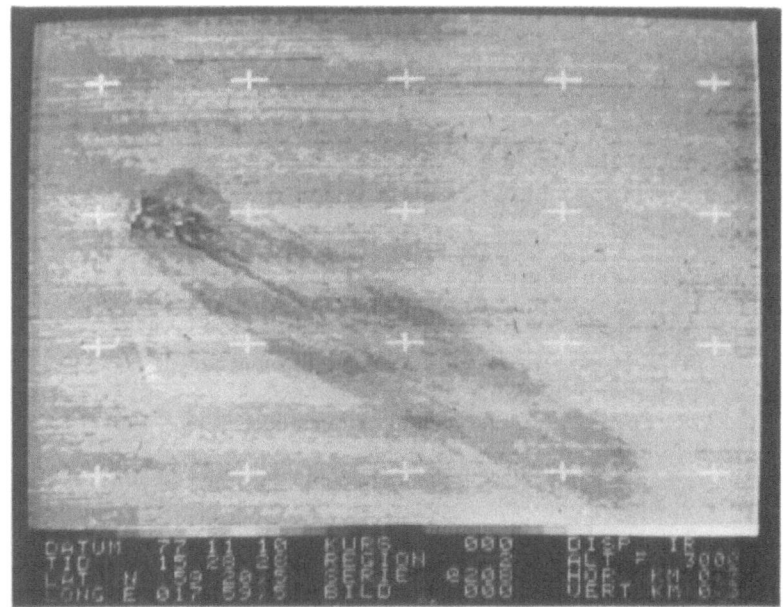

Figure 7.   Black-and-white photograph from the operator's
             monitor in the scanner system. The display here
             shows 400 litres of crude oil mapped in the IR
             channel. The oil slick is about two hours old.
             The size is about 300 m x 900 m. Thicker parts
             of the oil appear dark while thinner parts are
             depicted lighter. Thus, with the IR display,
             the core of more concentrated parts of an oil
             spill can be identified. The little spot under
             the oil slick is a research vessel.

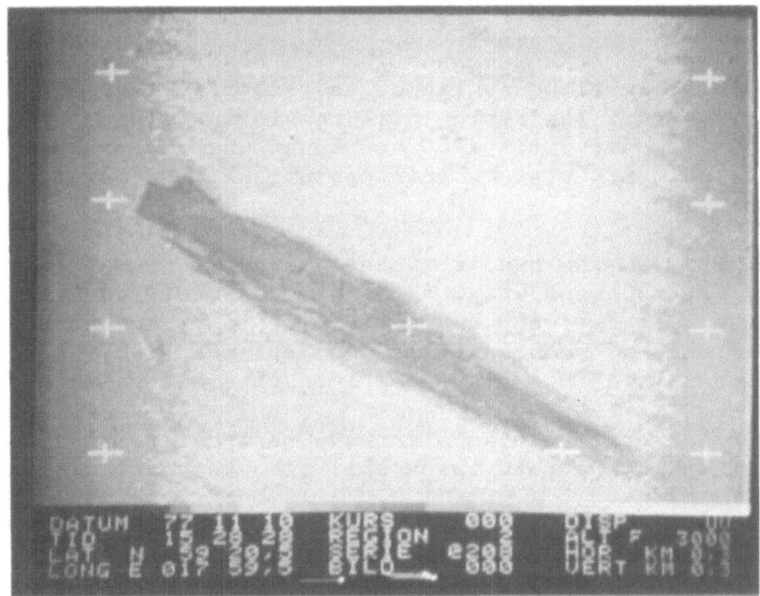

Figure 8. Black-and-white photograph from the operator's
monitor in the scanner system. The same slick
as in Fig. 7, but in the UV-channel. Detection
of oil in the UV relies on the fact that oil
has much higher reflectivity than water for UV
radiation. The UV-channel can only be used
during day-time when the UV radiation component of sunlight is reflected in the oil.

BUDGETS

The development of the camera, SLAR and IR/UV system was
done under rather tight economic and time restraints. It must
be pointed out that all systems are prototypes and comparatively low cost. Further, they are optimized to basic operational
requirements.

The camera system came first and was installed into the
aircraft in the beginning of 1977. Development cost per system
was about $ 35,000 US. Development of the SLAR began in 1976
and was completed in January 1978 with its installation in the
aircraft. Development cost including installation was about
$ 200,000 US. The IR/UV scanner system development was initi-
ated in early 1977 and completed in May 1978. Development and
installation cost was about $ 300,000 US.

PLANS FOR THE FUTURE

The on-going test programs for the SLAR and IR/UV line
scanner systems will be continued to gather further experience
from both systems. The system will remain installed in the two
Cessna 337 aircraft until 1980 and Coast Guard personnel gradually will put both systems into operational use for coastal
surveillance.

During 1979, the Swedish Coast Guard procured a third
bigger aircraft Cessna 402 C (Figure 9). A modified and upgraded
SLAR system with extended swath width up to 80 km and double-
sided coverage has been installed in September 1980. A new IR/UV
line scanner system has also gone into this aircraft (Fig. 10).

Other sensors, such as microwave radiometers, laser fluorosensor and low light television will also be considered in the

Figure 9.   The Swedish Coast Guard's Cessna 402C aircraft.

Figure 10.   The Swedish Coast Guard's Cessna 402C cabin
             installation. Upper left is the operator's colour
             TV monitor. On the right are SLAR, IR/UV, cameras,
             data recording and quick-look documentation systems
             control panel and display unit.

near future to obtain all-weather capabilities and more powerful sensor packages.

    The present prototype systems in the two small Cessna will
be exchanged for the new SLAR systems before the turn of the
year 1980.

ACKNOWLEDGEMENTS

    The two Swedish airborne oil spill surveillance systems
have been developed for the Swedish Coast Guard. Much of the
credit for starting these projects goes to the head of the
Swedish Coast Guard, Commodore R. Engdahl and his staff. Their
belief in new technologies and their active support have been
indispensable in transforming theoretical studies into operational remote sensing systems for coastal surveillance in
Sweden.

ADDENDUM*

   As seen previously, after some years of studies and prototype development a side-looking airborne radar (SLAR) and
an IR/UV-system were installed in two Coast Guard Cessna 337
aircraft. The first generation systems were flight tested and
then taken into operational use by the Swedish Coast Guard
during the years 1978-1981. A number of field experiments were
performed to verify the equipment. Much experience was also
gained from the daily operational use by the Coast Guard.

   Based on the experience and the requirements that this
work resulted in, a new fully operational second generation
system could be specified and developed. The second generation
system has been in use since 1981. It is installed in three
aircraft that are operated by the Swedish Coast Guard in daily
patrol over Swedish waters.

   One aircraft, a Cessna 402C (Figure 9) carries the complete remote sensing system, whereas the two Cessna 337 have
had the IR/UV excluded due to airframe installation and weight
limitations.

   From the start 1972, development of the Swedish Maritime
Surveillance System has taken place in very close co-operation
between the Swedish Coast Guard and the Swedish Space Corporation. The Swedish Coast Guard has contributed with its wide-
ranging experience by establishing the operational requirements
for the system and has also funded a large portion of the development cost. SCC has carried the overall responsibility for
the development program.

SYSTEM DESCRIPTION

   The operational experience gained from the first generation system made possible the definition and development of a
highly integrated second generation system designed exclusively
for maritime surveillance. Thus, many sophisticated features
mandatory for SLAR and IR/UV over land applications (like SLAR
antenna stabilization) could be omitted, and the development
work could be kept within a very reasonable budget.

*Editor's note: As agreed by the pilot-study steering committee,
the SSC has been asked for an up-dating of the previous contribution presented at the Washington workshop. This Addendum,
prepared by L. Backlund and O. Fäst, both from SSC, includes
the more recent development of the Swedish remote sensing sys-
system and presents the more recent research program which has
to be carried out by both SSC and the Swedish Coast Guard.

In the second generation system, a number of different sub-systems have been integrated to create a versatile and easy to handle system for surveillance of all the different activities on the sea surface (Figure 11):

* The SLAR gives all-weather capability and large area
  coverage, even against very small targets.

* The IR/UV scanner system is used for detailed mapping of
  oil spills, accident sites and clean-up operations.

* The camera system gives high quality, high resolution
  photographic evidence.

* A conventional forward-looking airborne radar facilitates
  in-flight mission planning.

* A microwave link permits immediate transmission of generated images to ships and command centers on the ground.

In the development work the aim has been to obtain a high degree of integration and simplicity of operation. Through the use of advanced technology the system also has low weight and low power consumption, permitting installation in a wide range of aircraft. Some of the features of the integrated system are:

* Annotation data is automatically inserted into all
  sensor and camera images immediately on recording. Images
  can never get into disorder, as date, time, position and
  mission identification are a part of the image itself.

* IR/UV and SLAR imagery is presented in real time on the
  operator's TV display, facilitating prompt action.

* Documentation of all sensor data on digital cassette tape
  which can be used for playback on the TV-screen or for
  producing Polaroid snapshots at any time.

* Overviews can be transferred in real time to surface
  units and land-based command centres using a microwave
  image link.

Side-looking Airborne Radar (SLAR)

The SLAR is used for sea traffic surveillance, oil spill detection, fishery protection, search and rescue and sea-ice mapping.

The SLAR has all-weather capability and is used for large area surveillance. It can be used to detect oil or life boats

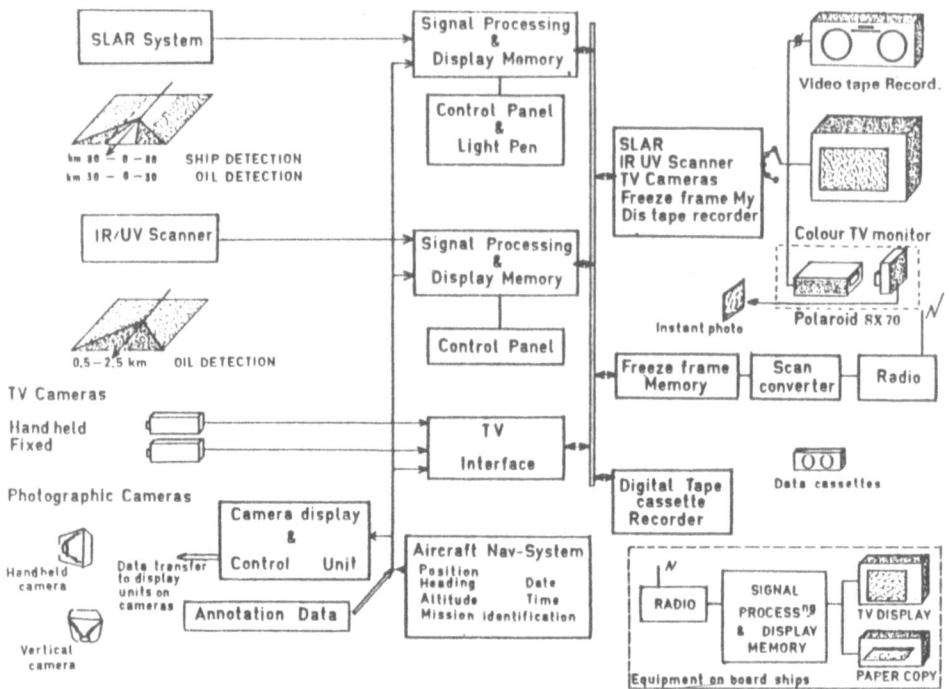

Figure 9. Block diagram of the Maritime Surveillance System.

out to 25-30 km on each side of the aircraft, and cargo ships out to 80 km on each side of the aircraft.

The SLAR image is presented as a rolling map in real time on a standard TV display. The TV lines are produced by a digital image processor, from the integration of radar pulses. The processed TV format facilitates system integration and gives the system operator a bright, flicker-free, easily interpreted presentation.

Automatic target positioning is one of the outstanding features of the SLAR. The operator can use a light pen to mark selected targets on the display. The geographic position of the target is then calculated and presented in the data block on the image.

This SLAR was developed by Ericsson in Sweden to meet SSC specifications. Some technical data have been given previously. Some other characteristics are listed below:

* Ground resolution: 75 m x 75 m
* Display ranges: 20-40-80 km both sides
* Weight: 70 kg
* Power consumption: 28 VDC, 11A

IR/UV Scanner System

The IR/UV system is used to get a high resolution mapping on the sea surface. It is used to monitor oil spills, to inspect ship's wakes for suspected oil discharges or to do a close survey of a water surface at night.

The IR can be used day or night to monitor the spreading of oil slick and to give information on the relative oil thickness within the slick. The latter is very valuable in clean-up operations, as usually 90 percent of the oil is located within 10 percent of the visual oil slick.

The UV channel is used only in daylight. It uses the fact that oil reflects UV light much better than water. Thus the UV channel adds confidence to the IR registration, resolving the ambiguity between pollution and natural phenomena, such as up-welling cold water.

The IR/UV imagery is also presented on the operator's TV monitor. False color coding can be added for image enhancement.

The scanner used is a Daedalus DS-1220. Other technical data of the IR/UV system are listed below:

* IR-channel: 8-14 microns (thermal infrared)
* UV-channel: 0.3-0.4 microns
* Resolution: 5 mrad or 2 meter ground resolution
                         at 1,000 ft flying altitude
* Field of view: 80 degrees
* Sensitivity: better than 0.5°K (IR)
* Weight: 58 kgs
* Power consumption: 28 VDC, 12 A

SLAR and IR/UV display monitor adds 30 kgs and 28 VDC, 4 A.

Camera System

As shown previously, the camera system comprises one vertical camera and one hand-held camera for oblique pictures. Both cameras use 36 exposures 24 x 36 mm miniature film and have a data box which superimposes all relevant nav-data on each frame.

Operational experience shows that the cameras are indispensable for routine surveillance, identification and documentation. In particular they are used to produce high-quality, high-resolution photographic evidence.

Data Storage and Documentation

A Polaroid camera has been integrated with a small TV
monitor in a quick-look documentation system. Instant photo
with the same image as on the operator's monitor can be readily
produced. The Polaroid photos create an easily accessible file
of events for each mission without delay for film processing,
etc.

SLAR and IR/UV are interfaced to a digital cassette recorder and a quick-look documentation system. Each cassette stores
up to 6 hours of SLAR or IR/UV information. Data can be replayed
either in flight or after landing for closer inspection.

In the present development, the image link and the TV
camera system have also been interfaced to the digital cassette
recorder. Recorded data can be replayed at any time and transmitted to surface units via the image link.

The technical data of the system are listed below:

* Data cassettes: DC300 type
* Capacity: 17 Mbite/cassette
* Polaroid: SX 70 type
* Weight: 16 kgs
* Power consumption: 28 VDC, 5 A

CONTINUING DEVELOPMENT

Two TV-cameras and a Microwave Image Link are presently
being flight tested.

TV-Camera System

The TV-camera system will provide immediately accessible
visual documentation to support the other sensors. TV-imagery
can be recorded on the digital cassette recorder, reproduced
by the Polaroid quick-look system, and transmitted via the image
link.

One hand-held TV-camera will be used by the operator to
produce general survey images. Furthermore, one fixed forward
looking TV-camera will also guide the pilot exactly on target
when using the vertical camera. Experiments will be conducted
to see whether it is possible to extend the capabilities of the
TV-system to night-time documentation of a ship's name.

Microwave Image Link

The microwave image link transmits a digitally recorded

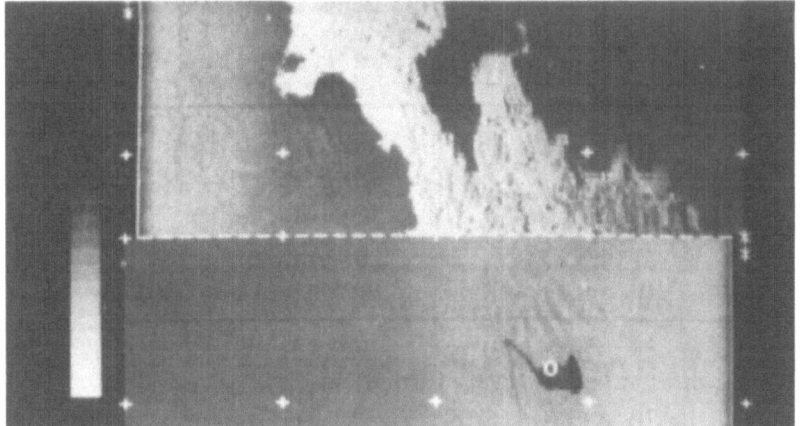

Figure 12.   SLAR image from a routine Coast Guard mission
It shows discharged oil from a tanker colli-
sion in the Baltic Sea. Upper portion of the
image displays 0-40 km to the right of the
aircraft (one can see Gotland), lower portion
0-40 km to the left. The oil slick has been
marked with the light pen ("O") and target
position is given at the bottom of the image.

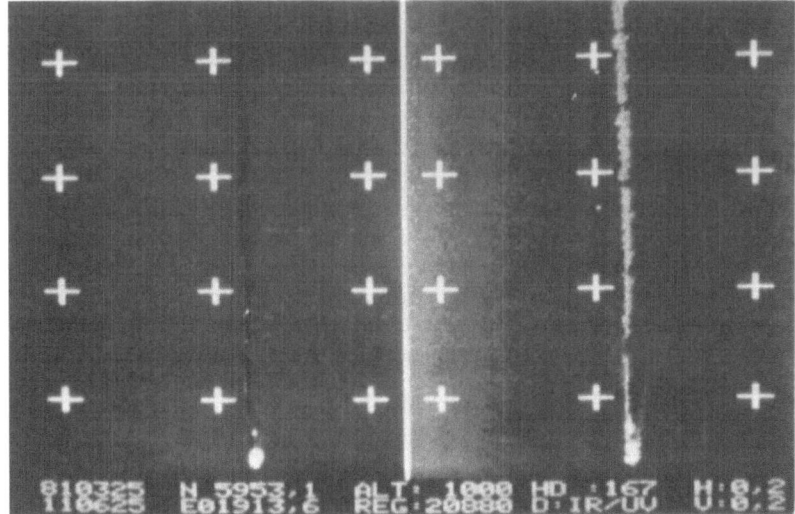

Fig. 13.   IR/UV display showing a 160 l/nautical mile discharge
in a ship's wake. The ship is the lower white spot in
IR (left) and UV (right). The little white spot just
above is a small work boat.
Interruptions in the slick are due to refilling of
the discharge container on board the ship.

image in less than one second (0.5 Mbit/s transmission rate).
All recorded images, whether they are SLAR, IR/UV or TV-camera
images, can be relayed to ships and land based units using the
microwave image link.

The receiving equipment has an imagery memory, a TV
display and a hard-copy unit that present exactly the same
images as in the aircraft. Thus the system can immediately
provide a ground command center with the aircraft's superior
overview. This will be useful on accident sites, when surface
units need to know whether booms and skimmers are at the
correct locations, which direction oil is spreading, and so on.
It may also prove useful in other situations, such as sea-ice
mapping or traffic surveillance in the fishing zones.

Microwave Radiometer

A prototype of an operational microwave radiometer is
under development. Flight tests are planned to begin in the
summer of 1984. A microwave radiometer offers the capability
to measure the thickness of an oil spill. By processing thickness data with an IR/UV scanner map of the area covered by oil,
a volume estimate can be obtained.

Volume data are important for deciding what clean-up activities are needed against an oil spill. Also, new regulations
are being enacted in Sweden and other countries, imposing
varying levies on the amount of oil spilled and the type of oil.

Laser Fluorescence

Laser fluorescence can be used for verification and a
coarse type classification of detected oil slicks, making
possible the selection of proper countermeasures. The instrument can also be used to detect neutrally buoyant oil just
below the sea surface. Studies are presently being carried out
to integrating a laser fluorescence instrument into the operational system.

International use

The Swedish Space Corporation is offering the integrated
maritime surveillance system for sale on the world market. The
Dutch Rijkswaterstaat has procured a complete system that has
been in daily operation from the beginning of 1983 and has
produced good results from the very first day.

In late 1983, the next system, procured by the British
Department of Trade, will be operational.

TECHNICAL RESULTS

OPTICAL MEASUREMENTS OF CRUDE OIL SAMPLES

UNDER SIMULATED NATURAL CONDITIONS

W.A. Hovis and J.S. Knoll

National Oceanic and Atmospheric Administration
National Earth Satellite Service
Washington D.C., United States

ABSTRACT

Spectrographic measurements of the solar reflectance from 0.4 to 2.4 µm and radiometric equivalent blackbody temperature measurements from 10.5 to 12.5 µm were made under simulated aging conditions of sun, wind, and salt water wave action on 10 different types of crude oil. These measurements were made at the USEPA OHMSETT site in Leonardo N.J. Four days of exposure produced little change in the reflectance from 0.4 to 2.4 µm, but oil-water temperature differences of up to 18°C were recorded in the 10.5 to 12.5 µm region. Large changes in viscosity were also measured.

INTRODUCTION

The spectral signature of crude oil is of interest in evaluating the potential of optical remote sensors for use in observing spills such as those that occurred after the accidents involving the TORREY CANYON, ARGO MERCHANT, and AMOCO CADIZ.

So, in order to understand the characteristics that crude oil might exhibit on the ocean when viewed by optical remote sensors from satellites, an experiment was carried out in August of 1978 to determine the reflected and emitted characteristics of crude oil specimens in those spectral regions now being examined by various satellite sensors. Some laboratory measurements were available; however, changes in the optical signature that might occur because of aging as crude oil floats on the surface of the ocean were not

understood. Deliberate oil spills on the ocean have been
limited so such small quantities that they could only be viewed
for a few hours before the oil had so widely dispersed as to be
undetectable. In order to overcome this, the U.S. Environmental
Protection Agency "Oil and Hazardous Material Simulated Environmental Test Tank Facility" (OHMSETT) was used with the generous
co-operation of the E.P.A. to examine ten specimens of crude
oil over a period of several days of aging.

EXPERIMENTAL PROCEDURE

    Special containers were constructed that would contain the
crude oil samples on the surface of the OHMSETT tank, but would
allow the oil to be subjected to the influx of sunlight and
contact with the water that would be encountered in the ocean

Figure 1. The U.S. E.P.A. Oil and Hazardous Material Simulated
          Environmental Test Tank Facility (OHMSETT) during the
          optical measurements of crude oil samples. Note the
          special containers at the surface of the tank.

Figure 2.   Optical measurements of a crude oil sample in the
            OHMSETT facility.

situation. In addition the oil was disturbed by using the wave-
making machinery of the OHMSETT facility to simulate the agita-
tion it would receive in the ocean.

        Ten floating cells were fabricated, each approximately
1.3 x 1.3 m and with sides projecting approximately 30 cm above
and below the water. The insides of the cells were painted with
a non-reflecting black to minimize reflections from the walls.
The cells were positioned by ropes from either side of the tank
to keep them centered in the tank and positioned underneath the
sensors (Figures 1 and 2).

SPECIMENS

        Ten specimens of crude oil were acquired for this study.
The crude oil specimens acquired were as follows:

Nigerian, Iranian Light, Lagunillas (Venezuela), La Rosa (Mexico), North Slope (Alaska), Brass River (Algeria), Arzew (Algeria), Suniland, Sahara and Gabon.

MEASUREMENTS

Measurements were conducted with a filter wedge spectrometer (FWS) measuring reflectance of solar energy in the nadir direction from 0.4 to 2.4 µm, and simultaneous measurements were made using a Barnes PRT-5 radiometer measuring equivalent black body temperature from 10.5 to 12.5 µm. A closed circuit

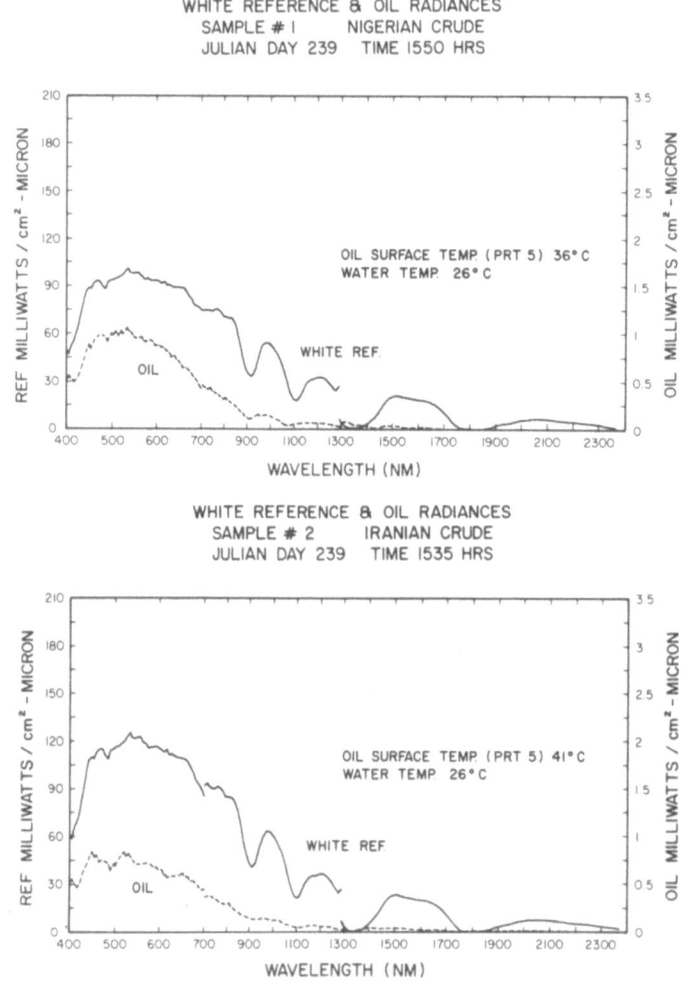

Figure 3.   The reflected solar energy for two crude oil specimens 1 and 2 and for the reference sample.

TV system was aligned with the sensors to view the sample cells in order to determine not only that the sensors viewed the oil surface, but that the field of view was not shadowed by the sides of the cell. A white reference material was provided for comparison for reflectance measurements while a blackened plate was inserted under the containers to prevent the spectrometer or radiometer from seeing through the oil to the tank itself.

The oil, when first placed on the water, exhibited a gray body reflectance from 0.4 to 2.4, with reflectance of approximately one to two percent in the nadir direction for all ten

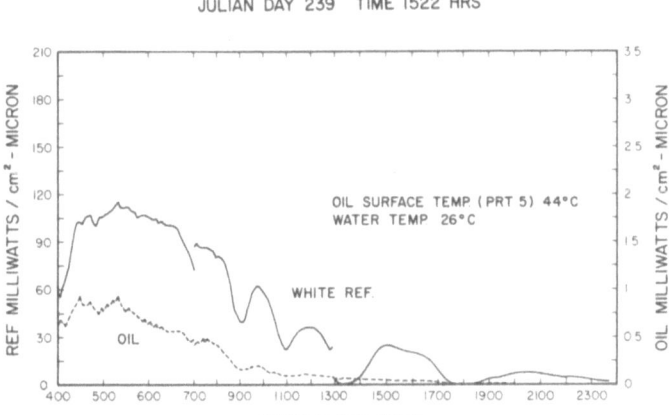

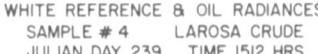

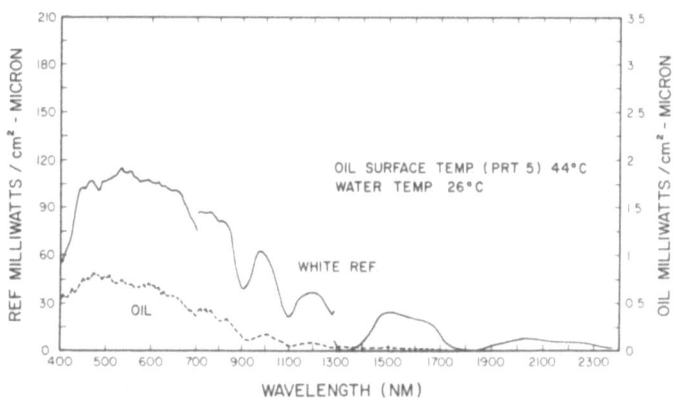

Figure 4. Reflected solar energy for two crude oil specimens and for a reference sample.

samples. After the oil had been on the surface of the water
and subjected to solar irradiation, evaporation and dissolution
for three days, a slight increase in reflectance was noticed
in the 400-700 nanometer region. This increase raised the reflectance by approximately one percent over the previous gray
body reflectance in the 400-700 nanometer region.

Figure 3 through 7 show the reflected solar energy in
units of spectral irradiance for the reference sample and the
10 specimens of crude oil taken on the last day of measurement.
It is clear that most of the samples exhibited no change in

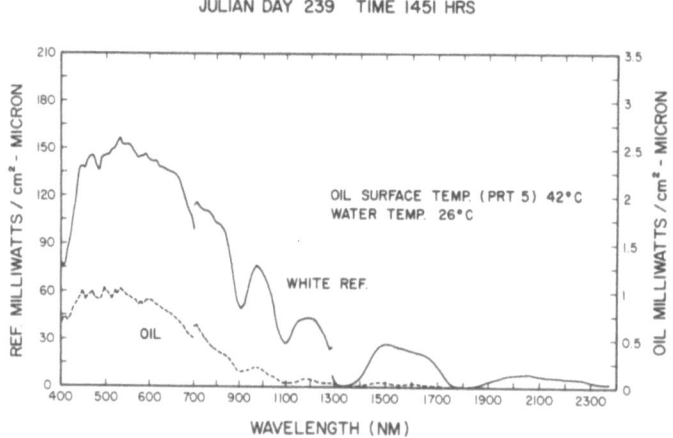

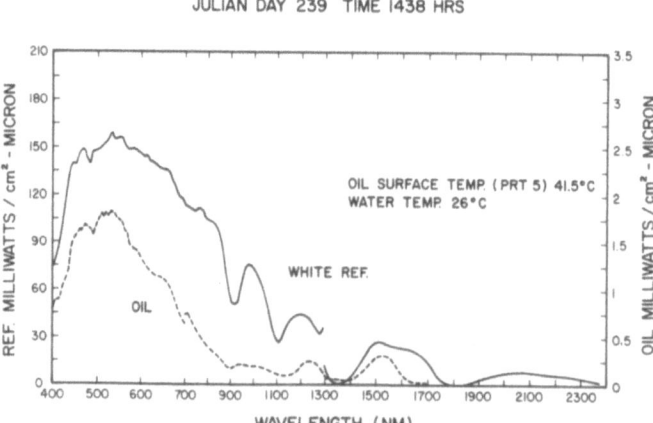

Figure 5. Reflected solar energy for two crude oil specimens
and for a reference sample.

reflective character with age that was detectable with our
instrumentation. Note that the scale for the reference sample
is given on the left side of the spectral plot and that for
the oil, on the right.

All the samples appeared to be reasonably gray from 0.4
to 1.1 μm (400 to 1100 nanometer region) with very low reflectance. In two cases, that of the Brass River Algeria (Fig. 7)
and the Arzew Algeria (Fig. 8) crude oils, there was an additional slight increase in reflectance between 900 and 1100
nanometers.

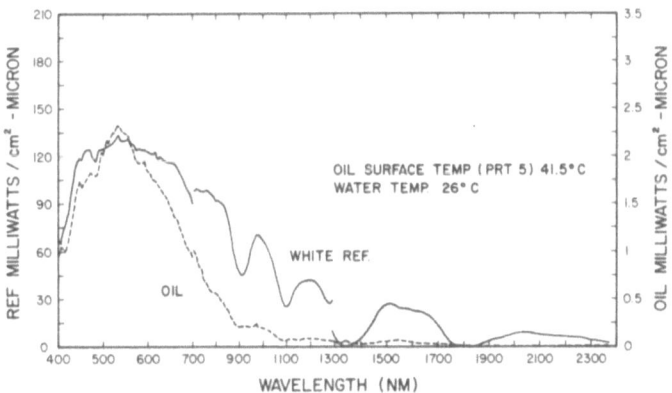

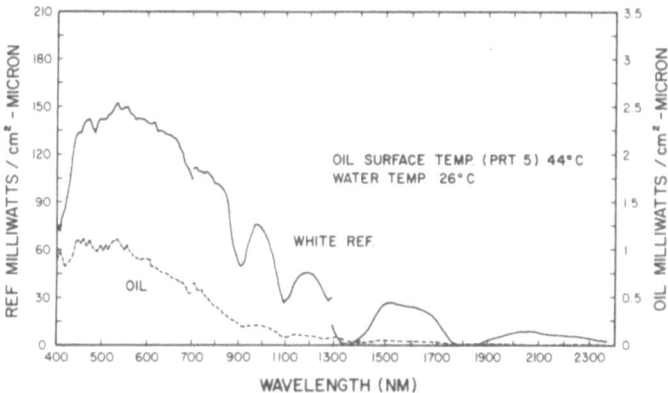

Figure 6.   Reflected solar energy from two crude oil specimens
            and for a reference sample.

Equivalent black body temperature measurements, measured as a function of the time of day, showed that the oil was always warmer than the surrounding water ranging from a temperature difference of 19°C at high noon to 10°C at 16:00 hours. It was also noted that the temperature difference between the water and the oil was reduced with increasing windspeed.

In addition to the optical measurements, measurements of specific gravity, viscosity, surface tension and interfacial tension were carried out by the Environmental Protection Agcy. during the test in which the oil was allowed to remain on the

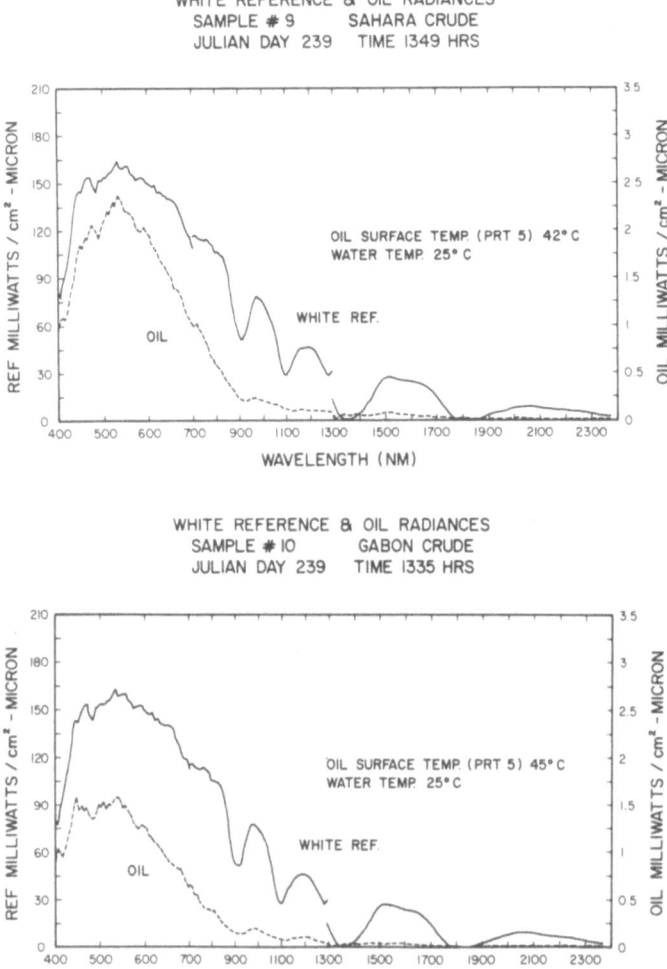

Figure 7.   Reflected solar energy from two crude oil specimens and for a reference sample.

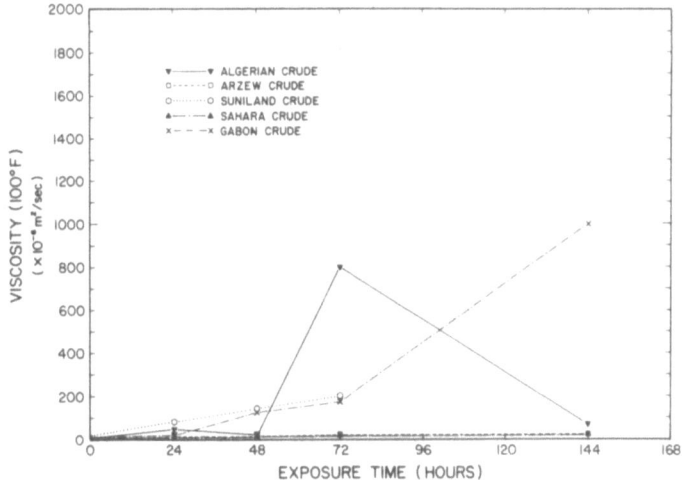

Fig. 8.   Viscosities of oil samples measured as a function of
          time after the oil was placed on the water surface.

surface for four continuous days. Figures 8 and 9 show the viscosities of the oil samples measured as a function of time after
the oil was placed on the EPA OHMSETT water surface. In general
the viscosities increased with time as volatiles evaporated
and solubles dissolved except for one case of Algerian crude

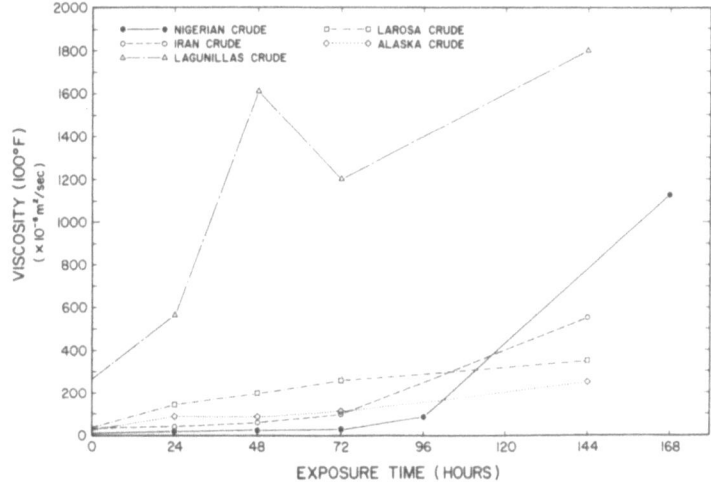

Fig. 9.   Viscosities of crude oil samples measured as a func-
          tion of time after the oil was placed on the water
          surface.

whose viscosity increased, then dropped sharply. The viscosity of Lagunillas crude dropped after 48 hours, but rose again after 72 hours to reach the highest viscosity of the samples measured.

CONCLUSIONS

These measurements indicate that crude oils of the type measured will be extremely difficult to detect from optical sensors on spacecraft sensing reflected solar energy, except in the case where they might drift over an area of shallow water with a highly reflecting bottom, in which case they would appear quite dark against that bottom. In cases where the water has higher reflectance, such as an area where there is a large concentration of suspended sediments or organic growth, oil might be also detected as a dark patch against a lighter background.

They also indicate that oil, under sunlight conditions on the ocean, would be expected to be warmer than the surrounding ocean water, and that a spacecraft sensor sensing equivalent black body temperature might be able to detect oil emanating from some point source such as leaking tanker by looking for the temperature difference.

These comments, of course, would also apply to remote sensing of oil from aircraft platforms.

REFERENCES

Deutsch, Morris, P. R., Strong, A. S., Estes, J. E., 1979, "Proc. 9th Annual Offshore Technology Conference".
Fantasia, J. F., Ingrao, H. C., 1975, "The Development of an Experimental Airborne Laser Remote Sensor for Oil Detection and Classification in Spills", U.S. Department of Transportation report CG-D-86-75, Washington, D.C.
Goldman, Gary, C., Horrath, R., 1975, "The Feasibility of Oil Pollution Detection and Monitoring from Space", Dept. of Transportantion report CG-D-117-75, Washington, D.C.
Hovis, W. A., Kley, W. A., Strange, M. G., 1967, Filter wedge spectrometer for field use, Applied Optics, 6:1057.
Hovis, W. A., Knoll, J. S., 1979, "Optical Measurements of Crude Oil Samples under Simulated Conditions", NOAA Technical Memorandum NESS 105, Washington, D.C.
Millard, J. P., Arvesen, J. C., 1973, Polarization: a key to an optical system for detection of oil on water, Science, 180:1170.

REMOTE SENSING ANALYSIS OF OIL POLLUTION

IN AUGUSTA BAY, SICILY, ITALY

A.L. Geraci and T. Caltabiano

Machinery Institute
Department of Engineering
University of Catania
Catania, Italy

ABSTRACT

The pollution of marine environment can be very harmful, because its effects on the ecological balance governing life. The level of danger depends on the level of pollution reached, the diversity of polluted discharges and the difficulty of en- forcing regulations. Remote sensing is of great importance in facilitating action designed to reduce or prevent pollution, providing an effective means of monitoring urban and industrial discharges.

The objective of this investigation was to use remotely sensed data for the assessment of oil pollution levels within Augusta Bay and the associated coastal area, in eastern Sicily.

The approach consisted of simultaneous acquisition of data from infrared photography and thermal infrared imagery. Most of the aerial explorations were conducted from helicopters, due to their capability to reach locations not accessible using wing aircraft, as well as their ability to maintain low altitudes more safely. The helicopter employed was an Agusta Bell 204/B ASW, supplied by an Italian Naval Air Station located in Catania.

Infrared photographs were taken to record in tonal or false color the near-visible infrared radiation from 0.7 to 0.9 micron. Conventional 35 mm cameras were used, because of their compactness, and ease of handling. They were equipped with 55 mm lenses. Also, a Kodak Wratten filter No. 12 (deep

yellow) was used over the camera lens to penetrate the haze
and absorb the blue light. An airborne AGA Thermovision system
(AATS) was used to detect and record infrared radiation from
2 to 5.6 micron. AATS was operated through the open door of
the helicopter. The thermal infrared images displayed on the
color monitor were photographed with a 35 mm camera.

The results included a series of color-coded maps, each
pertaining to both remote sensing techniques used (photographic and thermal infrared). Based on these results and the associated analysis that were made on the various kinds of imagery
the following conclusions were drawn up: highly polluted areas
and illegally discharging vessels were clearly discernible on
suitably enhanced imagery made from acquired data; infrared
photography and thermal infrared imagery must be used in conjunction with each other to provide accurate interpretation
of the data.

Figure 1. Map of the eastern Sicily's coast and data run.

INTRODUCTION

Sicily is the largest and one of the most densely populated island in the Mediterranean Sea, with several small adjacent islands. Its northern shore, site of the capital Palermo, is lapped by the Tyrrhenian Sea, and to the east it looks across the narrow Strait of Messina to the toe of the Italian peninsula. It lies about 160 km (100 miles) northeast of Tunisia, in North Africa. Sicily has a surface of approximately over 25,000 km^2 (9,600 sq mi) with approximately 5 million inhabitants, a coastline of 1,500 km (932 mi).

Sicily extends from about 36° 40' N to 38° 20' N latitude and from about 12° 30' E to 15° 30' E longitude. Augusta is a town north of the city of Syracuse on the eastern Sicilian coast (Figure 1). It lies on a long sandy island off the southeastern coast between Augusta Bay and the Jonian Sea. Long a naval station, Augusta has become a principal Sicilian trading port with industrial growth on its extensive waterfront, including oil refineries and chemical complexes. Shipping movement reach over to 32 million tons.

Augusta Bay is one of the Sicilian most heavily industrialized areas because of a number of process plants existing in the associated coastal area. Pollution resulting from increasing population and industrial activities has caused the need for protection of the marine environment. Marine pollution is one of the most dangerous kinds of pollution, because its effects on the ecological balance governing life.

Management is greatly facilitated if information about marine pollution is periodically available, and remote sensing might be used to facilitate the obtaining of such information[1]. Remote sensing techniques are considered as having the potential to provide a cost-effective method for detecting pollution features and water quality parameters. Large area synoptic views from aircraft and satellite platforms equipped with remote sensors provide information that is not readily available by the use of conventional measurement techniques. Such a system would benefit national, regional and local government agencies.

The objective of this investigation was to use remotely sensed data for the assessment of oil pollution levels within Augusta Bay and the associated coastal area, in order to demonstrate the usefulness and applicability of remote sensing to water quality monitoring and oil pollution analysis.

ANALYSIS METHODOLOGY

The approach involved simultaneous acquisition of data from infrared photography and thermal infrared imagery.

Remote sensing techniques are being used by the Machinery Institute, Department of Engineering, at the University of Catania in Sicily to determine levels of pollution in coastal waters. During 1978, infrared photography and thermal infrared imagery were used as part of an investigation on the Jonian Sea, in the coastal zone between S. Alessio and Murro di Porco capes from Taormina to Syracuse. Most of the aerial explorations were conducted from helicopters, due to their ability to reach locations not accessible using wing aircraft, as well as their capability of hovering and maintaining low altitudes more safely. The helicopter used was an Agusta Bell 204/B ASW. Helicopter and pilots were supplied by Maristaeli, an Italian Naval Air Station located in Catania, that has the best experience in flights over water.

For infrared photography, conventional 35 mm cameras equipped with 55 mm lenses were used, because of their compactness and ease of handling. Infrared photographs were taken to record in tonal or false color the near-visible infrared radiation from 0.7 to 0.9 µm. Kodak Wratten filters No. 12 (deep yellow) were used over the camera lenses to penetrate the haze and absorb the blue light. Films used were Kodak High Speed and Kodak Ektachrome infrared films, both of which were exposed 1/125 of a second at f/8 and processed by Kodak.

For thermal imagery, an airborne AGA thermovision system (AATS) was used to detect and record infrared radiation from 2 to 5.6 µm. Equipment included Thermovision camera, display

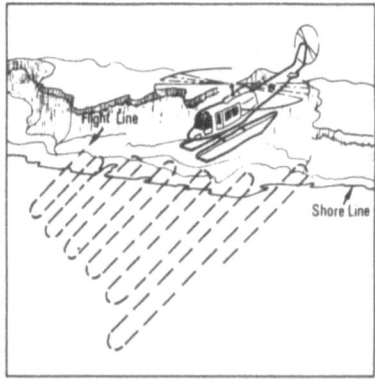

Figure 2.   Helicopter flight path.

unit, support frame, color monitor and camera adapter. AATS
was operated through the open door of the helicopter. The
thermal infrared images were photographed from the color
monitor with a 35 mm camera on Kodak Ektachrome 200 profes-
sional film (daylight). AATS is a real-time scanning system
with a 20° total field of view, a 3.4 mrad spatial resolution
and a resolving power of 100 elements per line. It works in
the temperature range from -20 to 900 °C (-4 to 1,1652 °F)
with a sensitivity of 0.2 °C (0.36 °F) at 30 °C (86 °F). The
color monitor of the system provided quantized images of gray-
tone thermal pictures, with individually color-coded isotherms
superimposed.

    Cameras for infrared photography were flown on the heli-
copter at an altitude of 1,800 m (5,900 ft) over the eastern
Sicilian coastal zone including Augusta Bay, on May 24, 1978.
Table 1 and Figure 1 show flight line data and data run respectively. Cameras were also flown at an altitude of 450 m over
Augusta Bay, back and forth across· the shoreline as indicated
in Figure 2 on May 25, 1978, and along the coast on May 26.

Table 1. Flight line data.

Check Points	Time GMT		Altitude ft/m	*Cloud cover/Remarks
	Start	End		
a - b	8:30	8:39	5900/1800	From S. Alessio cape to Minissale torrent
b - c	8:40	8:48	5900/1800	From Minissale torrent to Pozzillo
c - d	8:50	8:57	5900/1800	From Pozzillo to Cannizzaro
d - e	9:00	9:10	5900/1800	From Cannizzaro to the mouth of Simeto river
e - f	9:11	9:22	5900/1800	From the mouth of Simeto riber to Augusta harbor
f - g	9:23	9:30	5900/1800	From Augusta harbor to S. Panagia cape
1 - m	8:15	8:20	1450/450	
n - o	8:25	8:31	1450/450	
p - q	8:33	8:37	1450/450	
r - s	8:40	8:45	1450/450	
t - u	8:46	8:50	1450/450	
v - w	8:52	8:58	1450/450	
x - y	9:02	9:10	1450/450	

* clear

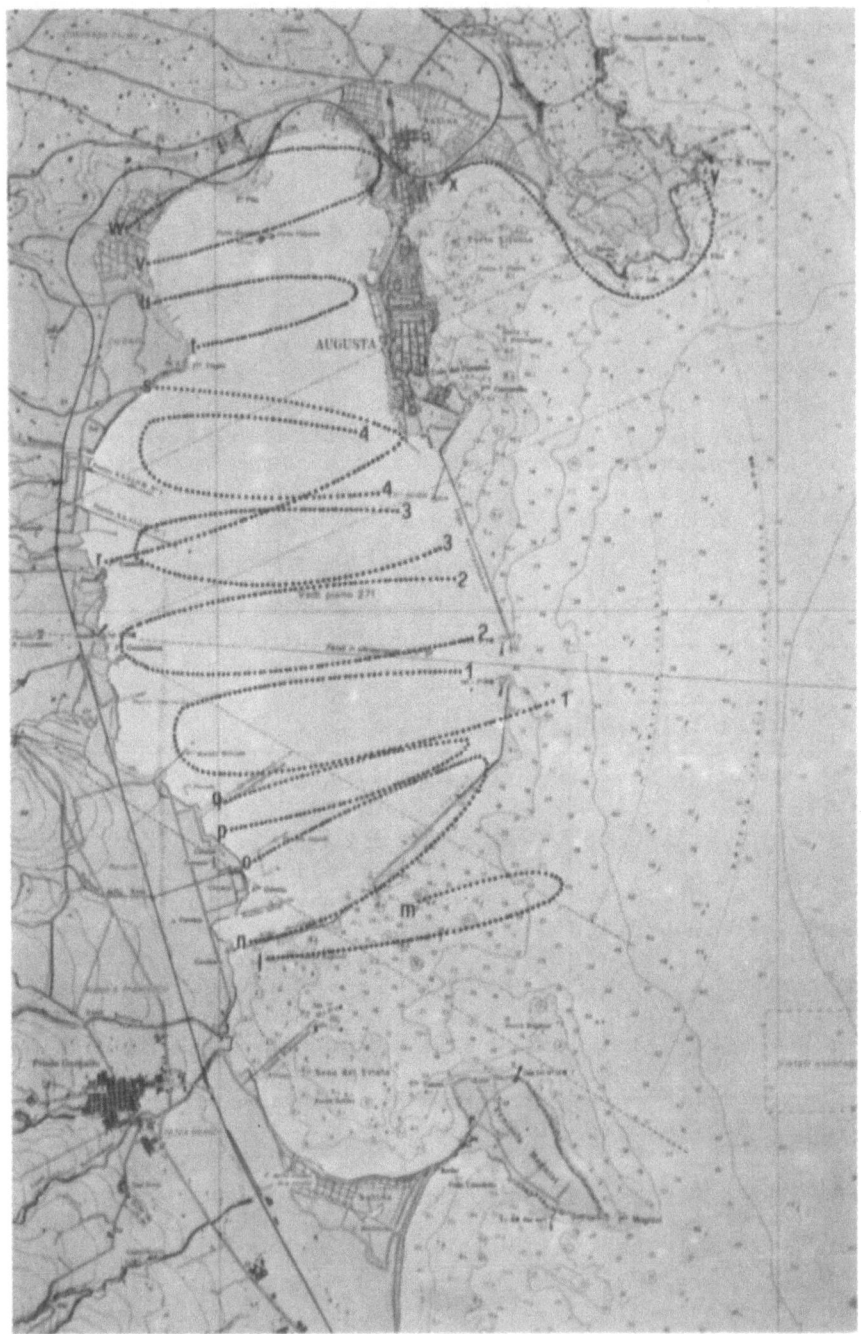

Figure 3. Map of Augusta Bay and data run.

Table 2. Flight line data.

Flight	Check Points	Time GMT		Altitude ft/m
		Start	End	
78-04	h - i	8:20	8:30	1450/450
78-07	1 - 1	8:30	8:40	650/200
	2 - 2	8:42	8:50	650/200
	3 - 3	8:55	9:06	650/200
	4 - 4	9:08	9:20	650/200

Flight line data and data run are shown respectively in Table 1 and Figure 3 and Table 2 and Figure 1. AATS was flown on the same helicopter at an altitude of 200 m (650 ft) over the study area on June 6, 1978, as indicated in Figure 2. Table 2 and Figure 3 show flight line data and data run respectively. During all flights the entire area was cloud free. No camera or processing malfunction vere noted, and the quality of the data was rated excellent.

RESULTS

Pollution in marine environment is caused by a wide range of factors, including waste disposal, oil spills, radioactivity, heat and pesticides, affecting various living organisms. The ever increasing number of oil spills , due to leaks from pipelines or accidental and illegal ship discharges makes more serious the resulting pollution. Oil pollution on the shore, in addition to the reduction of amenity, also affects marine and shore life and vegetation. As Reish[2] indicated, the effect of an oil spill depends not only of the quantity but also on the location and fraction of the oil. In fact, a major difficulty encountered in the setting of criteria for oils is that these are not definitive chemical categories, but include thousands of organic compounds with physical, chemical and toxicological properties. Before appropriate corrective action can be taken, a knowledge of the nature, thickness, areal extent, direction and rate of drift of the oil spill must be promptly established. Many investigators have repeatedly reported that the determination of an oil film thickness is important, because the film thickness along with areal extent allows the volume of the oil slick to be estimated . Hollinger[3] in 1974 indicated that in general the thick region of an oil spill contains more than 90% of the oil in less than 10% of the area of the slick. The volume of oil is essential for assessing the impact of the oil spill on marine life and environment.

In this investigation it was found that both types of imagery acquired (photographic and thermal infrared), used in various combinations, were often more effective than anyone alone for the purpose of water quality monitoring and oil pollution analysis. The results included a series of color-coded maps, each pertaining to both remote sensing techniques used.

Figure 4 shows a mosaic of color infrared photographs on

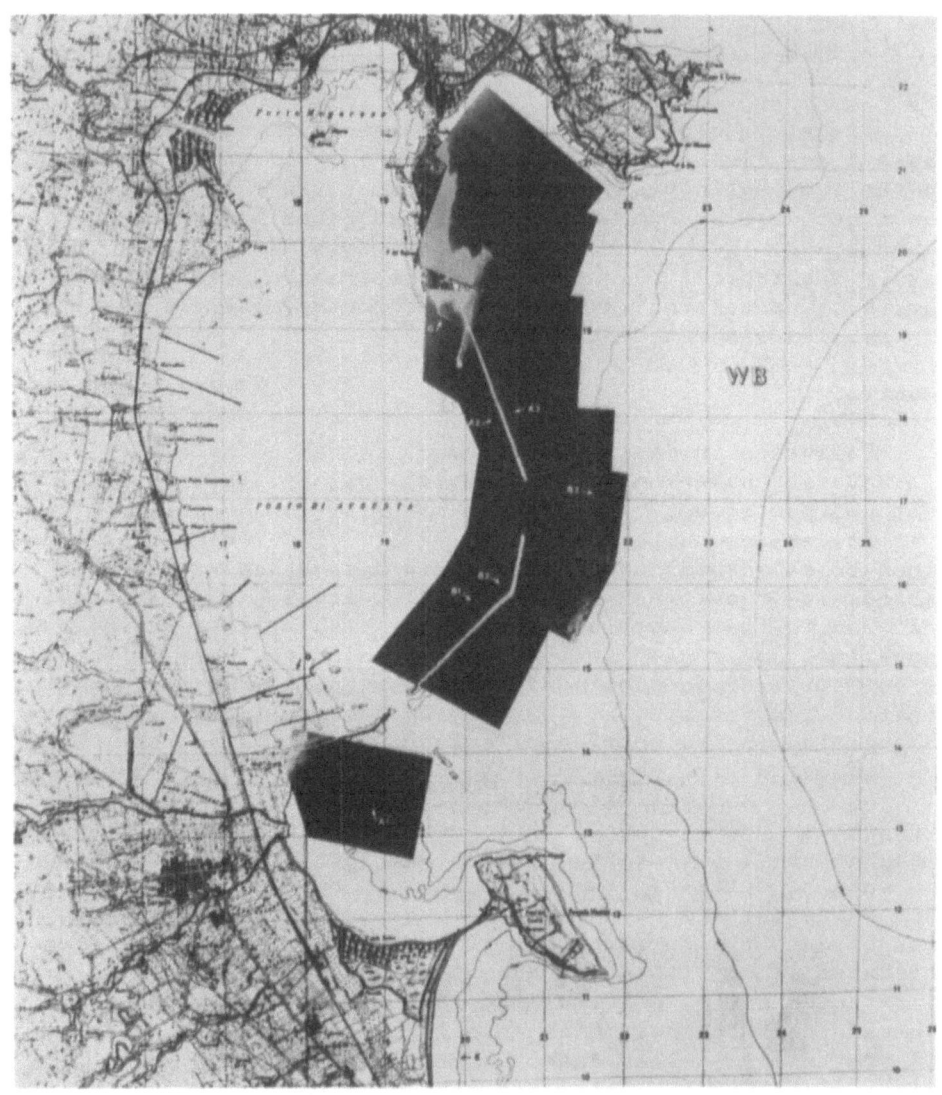

Figure 4.   Mosaic of color infrared photographs
on a map of the study area.

a map of the study area. Color anomalies related to water quality and oil pollution are indicated by white arrows. It is
known that the energy arriving at a camera is a function of the
suspended and dissolved material in the water and the material
floating on it[4]. A number of investigators have reported the
infrared photography as suitable for conducting water pollution
studies and detecting oil on water[5 to 10]. Turbid water generally has a higher reflectance than clear water[11], and the general pattern of infrared radiation from an oil slick demon-
strates a contouring effect with the highest emission from the
middle[12]. According to the studies done by these investigators,
the lighter areas at A were related to a higher turbidity, and
the signatures at B were identified as oil slicks.

Some researchers have investigated the use of computer-
assisted interpretation of digitized aerial imagery[13 14]. They
have reported benefits and problems associated with this technique. Following these investigators each color infrared photograph covering the study area was digitized with scanning
aperture of 50 μm, using an Optronics C-4500 drum microdensito-
meter. The scanned area was approximately 0.7 km^2 (0.27 sq mi)
with each pixel representing an area of 2 m^2 (21.5 sq ft) on
the ground. Analysis of various kinds of imagery and statistical calculations were made with a LogE/Spatial Data EyeCom II
system. Figures 5 and 6 show respectively the digital image
processing of the polluted areas at A1 and A2 in Figure 4.
Also the images of the oil slicks at B1 to B3 in the same
figure were digitally processed as in Figures 7 and 8. The
areal extent of the slick at B3 was calculated approximately
20,600 m^2 (221,700 sq ft).

Also during the survey a number of color infrared photographs, covering the study area from the lower altitude, were

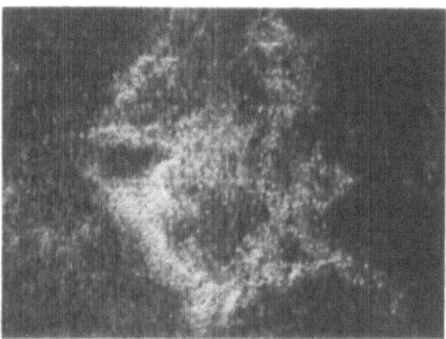

Figure 5. Digital image processing of the polluted area
           at A1 in Figure 4.

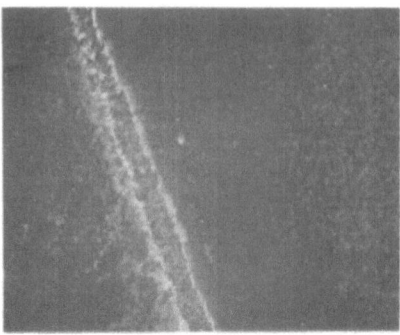

Figure 6.   Digital image processing of the polluted area
            at A2 in Figure 4.

collected. They were then digitized and the scanned area was
approximately of 0.12 km^2 (0.05 sq mi) with each pixel representing an area of 0.36 m^2 (3.9 sq ft) on the ground. Figure 9
shows a color infrared photograph of the oil slicks at C in
Figure 4 and the digital image processing of them. The areal
extent of the slicks was calculated approximately 9,200 m^2
(99,000 sq ft). The color infrared photograph of an industrial
discharge, located at D in Figure 4, and the digital image
processing of the associated polluted area are shown in Figure
10. An illegally discharging vessel is clearly discernible on
the color infrared photograph in Figure 11. The images of the
resulting oil slicks, having an areal extent of approximately
5,000 m^2 (53,800 sq ft), were digitally processed as in
Figure 11.

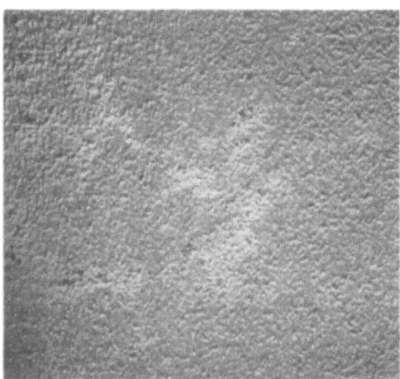

Figure 7.   Digital image processing of the oil slick
            at B2 in Figure 4.

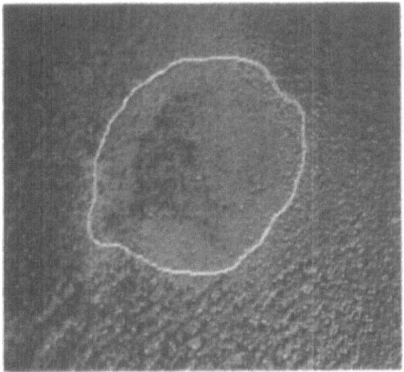

Figure 8.   Digital image processing of the oil slick at
B3 in Figure 4.

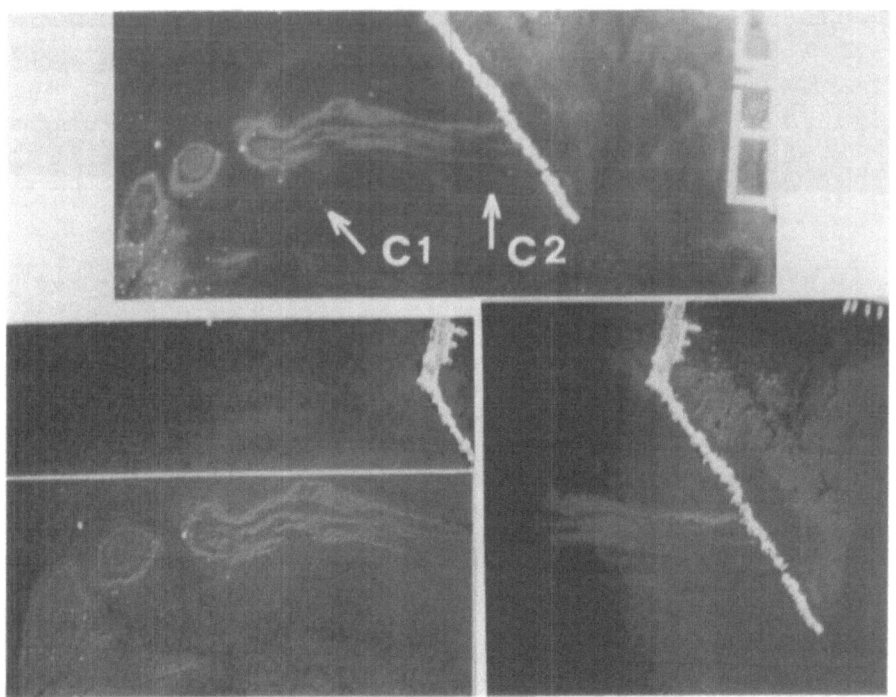

Figure 9.   Color infrared photograph of the oil slicks
at C (C1 & C2) in Figure 4 (top) and digital
image processing of the oil slicks at C1 and
C2 (bottom).

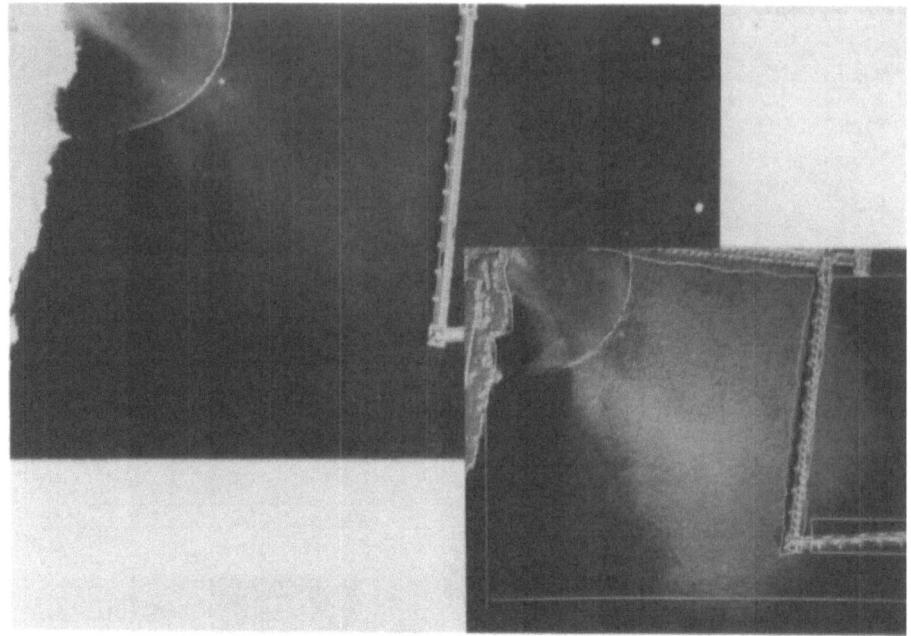

Fig. 10.   Color infrared photograph of the industrial discharge
           at D in Figure 4 (top, on the left) and digital image
           processing of the polluted area (botton, on the right).

        In addition to color infrared photography, several black-
and-white infrared photographs were collected from the lower
altitude during the survey.

        Thermal infrared imagery was found to be an effective tool
for the identification of oil on sea water. In fact, a number
of oil slicks were detected within the study area. Oil films
typically have a cool signature on thermal infrared images. An
oil film covering the surface of the sea produces a variation
in the emittance, due to the emission characteristics of the
oil-water surface. As reported by Estes[15] even if the thermome-
tric temperatures were the same, oil would record a lower radiometric temperature and exhibit higher reflectance characteristics than sea water. Therefore, refined and crude oils have
cooler signatures than the adjacent clean water. Buettner[16] in
1965 reported that the difference in radiant temperature between oil and water is approximately 1.4°C (2.5°F). However,
the oil radiometric temperature decreases as the film thickness
increases up to 50 μm[6]. Also a dependence of the oil emissivity
on the film thickness was indicated by Chandler[17]. This should
explain why in some cases infrared energy from the background

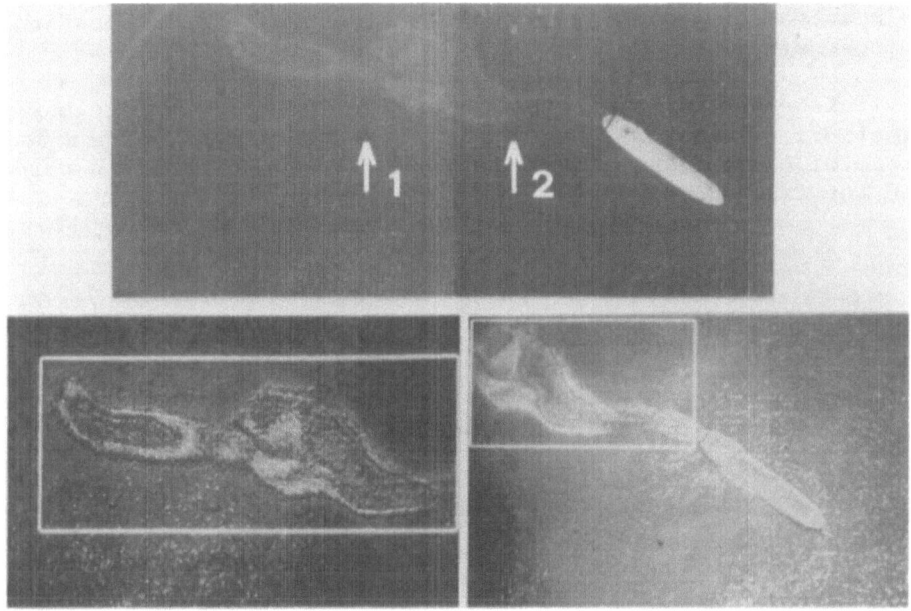

Figure 11.   Color IR photograph of an illegally discharging
             vessel (top) and digital image processing of the
             oil slicks at (1) and (2) (bottom)

is imaged through the oil slick, making warm streaks within
the typical cool signature of the oil. Figure 12 shows a mosaic
of color displays of calibrated thermal imagery on a map of the
study area. In these thermal images the values of temperature
increase in accordance with the following sequence of colors:
dark blue; violet; magenta; yellow; light blue, white (low
temperature ⟶ high temperature).

Figures on the right side of Figure 12 show the enlarge-
ments of the vessel at F illegally discharging oil into the
sea (top) and the oil covered sea surface at G close to the
loading berth of a refinery pier. The difference between radiometric temperatures of oil and water was calculated approximately 1.4°C (2.7°F), which is typical for oil-water surfaces.

As reported by Johnson[18], many investigators used quantitative analysis techniques to calibrate remotely sensed data,
and to map water quality parameters in the marine environment.
There is the need for calibrated measurements, by correlating
the remotely sensed data with the ground truth data. Based on
the results of the statistical analysis, Khorram[19] indicated the
regression models to represent the relationship between the

water quality measurements obtained from the sample sites and
the mean radiance value of remotely sensed data corresponding
to those sites.

Infrared photography and thermal infrared imagery clearly
detect oil slicks. It is generally easy to determine the areal
extent of a spilled oil on water, but is is difficult to determine the film thickness and the oil type on the imagery.

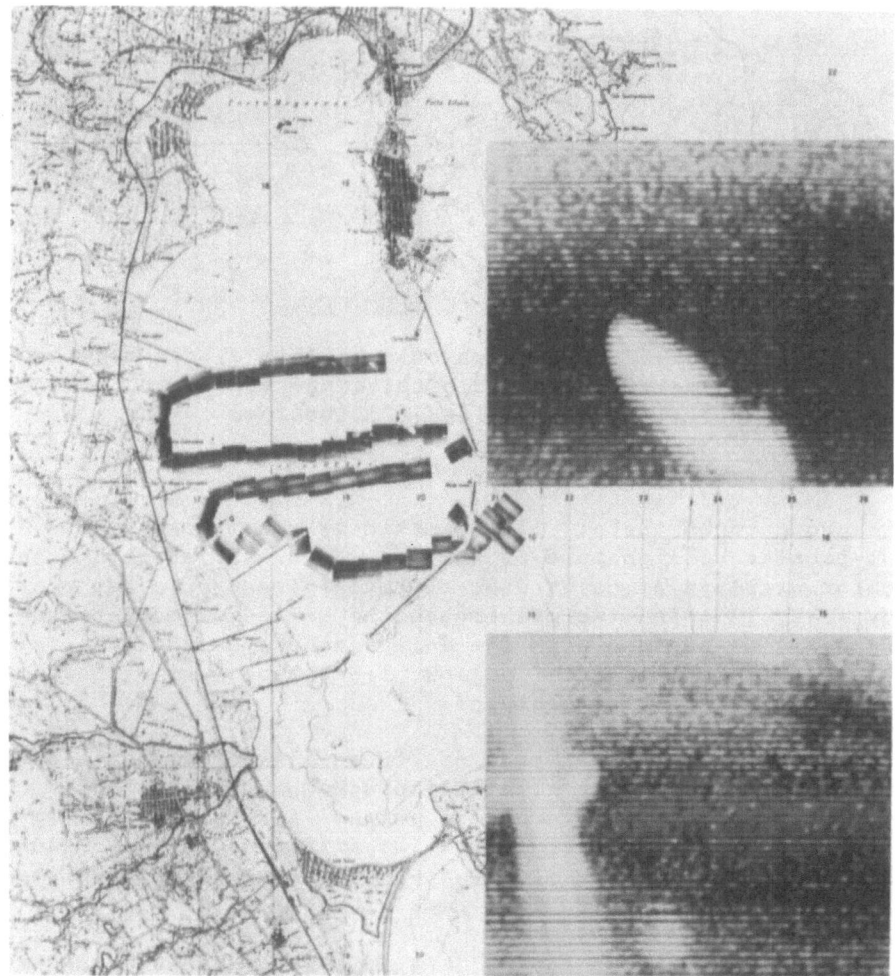

Figure 12. Mosaic of color displays of calibrated thermal
images on a map of the study area. On the right
side are color thermal infrared images of the
illegally discharging vessel at F (top) and of
the oil covered sea surface at G (bottom).

As suggested by Thaman[20], it is possible to map tonal or color levels within an oil slick, assigning different thicknesses to them. However, ground truth samples are required for calibration, in order to determine the correspondence between these levels and thicknesses, and between oil signatures and types. Also it has been suggested to correlate thermal infrared data with thickness and type of oil slicks.

Consistent with these investigators, in the present study it was virtually impossible to map water quality parameters, and determine thickness and type of oil slicks without calibrated measurements. The investigation did not include collection and laboratory analysis of water quality samples at predetermined sites in the study area for calibration of remotely sensed data. However, it is believed that the present study constitutes the first effort in this geographic area for water quality monitoring and oil pollution analysis using remote sensing techniques.

Further, it was proved the significant contribution of remote sensing to water quality monitoring and.oil pollution analysis. Particularly, it was proved that in terms of cost both techniques used were effective. However, the possibility of finding a meaningful correlation between remotely sensed measurements in coastal zones and the urban and industrial agglomerations bordering them, should be explored more. Water pollution remotely sensed measurements, and information coming from process plants and urban agglomerations in coastal zones should be stored together to establish a bank of data. With the help of this bank of data, those measurements and information should be processed together in order to make correlations between them. The final resulting information could be used to develop appropriate regulations and procedures to control pollution sources in the marine environment. Information derived from remote sensing data processed in this way could be very useful in the making of management decisions.

CONCLUSIONS

The early detection and monitoring of oil in the marine environment will significantly improve predictions of its location, nature, areal extent, and behavior over subsequent time periods. The treatment required and a determination of the responsibility of any damage will depend on the reliability with which the oil can be monitored under the prevailing weather conditions.

Remote sensing is of great importance in facilitating action designed to reduce and prevent pollution, providing an effective means of monitoring urban and industrial discharges.

Particularly, its capability to identify critical areas in coastal zones by monitoring pollution sources can greatly increase the effectiveness of a governmental agency that is responsible for pollution control.

The results presented in this paper showed how remotely sensed measurements and related observations can provide useful information on water quality and oil pollution in the marine environment. Based on these results, the following conclusions were indicated:

* Highly polluted areas and illegally discharging vessels were clearly discernible on suitably enhanced imagery made from acquired data.

* Infrared photography and thermal infrared imagery must be used in conjunction with each other to provide accurate interpretation of the data.

* According to the studies done by many investigators from different parts of the world, it was virtually impossible to map water quality parameters, and determine thickness and type of oil slicks without calibrated measurements.

* It is believed that the present study constitutes the first effort in this geographic area for remote sensing analysis of oil pollution in the marine environment.

* However, the possibility of finding a meaningful correlation between remotely sensed data in coastal zones and the agglomerations bordering them should be explored more.

FUTURE WORK

In the perspective of time and viewed in a multidisciplinary approach, remote sensing has already given significant contributions to water pollution control, and it will very likely give more in the future.

The results presented in this paper showed how infrared photography and thermal infrared imagery can provide useful information on water quality and oil pollution in the marine environment. However, additional studies of those phenomena associated with the photographic and thermal infrared responses of the major features of water and oil are required. These studies involve laboratory researches for calibration of remotely sensed data, by correlating them with water quality measurements, and thickness and type of oil slicks.

The authors plan to do work in this area in the very near

future. Also efforts are planned with Agusta, Helicopter Division, in Milano, Italy, in order to test a low-cost multi sensor system for use in helicopter marine environmental control operations. The system is expected to be used for other missions such as search and rescue, law enforcement and disaster relief.

ACKNOWLEDGEMENTS

The authors would like to thank Italian Navy for providing helicopter and pilots for the survey, and especially the personnel of the Naval Air Station (Maristaeli) at Catania, Italy, for assistance and support during the flights.

REFERENCES

1. A. L. Geraci, Remote sensing versus marine pollution, in: "The First Week of Photography", Kodak, Terrasini (1980).
2. D. J. Reish and J. K. Thomas, Marine and estuarine pollution, Wat. Poll. Control Fed., 6:1437 (1974)
3. J. P. Hollinger, Passive microwave sensing of slicks, in: "Proc. 9th Intl. Symposium on Remote Sensing of Environment, Ann Arbor" (1974).
4. J. P. Scherz, Remote sensing considerations for water quality monitoring, in: "Proc. 7th Intl. Symposium on Remote Sensing of Environment, Ann Arbor" (1971).
5. A. A. Allen and J. E. Estes, Detection and measurement of oil film, in: "Proc. Santa Barbara Oil Symposium", University of California, Santa Barbara (1970).
6. J. C. Munday, W. G. McIntyre and M. E. Penny, Oil slick studies using photographic and multispectral scanner data, in: "Proc. 7th Intl. Symposium on Remote Sensing of the Environment, Ann Arbor" (1971).
7. J. P. Millard and J. C. Arvesen, Airborne optical detection of oil on water, Applied Physics, 1:102 (1972)
8. K. P. B. Thomson, S. L. Ross and H. Howard-Lock, "Remote Sensing of Oil Spills", Environmental Impact Control Directorate, Economical and Technical Review, Report EPS 3-EE-74-2 (1974).
9. J. E. Estes and S. P. Kraus, "Airborne Remote Sensing Applications for the Detection and Monitoring of Oil from Natural Seeps and other Sources", California State Lands Division, Contract No. LC-6068 (1976).
10. A. L. Geraci, "Jonian Sea Water Exploration and Evaluation", Interim report, Machinery Institute, Department of Engineering, University of Catania (1978)
11. L. A. Bartolucci, Field measurements of the spectral response of natural waters, Photogrammetric Engineering and Remote Sensing, 5:595 (1977)

12. R. M. Watson and C. Y. Tippet, A report on some studies
     of marine oil pollution in the North Sea, in: "Resource
     Management and Research", London,(1978).

13. R. Hoffer, P. Anuta and T. Phillips, ADP, multiband and
     multiemulsion digitized photos, Photogrammetric Engi-
     neering and Remote Sensing, 10:885 (1971)

14. F. L. Scarpace, B. K. Quirk, R. W. Kiefer and S. L. Wynn,
     Wetland mapping from digitized aerial photography,
     Photogrammetric Engineering and Remote Sensing, 6:829
     (1981).

15. J. E. Estes, L. W. Senger and P. R. Fortune, Potential
     applications of remote sensing techniques for the study
     of marine oil pollution, Geoforum, 9:69 (1972).

16. K. J. K. Buettner, C. D. Kern and J. F. Cronin, The conse-
     quences of terrestrial surface infrared emissivity, in:
     "Proc. 3rd Intl. Symposium on Remote Sensing of Environ-
     ment, Ann Arbor" (1965)

17. P. B. Chandler, Remote sensing of oil polluted sea water,
     in: "Proc. 16th Annual Technical Meeting on the Environ-
     mental Challenges of the 70's", Institute of Environ-
     mental Sciences, Boston (1970).

18. R. W. Johnson and C. W. Ohlhorst, Application of remote
     sensing to monitoring and studying dispersion in ocean
     dumping, in: "Proc. 1st Intl. Ocean Dumping Symposium,
     Kingstone, Rhode Island" (1978).

19. S. Khorram, Use of ocean colour scanner data in water
     quality mapping, Photogrammetric Engineering and Remote
     Sensing, 5:667 (1981).

20. R. R. Thaman, J. E. Estes, R. W. Butler and J. M. Ryerson,
     The use of airborne imagery for the estimation of area
     and thickness of marine oil spills, in: "Proc. 8th
     Intl. Symposium on Remote Sensing of Environment, Ann
     Arbor" (1972).

(Editor's note: The authors were unable to present this paper
at the Paris workshop. It is included as an addendum to the
Washington and Paris proceedings because of its pertinence and
interest, particularly with regard to the Mediterranean countries which might be reluctant to undertake oil pollution
monitoring.

HIGH CONTRAST IMAGING OF AN OIL SLICK

BY MEANS OF A LOW LIGHT LEVEL TELEVISION

R.A. Neville[1], V. Thomson[1] and R.A. O'Neil[2]

[1]Intera Environmental Consultants Ltd.
Ottawa, Ontario, Canada
[2]Canada Centre for Remote Sensing
Ottawa, Ontario, Canada

ABSTRACT

Because the reflectance of oil films is higher in the near ultraviolet and near infrared, it is possible to enhance the contrast of oil slicks on the water surface by observing them through a Corning type 7-51 filter which transmits in the 320-420 nm and the 650-850 nm regions of the spectrum. The radiance arising from volume scattering in the water column can be reduced by viewing the surface at Brewster's angle (an angle of incidence approximately 53° for water) through a horizontal polarizing filter.

A silicon intensified target (SIT) television camera (RCA model TC1030 H) was fitted with a 12.5 mm lens to give a 42° x 54° field of view. To enhance the contrast of oil slicks on the water surface the filters described above were used. This configuration then accentuates sky light reflected from surface films. Other materials, such as dye, dissolved in the water column are not seen by the television camera.

High contrast images have been obtained with such a television system when flown over a natural oil seep in Baffin Bay and test spills in the Atlantic Ocean.

Though this is a relatively inexpensive system, it is believed to be useful both for surveillance and tactical purposes in oil spill countermeasures activities.

INTRODUCTION

A low light level television system can be an excellent sensor for viewing oil on water[1] for several reasons. First, the presently available Silicon Intensified Target (SIT) television tubes offer high sensitivity and low blooming, as well as good picture quality. Second, television systems offer real-time display of imagery suitable for both surveillance and tactical purposes. Line scanners and real aperture side looking radars acquire a single line of data at a time, whereas television systems present a dynamic image. The target remains in the field of view for several seconds for a rigidly mounted camera and for gimbal mounted or hand-held cameras, images of interest may be viewed for considerable lengths of time. Finally, television systems are inexpensive and easy to operate; they provide a considerable remote sensing capability for very little capital outlay.

Low light level televisions can have a spectral response in the region from 250 to 1,200 nm (different television tubes use different portions of this region). In this spectral region, oil has a greater surface reflectance than water. By careful choice of spectral region and polarization component, the contrast between oil and water can be maximized.

POLARIZATION

As discussed elsewhere[2] the total radiance reaching a remote sensor consists of a surface reflected component, a volume reflected component, and a path radiance scattered by the intervening atmosphere into the field of view of the sensor. The desired information on the presence of oil on the water is contained, for the most part, in the surface reflected component; the other two components then act as a contrast background radiance.

To maximize the oil-water contrast one should maximize the surface component to background component ratio. This can be accomplished by employing a polarizer.

For non-zero incident angles i.e. for look directions other than nadir, the surface reflected light is linearly polarized to a degree dependent on the incident angle. At Brewster's angle, which for water is approximately 53° from the vertical, the light is completely linearly polarized parallel to the water surface. On the other hand, the path radiance component, under most operational conditions, is less than 20% horizontally polarized. At the same time, the volume reflected component is vertically polarized to a slight degree. Hence, one can reduce the background radiation by approximately

50%, while preserving the component with the information content by looking at or near the Brewster's angle and filtering
the received radiation with a polarizer adjusted to pass the
horizontally polarizer component.

An additional factor to be considered is the degree of
polarization of the sky[3] as it is the sky light that is
reflected from the water or oil surface. The sky polarization
depends on the angle between the look direction and the sun's
position. In practice, one would normally direct the sensor
away from the sun to avoid glitter. It happens that this is
also the correct procedure to increase the oil-water contrast
when viewing at or near the Brewster's angle.

SPECTRAL REGION

Continuing with the objective of maximizing the oil-water
contrast one can achieve improved results by restricting the
observed spectral range to that in which the contrast is
greatest. There are a number of factors to be considered: the
spectral variation of oil-water contrast in the surface
reflected light; the spectral absorption properties of water
and oil as it affects the volume reflected light; the spectral
qualities of skylight vs sunlight, the former contributing to
the surface reflected light, the latter to the volume and
atmospheric components, the sensitivity of the detector being
used.

Index of refraction measurements[4] indicate that the
contrast in the surface reflected light is highest in the UV.
From this and the fact that for a clear sky the sky light to
sunlight ratio is highest at the blue end of the spectrum, one
concluded that the near ultraviolet/violet region should be
good under clear sky conditions. This must be tempered somewhat
by the declining light levels and television sensitivity as the
wavelength decreases. In addition the absorptivity of most oils
increases with shorter wavelengths[4].Thus, one might expect
that for optically thick layers of oil, the increased surface
reflectance from the oil would be offset by a decreased volume
reflected component from the water below the oil.

For overcast skies, although the global (sky plus sun)
radiance is reduced, the sky light to global radiance ratio is
increased; as a result, the surface reflected component is
enhanced compared to the volume and atmospheric components. At
the same time, the spectrum is shifted into the red, the clouds
appearing white as opposed to the blue of the clear sky. Added
to this is the increased absorptivity of water in the red,
giving a lower volume reflected component, and the reduced
atmospheric scattering for the longer wavelength. Therefore,

although the oil-water contrast in the surface component may
be lower in the red part of the spectrum, the reduced background components, volume reflected and atmospheric, especially for overcast skies, favours the red-near IR region.

TELEVISION CAMERA AND FILTERS

In the experiments being reported here, a film polarizer
that is effective in the near UV down to 275 nm (Polaroid
HNP'B), as well as in the visible, was fitted to the television camera lens. It is noted that most commonly available
film polarizers are not effective in the UV.

The spectral range was narrowed down to cover the two
regions, near ultraviolet/violet and red/near infrared,
expected to give optimal results. This was accomplished by
using a Corning 7-51 filter which passes light in the bands
310-420 nm and 700-1,100 nm (Figure 1).

Specifications are given in Table 1 for the low light
level television as flown with the polarizing and spectral
filters. Figure 1 displays the spectral sensitivity of the
television camera as fitted with the prescribed filters. The
units are NEN (Noise Equivalent Radiance) per $W \cdot m^{-2} \, sr^{-1} \, 10^{-5}$.
It is noted that the NEN value was measured to be that radiance

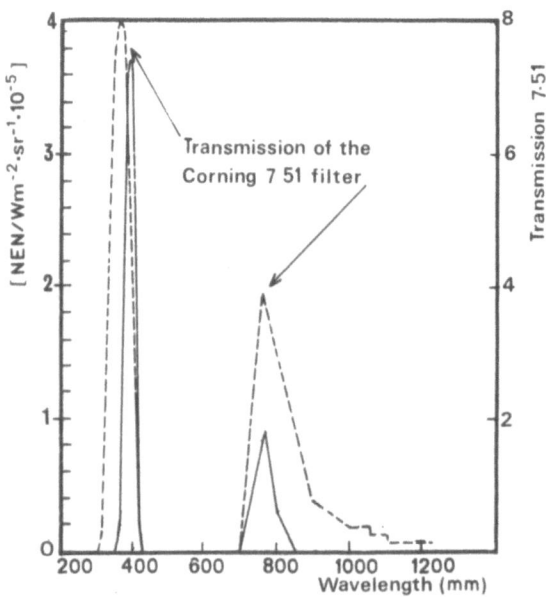

Figure 1. Sensitivity of the RCA TC1030H camera with
           horizontal polarizer and Corning 7-51 filter.

Table 1. Specifications for the low light level television system as flown on a CCRS aircraft.

CAMERA	RCA TC1030H
LENS	Fujinon TV LH15A02 f1.4 12.5 mm focal length 42.4° x 54.4° field of view
SENSITIVITY	10 NEN/W $m^{-2}sr^{-1}$ (350-740 nm full width)
POLARIZER	HNP'B horizontal (240-740 nm full width)
FILTER	Corning 7-51 Transmission: 310-420 nm 700-1100 nm full width
LOOK ANGLE	53° off nadir, forward
SENSITIVITY as outfitted	9.5 x 10 NEN/W m^{-2} sr^{-1} (370-420 nm full width) 6.6 x 10 NEN/W m^{-2} sr^{-1} (690-840 nm full width)

required for the camera to produce a just discernable image of a half-white/half-black target. These results are for the camera fitted with a Fujinon-TV LH15A02 lens; the system response could be extended another 20 nm into the UV if a special quartz glass lens was substituted.

RESULTS

Data were obtained over a natural oil seep in the Scott/ Inlet area of Baffin Bay and over a test oil spill at Dump site 106 off the coast of New Jersey. Imagery from these missions is shown in Figures 2 to 5. Figures 2 and 4 are photographs of a television monitor displaying video-taped data. Figures 3 and 5 were produced on a scan printer (Honeywell Visicorder 1856A) by exposing the same scan line from subsequent television frames; thus, the television system is used as if it was a 30 lines/s line scanner.

It is noted that by photographing the television monitor, one loses the advantage of a dynamic picture. Hence the effective quality of the imagery in Figures 2 and 4 is reduced from that of the original television picture. In Figures 3 and 5, the content of the oil spill image is regained to some extent although the dynamic quality is still missing.

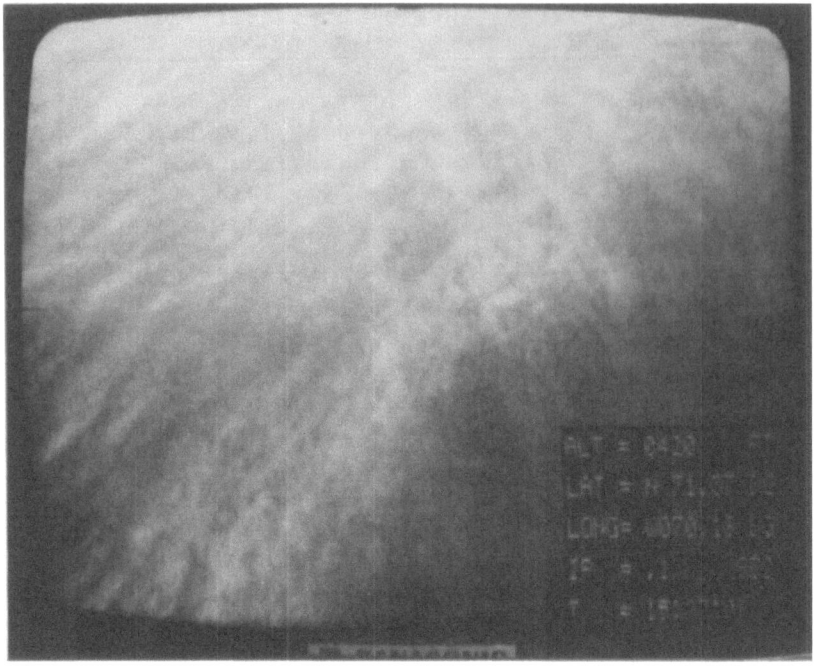

Figure 2. Picture of television monitor showing an oil seep
           off Baffin Island, September 1978. Annotation is
           video mixed on display; it originates from another
           sensor.

     In short, it is difficult to equal the effect of the orig-
inal television picture by means of a static representation;
this simply points out one of the main advantages of the television system.

DISCUSSION

     In general, the low light level television system,
filtered as discussed in section above, provides high contrast
dynamic imagery of oil slicks under various sky conditions. It
has proven to be a valuable tactical sensor as it provides the
operator with real-time imagery. Using this system, an operator
can easily and readily distinguish an oil slick from, for
example, a wind slick or any slick not caused by a surface oil
film having optical properties similar to those of an oil film.

     As discussed in section "Polarization" and in section
"Spectral region", much of the improved contrast, as compared
to visual sightings, can be attributed to the polarizing and

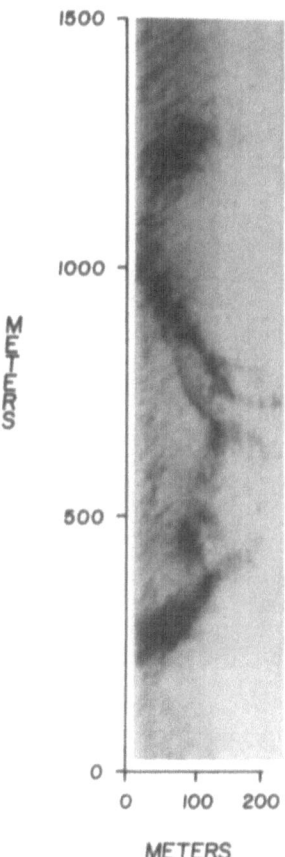

Figure 3. Low light level television picture formed on a scan
           printer by exposing one scan line per television
           frame (Oil seep off Baffin Island, Sept.19, 1978,
           flight line 16).

spectral filters. It is because these filters severely attenuate
the radiance from an already dark sea surface that the high sensitivity of the low light level television system is required.

     There are situations, however, in which the LLLTV system
has failed to provide adequate imagery. In one instance the
system was flown at night over an oil spill in Montreal harbour.
It was impossible to determine if the ambient skylight was sufficient to produce an image because the presence of even one
harbour or ship's light in the camera field of view caused the
automatic gain control and the auto-iris to reduce the sensitivity drastically. This problem could possibly be circumvented
by the use of an automatic iris/eclipser control system at the

Figure 4. Picture of a TV monitor showing oil spill
           at Dump site 106 off the New Jersey coast.

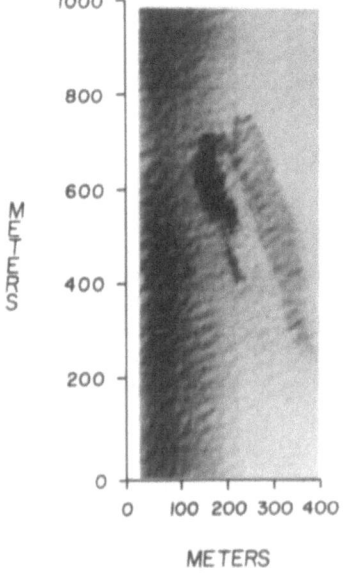

Wallops Island
Sortie 3
Flight line 6
3 Nov., 1978

Figure 5. Scan print of a LLLTV; oil spill at Dump site 106.

risk of damaging the camera if focused on a stationary light
source for an extended period of time.

The LLLTV system has encountered sea surface conditions
under which it was unable to provide unambiguous imagery.
These include high sea states where foam is indistinguishable
from oil. In another case, slush and patches of mixtures of
water and small bits of ice were not discriminated from oil.
To solve these problems, one might consider using a polarizer
that can be switched quickly between horizontal and vertical
polarization modes synchronously with the 30 Hz picture repetition rate. The system should display oil as a flashing signal
whereas the remainder of the scene, including foam and ice,
would remain constant.

REFERENCES

1. J. Millard and J. Arvesen, "Development and Test of Video
    Systems for Airborne Surveillance of Oil Spills", Dept.
    of Transportation report CG-D-95-75, Washington, D.C.
    (1975).
2. R. A. Neville, V. Thomson, K. Dagg and R. A. O'Neil, An
    analysis of multispectral line scanner imagery from two
    test oil spills in: "Proc. Washington Workshop of the
    NATO/CCMS Pilot-study on the Use of Remote Sensing for
    the Control of Marine Pollution", United States Coast
    Guard, Washington, D.C. (1979).
3. J. Millard and J. Arvesen, Effects of skylight polarization,
    cloudiness, and view angle on the detection of oil and
    water, in: "Proc. Joint Conf. on Sensing of Environmental
    Pollutants, Palo Alto, California" (1971).
4. R. Horwath, W. Morgan and S. Stewart, "Optical Remote
    Sensing of Oil Slicks: Signatures Analysis and Systems
    Evaluation", Dept. of Transportation report DOT-CG-92
    580-A, Washington, D.C. (1971).

OIL ON THE SEA: APPLICATION OF PATTERN RECOGNITION TECHNIQUES

TO THERMAL INFRARED AND SPOT IMAGES

M. Brussieux and G. Massart

Centre National pour l'Exploitation des Océans
Centre Océanologique de Bretagne
Brest, France

ABSTRACT

Evaluation of amounts of oil spilled at sea after any kind
of accident (tanker or oil rig) which would occur by the shore
or in the open sea: the most interesting results are obtained
through analysis of digital informations given by scanning radiometry, the image appearing on a video screen while informations are treated. Direct mapping is possible. As slight differences in contrast can be translated into thicknesses, it is
also possible to evaluate the amount of oil contained in a
slick.

INTRODUCTION

In case of a large oil spill at sea, it is necessary to
provide the on-scene commander with the useful informations
about the spill, which are:

* surface of the slick, and areas of the thin and thick
  oil layers;
* amount of oil contained into the slicks or, more easily,
  an estimation of this amount with particular attention
  paid to the thickest areas;
* mapping of the slicks, showing the position of the thin
  and thick areas.

Remote sensing and automatic processing of the data seem
to be the only mean of getting these informations early enough
to allow their efficient use.

So, we are going to explain what can be done using the
data given by a single channel thermal infrared scanning radiometer and with SPOT satellite images.

THERMAL INFRARED IMAGE PROCESSING

When oil is poured at sea, observation of the slick through
a thermal infrared radiometer generally shows hydrocarbons to
be "colder" than the sea. When the amount of oil is large enough,
one can sometimes see "hot" areas appearing in the slick, and
in-situ measurements show that these areas correspond to thick
oil layers.

Experiments conducted in the past, such as Polumer 75,
organised by the Centre National pour l'Exploitation des Océans
(CNEXO) and the French Petroleum Institute (IFP) allowed observations of spills using radiometry either in the 10 to 12 µm
or in the 0.7 to 1.1 µm wavelengths ranges. A comparison of
the two sets of data showed that the short wavelengths (visible
to close infrared) allow detection of very thin layers (sheens)
while thermal infrared detects layers the thickness of which
is at least equal to the wavelength (i.e. 10 µm).

In that particular case, "hot" areas did not appear in the
slick, and calculation established a mean thickness of 52 µm
using as a data the known volume of oil spilled to create the
slick. This result is in good agreement with that is generally
accepted when using thermal infrared observation -i.e. cold
oiled areas are covered with a 10 to 100 micron thick oil
layer-.

More recent measurements indicated a lowest detection
limit for infrared radiometry at 25 to 50 micron layer thicknesses, and mean thicknesses of up to 1,000 micron for the hot
areas[1].

Taking these data into account, a methodology has been
studied in Brest for automatic estimation of areas and amounts
of oil on thermal infrared images.

Figure 1 shows a radiometric profile obtained above pure
water and Figure 2 another profile above water and oil (after
low-pass filtering). The distorsions observed on the first
signal have various origins: radiometer drifts, position of the
sun, etc. On the second signal, those parasitic effects are
superimposed to positive and negative edges of contrast corresponding to cold and warm areas of the image.

Substracting the first signal point by point from the
second thus gives direct information upon presence or absence

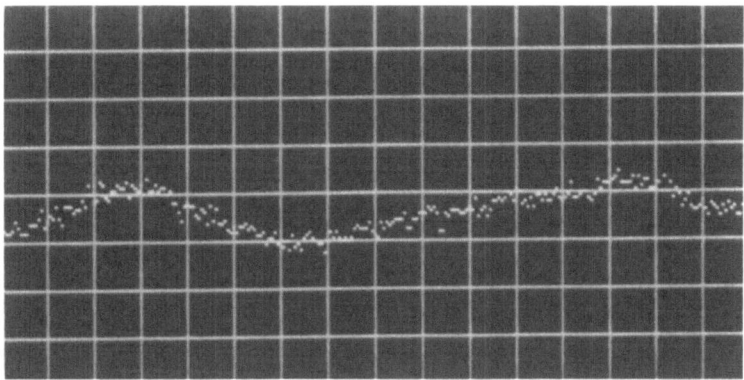

Figure 1:   Distortion of a thermal profile
            on non-polluted sea-water.

of oil on the surface. In practice, the detection is implemented
as following:

   * During a first step, the profile corresponding to unpol-
     luted areas is "learnt" by the machine (stored in memory)

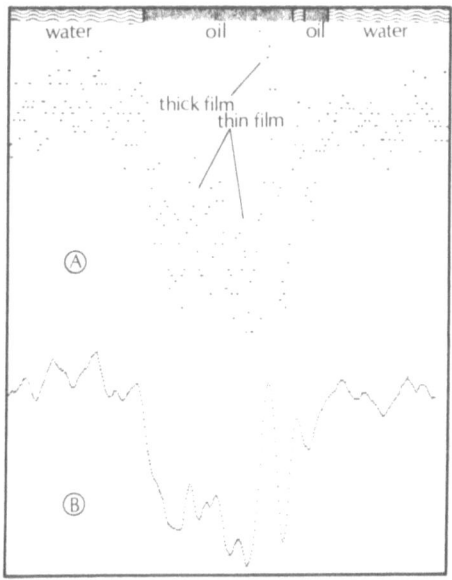

Figure 2:   Improvement of the signal to noise ratio by using
            triangulary smoothing of imagery.

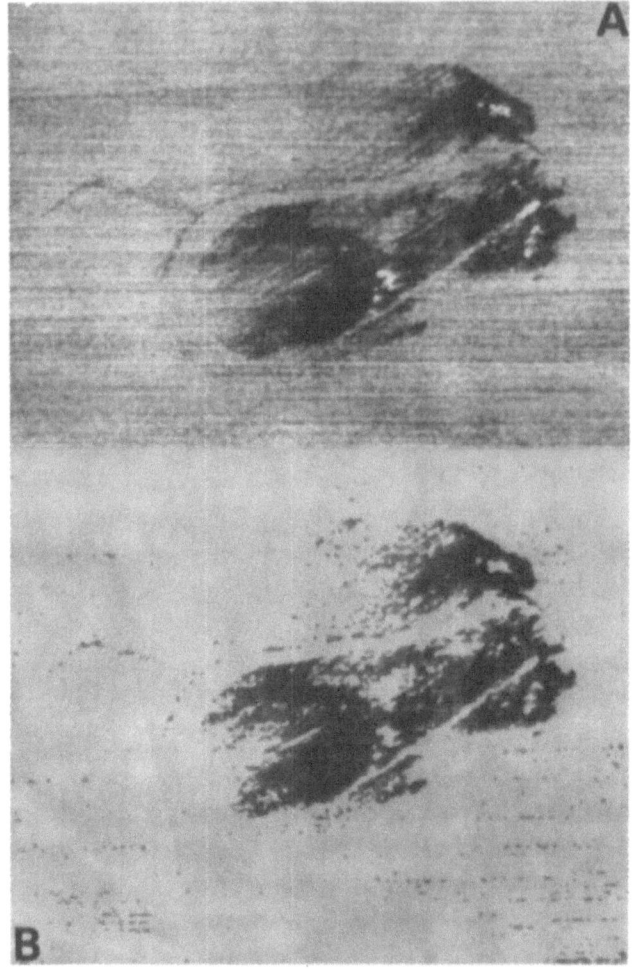

Figure 3.   Top: A thermal infrared imagery of a test oil spill
                 (Protecmar II exercise)
            Bottom: The same imagery after computerized treat-
                 ment. The oil is automatically distinguished
                 in real time on the noisy imagery.

* This typical profile is then substracted from the subsequent lines of image coming out from the scanner.
* An alarm is set each time a positive or negative change
  in level is recognized on the outcoming signal. Hinckley
  detectors have been proved to give optimal results for
  this operation when signal/noise ratio is weack[2,3].
* Between these alarms, a finite-state automaton is pro-
  grammed for recognizing which parts of the signal are

"containing" oil, according to the context: for example, sets of warm points, like ships, are not detected.
* Following the detection, computed corrections are brought to estimate dimensions, areas and amounts.

The black-and-white image of the extracted slicks (Figure 3) is displayed through pseudo-colour memories on a video-colour monitor.

The methodology has been tried and gave good results on thermal infrared images coming from Protecmar II experiments. Errors did not exceed 20% in the estimation of amounts for this training set.

A hardwired version of the algorithm is being studied in Brest for real-time oil slicks detection and mapping, on board of the plane carrying the infrared radiometer.

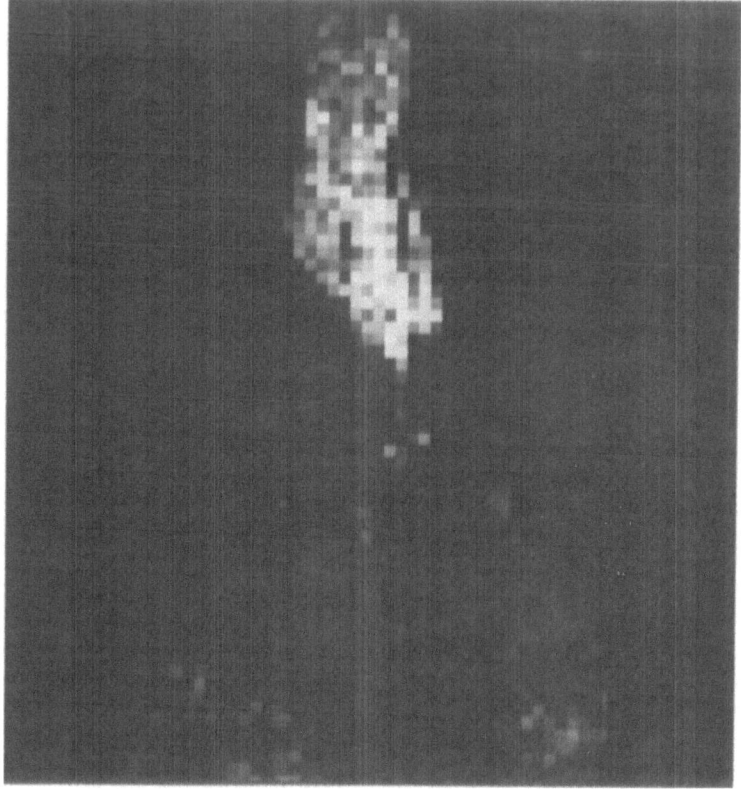

Figure 4. A test oil spill as seen from satellite SPOT (SPOT simulation, Protecmar II exercise, 1981).

ANALYSIS OF SPOT IMAGERY

SPOT channels give information in the visible and near infrared bands. The pixel size is 20 meters, which has been enough for observing the Protecmar slicks and developing the following methodology:

* A principal components analysis on the three channels shows that more than 98% of the total amount of information can be condensed into one image plane (Figure 4). This result was obtained because of the strong initial intercorrelation between the three channels.

* Thresholding the gray level histogram on the first principal plane allows the separation between hydrocarbons and water.

* A new principal components analysis on the isolated slick now condenses more than 99% of information in the two principal planes.

* Drawing a bidimensional histogram of those two sets of pixels leads to observe "clusters" among them, corresponding to:
   - thin oil layers;
   - thick oil layers;
   - boats and other floating objects.

If one assigns a different colour to each cluster, it is possible to obtain thematic maps of pollutions.

CONCLUSION

Image processing on views of Protecmar slicks gave excellent results. It may be noticed that SPOT satellite would easily detect a few cubic meters of oil on the sea, and moreover give informations on the thicknesses of the layer. New experimentations should allow, by simultaneous recordings of the sea-truth, to increase precision in the results obtained.

REFERENCES

1. H. Parker and D. Cormack, Evaluation of infrared line scan and side-looking airborne radar over controlled oil spills in the North Sea, in "Airborne Remote Sensing of Oil Spills in Coastal Waters", Proceedings of the first workshop sponsored by Working group I of the French pilot-study for the NATO Committee on the Challenges of Modern Society, held at Washington, D.C., U.S.A. (1979).

2. Gasnier, "Detection et Suivi des Contours: Etude Comparati-
       ve de Méthodes Séquentielles et Application à des Radio-
       graphies et des Images T.V.", Thèse de 3ème cycle, Uni-
       versité de Rennes I (1980).
3. Basseville, Edge detection using sequential methods for
       change in level, Inst. Electrical & Electronics Engineers
       Publication, trans on Accoustics Speach and Signal pro-
       cessing, Vol. 29, No. 1 (1981).

AN ANALYSIS OF MULTISPECTRAL LINE SCANNER IMAGERY

FROM TWO TEST OIL SPILLS

R.A. Neville[1], V. Thomson[1], K. Dagg[2]
and R.A. O'Neil[3]

[1]Intera Environmental Consultants Ltd.
Ottawa, Canada
[2]The Genesys Group
Ottawa, Canada
[3]Canada Centre For Remote Sensing
Ottawa, Canada

ABSTRACT

Results of flights of a Daedalus model 1260 multipectral
line scanner and a Daedalus model 1230 infrared/ultraviolet
line scanner over two test oil spills off the New Jersey coast
are reported.

The Daedalus multispectral scanner has 10 channels in the
spectral range of 0.38 to 1.1 μm and an eleventh in the thermal
infrared from 8.5 to 12.5 micron. A rotating mirror covers a
scan angle of 85.9° with an instantaneous field of view of
2.5 milliradian. High contrast imagery of the oil slick was
obtained in the thermal infrared channel. Images of oil spills
in the near-infrared channels showed only the thicker regions
of the oil slicks and these with less contrast than was apparent in the thermal infrared channel. Imagery from succesively
shorter wavelength channels showed a larger area slick corresponding to the thinner.outer boundaries. The area of the
slick observed in the 0.70-0.79 micron range was nearly equal
to that in the 8.5-12.5 channel. Sun glitter from the sea
surface can degrade the imagery in all but the thermal channel
though it was observed to be slightly reduced in the blue
channel (wavelength less than 0.5 micron).

The Daedalus model 1230 dual channel line scanner was also

flown over the test oil spills. It was fitted with a thermal
infrared detector (8.5 - 12.5 μm) and an ultraviolet detector
(0.27 - 0.37 μm). The scan angle is 77.3°. The UV/IR instantaneous field of view is 5.5/2.5 milliradian. The thermal infrared imagery obtained with this scanner was identical to that
obtained with the multispectral scanner. Two techniques were
used to display the data from the UV and IR channels. The
first was simply to colour density slice the data from each
channel separately and to overlay the two images. In the
second approach, for each pixel in the image, the pair of
intensity values (one the UV, the other the IR) plotted in a
two dimensional vector space. The points in this spectral space
are grouped into clusters containing nearest neighbours, and to
each cluster a colour is assigned. Each image pixel represented
by a point in a given cluster acquires the cluster's colour,
thereby producing a thematic image in which the themes
correspond to oil films of different thicknesses. Though this
produced essentially the same image as was obtained by the
simple photographic overlay technique, it may be of value in
the development of data acquisition systems or real-time
displays for this type of data.

INTRODUCTION

It has been shown[1] that oil slicks on water are detectable
by observing the reflected light in the near ultraviolet,
visible, and near infrared regions of the electromagnetic
spectrum. It is also known that oil slicks are "visible" in
the thermal infrared because of their effects on the apparent
temperature of the surface. The object of this study is to
determine the best spectral regions for the detection of oil.
This report presents scanner data taken in the thermal IR
(8.5 - 12.5 μm) and in 8 bands in the 270 nm/1,100 nm range.

EFFECT ON REFLECTANCE

Indices of refraction of representative crude oils have
been measured in the visible and near IR region. From these it
can be calculated that the surface reflectance of oil is
greater than that of water by approximate factors of 2.1 at
400 nm and 1.6 at 1,200 nm. One can speculate, therefore, the
oil-water contrasts are greater for shorter wavelengths.
However, there are other factors to be considered. The total
radiance reaching a remote sensor consists of a surface
reflected component, a volume reflected component, and a path
radiance scattered by the intervening atmosphere into the
field of view of the sensor. An increase in one of the latter
two background radiances reduces the contrast available in the
surface component. The volume reflected radiance for water has
a maximum in the 450 - 650 range with a magnitude that may be

comparable to the surface reflected component. The path
radiance has a λ^{-4} dependence resulting in greater intensities
in the UV-blue range. This latter contribution varies as the
path length and at higher altitudes can dominate the total
radiance. An additional factor is the magnitude and spectral
distribution of the sky radiance as compared with the global
radiance; the former being reflected by the surface, the latter
reflected by the water volume and the atmosphere. Therefore,
the cloud cover can affect the contrast. The contrast may not
arise solely from the surface reflected light. Different oils
and/or different states of emulsification may increase or
decrease the volume reflected component as compared with that
from water. Optically thick layers of oil, since oils absorb
preferentially in the UV-blue-green, may appear darker in
these regions in spite of the increased surface reflectance.

Another important consideration, adressed elsewhere[2], is
the effect on polarization introduced by surface reflection at
non-zero incidence angles and by skylight polarization properties.

The results reported on here apply primarily to the
problem of the variation of contrasts with wavelength in the
near-UV to near-IR range. Two test spills of different crude
oil types were observed under similar sky conditions, sea
states and water types. Flights were made at different altitudes
but the effect of this variation will not be addressed here.
Polarization effects cannot be dealt with since the scanners
do not employ polarizers.

THERMAL EFFECTS

The apparent temperature measured by a thermal radiance
measuring device is affected by both the physical (real)
temperature and the thermal emissivity. An oil slick can cause
changes in both. The presence of an oil layer on the surface
of the sea can have one, or any combination of a number of
effects on the physical temperature of the surface:

* Optically thick oil can absorb sufficient solar
  irradiation to heat it above the water temperature;
* Oil near the source of the spill may retain the source
  temperature which could be different from that of the
  water;
* Volatile components from freshly spilled oil may provide
  cooling by increased evaporation;
* Less volatile oils may impede evaporation of the water,
  thereby producing a warming effect;
* An oil film can impede the flow of heat at the water
  surface, resulting in an increased or decreased surface

temperature depending on whether the air temperature
is above or below the water temperature.

The thermal emissivity (efficiency as a radiator of thermal radiation) of oil is generally less than that of water.
Thus, oil at the same physical temperature will appear colder
than water. For emissivities of .95-.99 and temperatures in
the range 275-295°K, a radiance measurement at a wavelength of
10.5 micron gives apparent temperatures that decrease by
0.6°K for a 0.01 decrease in emissivity. Oil is variously reported[3] to have an emissivity 0.02-0.05 less than that of
water giving a 1-3°K temperature contrast.

LINE SCANNERS

Two imagers, one a Daedalus model 1230 dual channel line
scanner (DCLS), the other a Daedalus model 1260 multispectral
scanner (MSS) were flown on two separate CCRS aircraft over
the same test oil spills.

The dual channel scanner is fitted with an ultraviolet
detector (silicon) sensitive in the 0.27-0.37 micron region,
and a thermal infrared detector covering the range 8.5-12.5
micron. This scanner scans at 77.3° scan angle at 60 lines per
second and has an instantaneous field of view of 5.5 mrad for
the UV detector and an instantaneous field of view of 2.5 mrad
for the IR detector.

The multispectral scanner has, in one part, a thermal IR
(8.5-12.5 μm) detector identical to that in the dual channel
scanner. In the other part is a silicon diode array spectrometer that divides the near UV to near IR range into 10 bands
as listed below:

MSS Band	Wavelength range
1	0.38 - 0.42 μm
2	0.42 - 0.45
3	0.45 - 0.50
4	0.50 - 0.55
5	0.55 - 0.60
6	0.60 - 0.65
7	0.65 - 0.69
8	0.70 - 0.79
9	0.80 - 0.89
10	0.92 - 1.10
IR	8.5 - 12.5

This scanner scans an 85.9° angle at 12.5, 25, 50 or

100 lines per second and has an instantaneous field of view of
2.5 milliradian.

TEST OIL SPILLS

The data for this report were collected over two intentional oil spills conducted by JBF Scientific for the American
Petroleum Institute. Each spill contained 10 bbl of oil, Merban
crude on November 2, 1978, La Rosa crude on November 3, 1978.
Remote sensing aircraft and/or personnel from NASA, U.S. Coast
Guard, ERIM and CCRS participated in the remote sensing aspects
of the experiment. Data have been shared and ultimately will be
correlated to determine the relative performance of the various
sensors. The slicks were imaged by the MSS on 47 passes and by
the DCLS on 11 passes.

RESULTS AND DISCUSSION

Thermal Infrared

The thermal infrared band is found to distinguish two
distinct regions within the oil slick, one region brighter
(hotter) than the water, the other darker (colder), as shown
in Figure 1. Each of these main regions can be resolved into
sub-regions corresponding to smaller temperature intervals.

These effects are observed in data from both the MSS and

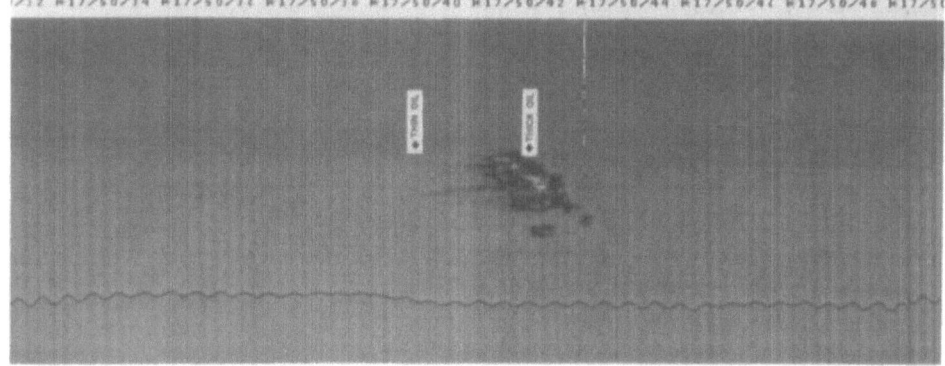

Figure 1. MSS thermal IR unenhanced image showing the effects
           of an oil slick on the apparent temperature of the
           surface. Cold appears black, hot, white. The temper-
           ature range covered is 280.3-294.9°K. Data was col-
           lected over the Wallops island oil spill at 17:50
           GMT, November 2, 1978.

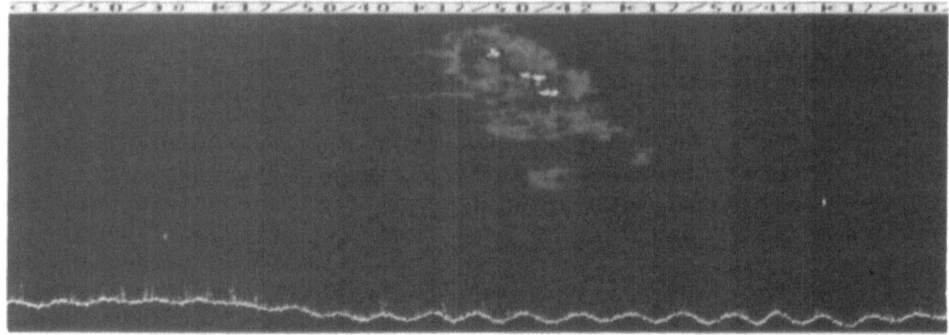

Figure 2. **MSS** thermal IR image as in Fig. **1** is here grey level
            sliced and colour-coded to enhance the contrasts
            between the various regions in the oil and the water;
            white corresponds to apparent temperatures in the
            range 289.2-290.3°K, the immediate shade of yellow
            to the range 285.8-286.4°K, and red to the range
            287.0-287.7°K.

the DCLS IR channel for both crude types and for the duration
of the observation periods which extended up to three hours
after the spills. It is therefore assumed that the oil had
reached an equilibrium temperature and that these effects were
not transient phenomena associated with a different source
temperature or additional evaporative cooling resulting from
the volatile crude components. It is concluded that the hotter
region is the result of solar heating of an optically thick
oil layer and that the apparently cooler region is the result
of the lower emissivity of oil.

     Radiance calculations using the data from the Merban (La
Rosa) crude oil indicate that the apparent temperature of the
hottest region is 2.5°K (6.4°K) above that of the water while
the coldest region is 2.4°K (4.2°K) below. The emissivity of
the oil is calculated to be .949 (.921) assuming that of water
is .988. The hottest region of the oil would then have had a
physical temperature of 4.9°K (10.9°K) above the water temperature.

     In addition to the preceding distinctive effects there
are subtle differences in the apparent temperature observed in
the thin slick areas. In the main body of the thin slick,
there is a region slightly warmer than the water surrounded by
a slightly cooler fringe. The warming could be explained by
assuming that the oil is too thin for its lower emissivity to
have an effect, but is sufficient to impede evaporation from

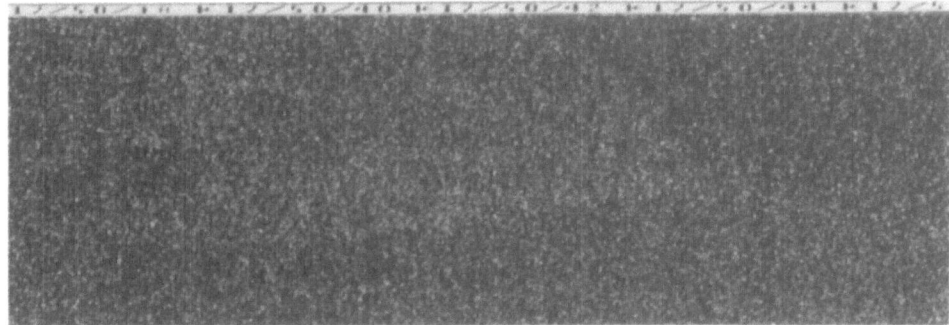

Figure 3.   **MSS** channel 2 colour enhanced image showing the
            extent of the thin oil film. Contrast is poor
            because of the detector's low signal to noise level
            in the channel 2. (Wallops island oil spill, 17:50
            GMT, November 2, 1978, C-GRSA flight line 13).

the water. The apparently cool fringe might be caused by interference effects.

    As shown in Figure 2, the IR image has been grey level
colour coded to accentuate the different slick regions, white
corresponding to the thickest region.

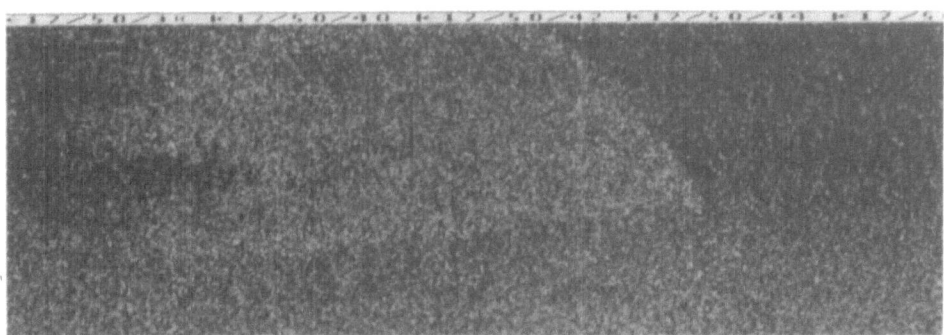

Figure 4.   **MSS** channel 3 colour enhanced image showing the
            extent of the thin oil film. Contrast improvement
            over the channel 2 is attributed to the better
            signal to noise level in this band. Note that the
            thick oil region is not as bright as the thin.
            Data was collected over the Wallops island oil
            spill at 17:50 GMT, November 2, 1978, on C-GRSA
            flight line 13.

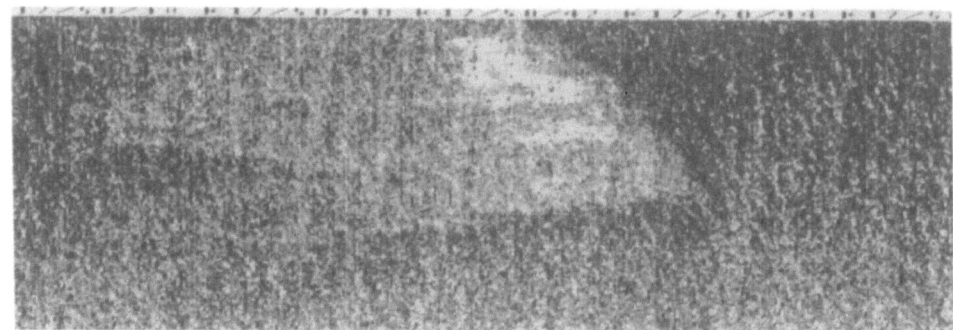

Figure 5.   MSS channel 4 colour enhanced image showing the thin
             oil film. Note that the thick oil shows brighter
             here than in channel 3. (Wallops island oil spill,
             17:50 GMT, November 2, 1978, C-GRSA flight line 13).

Visible to Near-infrared

        MSS images of the Merban crude oil slick show that the
thinnest areas of the slick are detected in MSS bands 2 and 4,
but as the MSS band number or wavelength region increases, the
oil thicknesses required for detection also increase. Bands 9
and 10 see only the thickest regions of the oil slicks. This
effect is apparent in Figures 3 to 9. These images have been
colour enhanced to accentuate the oil slick. In MSS channels 6
to 9, two oil thicknesses can be distinguished, thick coded
white, thin yellow in these images, the radiance levels being

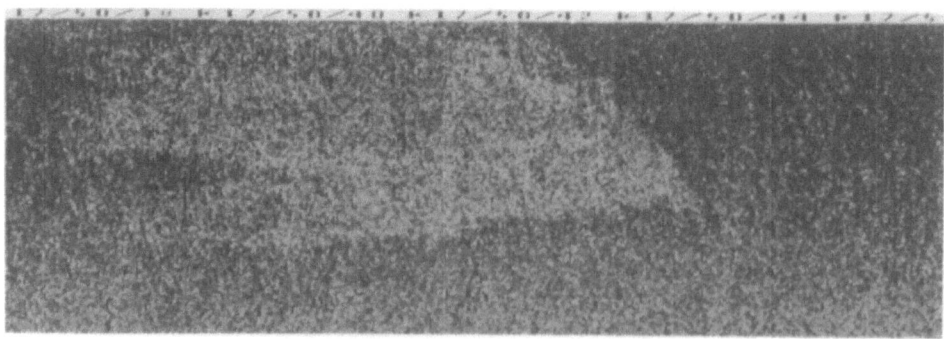

Figure 6. MSS channel 6 colour enhanced image showing two
          intensity levels. The brighter (shown white) is the
          thicker oil; the less bright (shown grey) is the thin
          oil film (Wallops Island oil spill, 15:50 GMT, Nov. 2,
          1978).

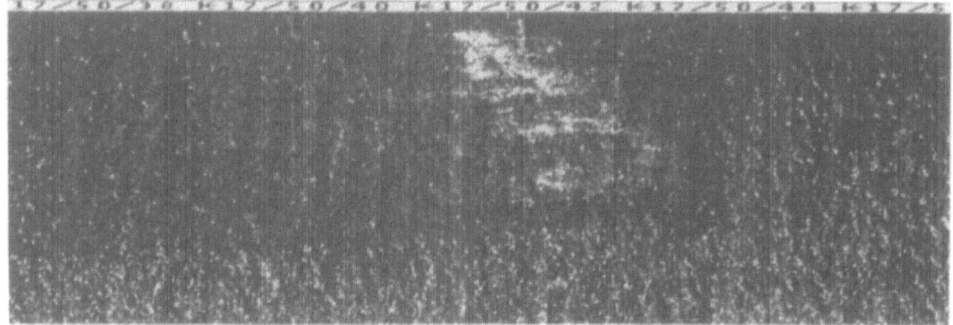

Figure 7.   MSS channel 8 colour enhanced image showing two in-
            tensity levels. The thick oil is brightest (shown
            white); the thin is not so bright (shown grey). Note
            that the thin oil film-water contrast is not so good
            as in channel 6.(Wallops Island oil spill, 17:50 GMT
            November 2, 1978, C-GRSA Flight line 13).

higher for the thick regions. In MSS channels 8-10, the outer
edges of the thin film region of the slick appear as areas of
reduced reflectance, possibly caused by the suppression of the
higher wave slopes.

    It must be pointed out, however, that the visibility of
the thick layers of oil in these channels depends on the oil
type and/or its state of emulsification. For La Rosa crude oil,

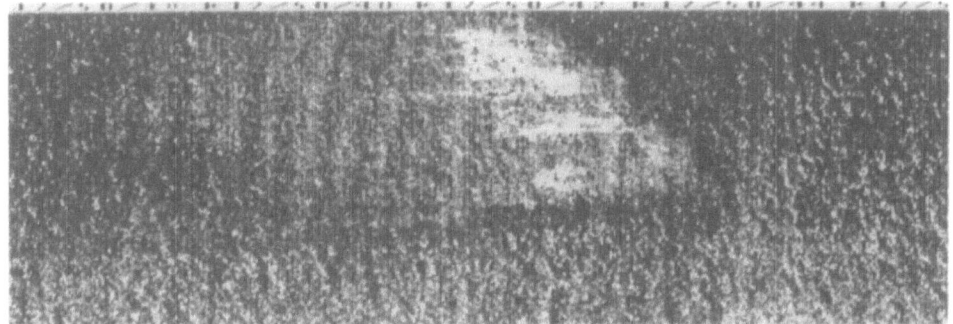

Figure 8.   MSS channel 9 colour enhanced image showing two in-
            tensity levels. The thick oil is brightest (shown
            white); the thin is not so bright (shown grey). Note
            the poor oil film-water contrast. The data was col-
            lected over the Wallops Island oil spill,at 17:50
            GMT, Nov. 2, 1978, C-GRSA flight line 13.

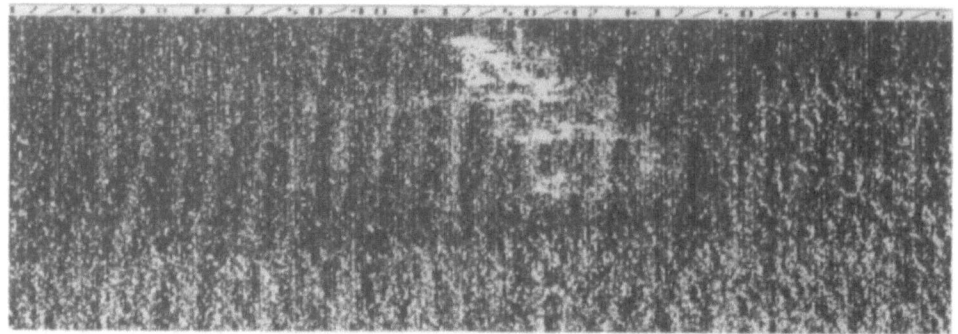

Figure 9.   MSS channel 10 colour enhanced image showing two
            intensity levels. Thick oil is bright (shown white);
            thin oil, barely visible, is shown grey. Data from
            Wallops oil spill, 17:50 GMT, Nov. 2, 1978, C-GRSA
            flight line 13.

a heavier oil type, the thick regions are not visible and in
fact appear as dark areas in MSS channels 3-6. MSS channels
8 and 9 do not discriminate the thick from the thin regions and
MSS 10 provides only marginal detection of oil.

     One must conclude that the radiance levels from the thick
oil is determined primarily by the volume reflected component.
If the oil is highly absorptive, as is the La Rosa crude, the
volume reflected radiance from the oil and/or the water beneath
the oil is severely attenuated. If, on the other hand, the oil
is less absorptive, as is the Merban crude, and if it is emul-
sified giving a more highly scattering medium, the volume reflected radiance will be increased.

Ultraviolet

     The UV imagery from the DCLS gives good contrast for the
thin regions of the oil slick (Figure 10). It is noted that
the thick parts of the oil slick exhibit lower reflectances
than do the thin. The reason for this is that whereas the thin
film of oil increases the surface reflectance, leaving the
volume reflected component lightly mainly attenuated, the thick
layers absorb the volume reflected component while increasing
the surface component. The improvement of the UV imagery from
the DCLS over the MSS 2 imagery is due in large part to the
lower noise equivalent radiance value for the UV detector.

     It should be noted that these results, obtained in the
UV-visible-near IR, apply only to a detector with sensitivity

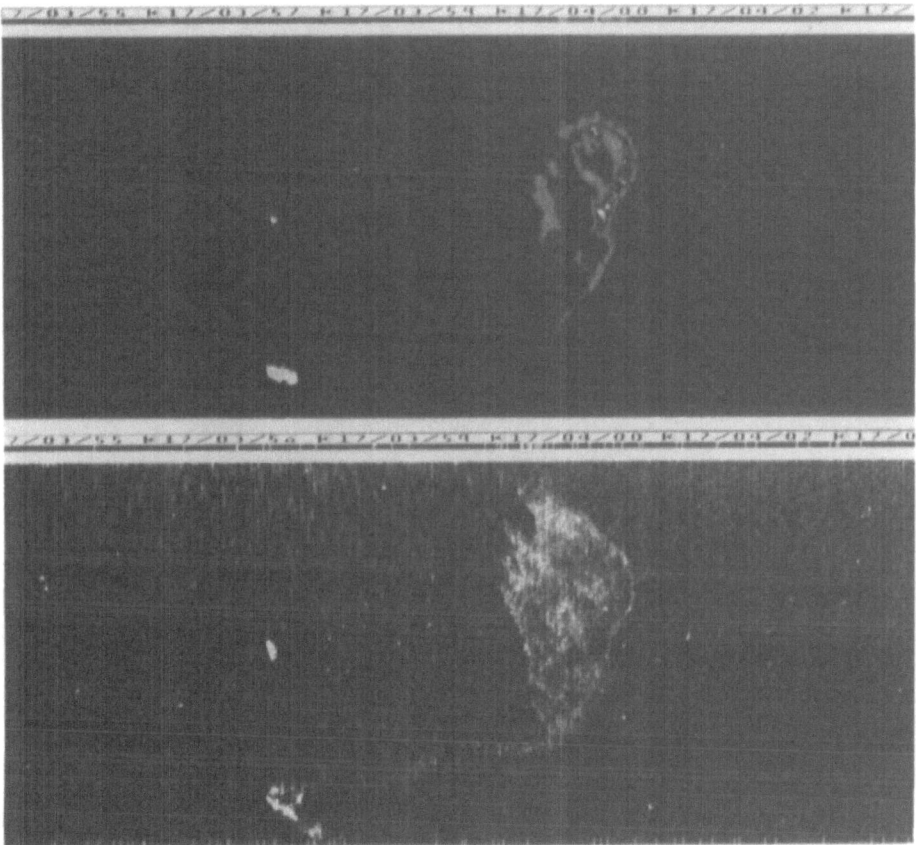

Figure 10.   DCLS-UV colour enhanced image showing the thin oil
             film (down) and DCLS-thermal IR colour enhanced
             image (up) showing the apparent thermal variations
             in the oil slick. The better contrast observed on
             the bottom as compared with the MSS 2 channel image
             is attributed partly to the better signal to noise
             level for the UV detector, partly to the higher
             intrinsic contrast in the UV (Wallops oil spill,
             17:04 GMT; Nov. 2, 1978, C-GRSC flight line No. 3)

characteristics similar to the detectors used here. The reason
is that the oil-water contrasts apparent in the imagery are
determined by the signal-to-noise level to an increasing degree
when the sensitivity limit is approached. This is the case
for the lower MSS channels.

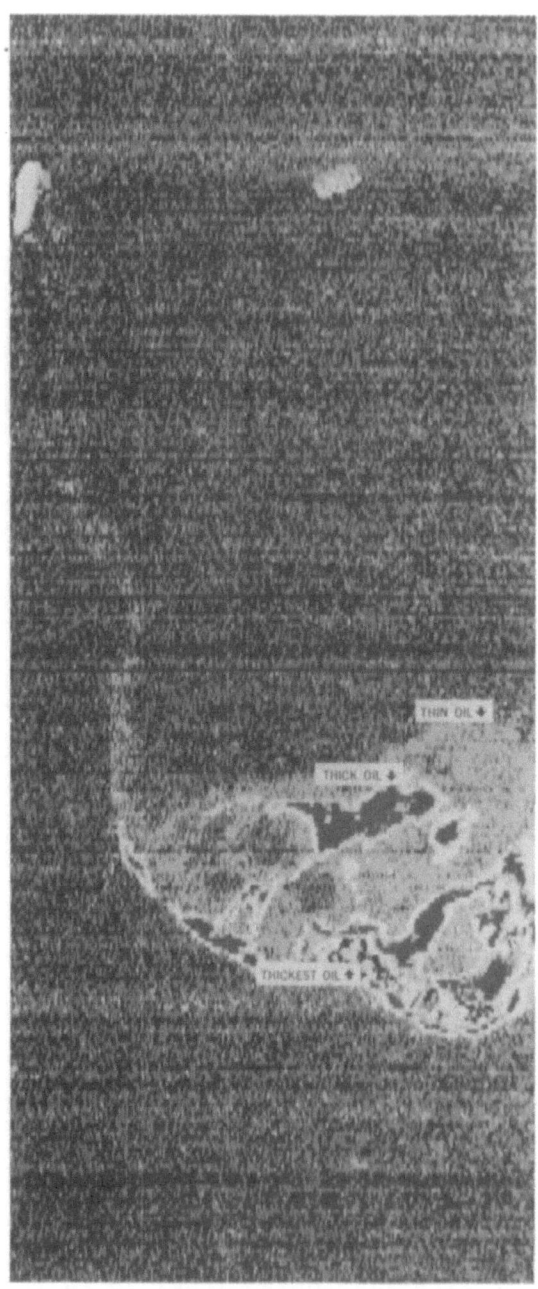

Figure 11.   DCLS-UV and IR images digitally superimposed and
             subjected to an clustering procedure in the two
             dimensional UV-IR space. The result is a coulour
             coded thematic image corresponding to the various
             UV-IR signatures exhibited by the target area.

Figure 12.   DCLS-UV and IR images individually colour enhanced
             and then photographically superimposed proving an
             effect similar to that in Figure 11. The data for
             this image is the same as for that in Fig. 11 but
             has not undergone aspect ratio correction (Wallops
             oil spill, 17:00 GMT, Nov. 2, 1978, flight line 3.

Ultraviolet - Infrared Combination

     Comparison of the UV image (Figure 10) with its infrared
equivalent shows that the UV and IR complement each other. The
IR detects the thicker regions of the oil slick, the UV the
thin regions. This feature has been exploited in constructing
the dual-channel UV-IR image shown in Figure 11. The two images

have been digitally superimposed with the CCRS Image Analysis System (CIAS). For each pixel in the composite image, the pair UV, IR of intensity values is plotted in two-dimensional UV-IR vector space. These points (vectors) are then grouped into clusters of nearest neighbours, and each cluster is assigned a specific colour. Each pixel corresponding to a vector in a given cluster then acquires that cluster's colour. The result is a thematic image in which the various colours indicate relative oil thickness. In Figure 11, there are 8 classes corresponding to water, thin oil that is bright in the UV, several gradations of thick oil that is dark in the IR, thicker oil that is bright in the IR, and a hot spot on the ship.

An almost equivalent result can be achieved (Figure 12) by colour coding the individual UV and IR images and photo-graphically superimposing the two resultant images.

CONCLUSION

It must be concluded that as far as the extent and the relative thickness of the slick is concerned, the UV and thermal IR together provide virtually all the information available in the whole spectral region including UV, visible, near IR, and thermal IR. Hence, for airborne remote sensing, a UV-IR line scanner provides an attractive method for oil detection. For higher altitudes and satellite platforms, the UV would be seriously plagued by atmospheric scattering and path radiance. Under these circumstances, a longer wavelength, between 400 nm and 650 nm would be preferable.

REFERENCES

Axelsson, S., 1975, "Remote Sensing of Oil Slicks: Results from a Field Experiment in the Baltic Sea, September 1974", SAAB-SCANIA AB RL-O-3 R27.

De Villiers, J., 1973, "Airborne Detection and Mapping of Oil Spills, Grand Bahamas, February 1973", EMR CCRS Data report 73-7

Edgerton, A., Bommarito, J., Schwantje, R., Meeks, D., "Development of a Prototype Oil Surveillance System", U.S. Department of Transportation report CG-D-90-75, May 1975.

Estes, J., Senger, L., Fortune, P., Potential applications of remote sensing techniques for the study of Maine oil pollution, Geoforum, 9:69

Fontanel, A., Roussel, A., 1976, La détection des nappes d'hydrocarbures en mer, in: "Proc. Journées de Télédétection, Toulouse, Octobre 1976".

Hornstein, B., "The Appearance and Visibility of Thin Oil Films on Water", Environmental Protection Agency report R2-72-039.

Horvath, R., Morgan, W., Spellicy, R., 1970, "Measurements Program for Oil Slick Characteristics", U.S. Department of Transportation report CG-92580-A, February 1970.

Millard, J., Arvesen, J., Lewis, P., 1975, "Development and Test of Video Systems for Airborne Surveillance of Oil Spills", U.S. Department of Transportation report CG-D-95-75.

Munday, J., McIntyre, W., Penney, M., Oberholtzer, J., 1971, Oil slick studies using photographic and multispectral scanner data, in: "Proc. 7th Intl Symposium on Remote Sensing of the Environment, ERIM, Ann Arbor".

Neville, R. A., Thomson, V., O'Neil, R. A., Buja-Bijunas, L., Gray, L., Hawkins, B., Dagg, K., 1979, Remote sensing of oil spills, Spill Technology Newsletter.

Thomson, K., McColl, W., 1972, A remote sensing survey of the Chedabucto Bay oil spill, Environment Canada Scientific Series, No. 26.

Thomson, V., Neville, R. A., O'Neil, R. A., 1979, High contrast imaging of an oil slick by means of a low light level television, in: "Proc. of NATO-CCMS Workshop on the Use of Remote Sensing for the Control of Marine Pollution, held in Washington, D.C., April 18-20".

Wolfe, W., "Handbook of Military Infrared Technology", Dept. of the Navy, Washington, D.C.

DETECTION OF OIL SLICKS USING REAL APERTURE AND

SYNTHETIC APERTURE IMAGING RADARS - EXPERIMENTAL RESULTS

A. Fontanel

Institut Français du Pétrole
Rueil Malmaison, France

ABSTRACT

Experiments were conducted at sea to compare the responses of a real aperture horizontally polarized radar and a synthetic aperture vertically polarized radar (X-band).

The altitudes of the planes varied from 300 m to 3,000 m as well as the direction of flight with respect to the wind direction. Aerial photographs were taken at the same time.

The best detection of oil was obtained with the synthetic aperture radar because, independently of polarization conditions, the higher resolution of SAR seems to favor oil detection.

In addition, measurements were made at sea using a fixed four-frequency scatterometer in order to quantitatively evaluate the attenuation of the backscattered signal by the oil films.

INTRODUCTION

Several experiments have been carried out jointly at sea since 1973 by the Centre National pour l'Exploitation des Océans (CNEXO) and the Institut Français du Pétrole (IFP) in order to test and, if possible, to develop various remote sensing techniques for the detection and observation of oil slicks on the sea.

The results show that in order to detect oil spills with maximum efficiency it is necessary not only to operate an all weather remote sensing system but also to obtain an unambiguous

response denoting oil without requiring in-situ verifications.
Various results in France and abroad (Maurer, 1976) seem to
indicate that such a system could make use of sensors in the
blue and thermal bands of the spectrum as well as in the microwave range.

Microwave sensors (wavelengths from 0.1 up to 100 cm) can
operate either in a passive mode (radiometers measuring
differences in brightness temperature between oil and water at
a given wavelength) or in active mode (side-looking radar).
Only side-looking radar images will be discussed here.

EXPERIMENTAL CONDITIONS

The POLUMER 1976 campaign was carried out in the Mediterranean Sea from 30. November to 1 December 1976, 100 kilometers
off Toulon. There was intentional spillage by a ship travelling
at a speed of 1.5 knots in a direction perpendicular to the
wind direction. The wind was blowing from sector 230° with a
mean speed of 8 m s^{-1}. The sea was calm with a H 1/3 of about
50 cm.

The performances of two imaging radars were compared: a
synthetic aperture radar (SAR) operated by Centre d'Essais en
Vol (Brétigny, France), made by Thomson CSF and an EMI real
aperture radar (SLAR) operated by National Aerospace Laboratory,
Amsterdam, The Netherlands. The altitudes of the planes varied
from 300 m up to 3,000 m as did the direction of flight with

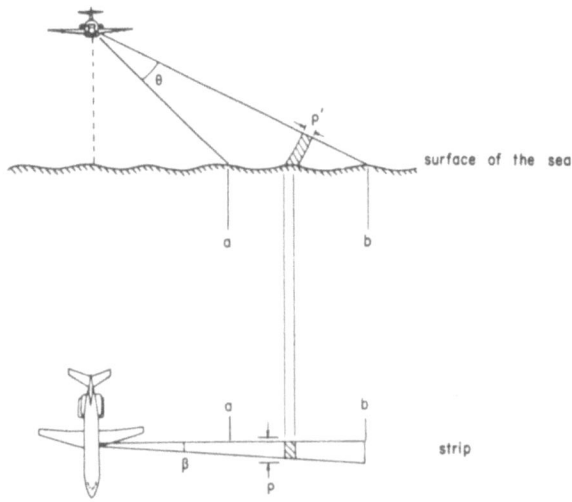

Figure 1. Principle of imaging radar.

respect to the wind direction. The characteristics of radars
used in the experiment were as follows:

Real Aperture Radar (EMI type)

* Frequency: 9500 to 9700 MHz (X-band)
* Pulse length: 200 nanoseconds
* Pulse rate: 1260 to 2500 s
* Antenna length: 2.25 m
* Beam size: 36° in the vertical plane
                    1° in the horizontal plane
* Angular along-track resolution: 13 x 10^{-3} radian
* Across-track resolution: 26 m at the range of 2 km
                           52 m at the range of 4 km
* Polarisation: horizontal
* Cathode-ray tube display with an image on a 5" film
* Aircraft: BEECHCRAFT QUEEN AIR Model 80
* Incident angle: 80 to 87° (altitude 1,000 feet)
                  61 to 83° (altitude 10,000 feet)

Synthetic Aperture Radar (Radar RAPHAEL, Thomson CSF)

* Frequency: X-band
* Polarisation: vertical
* Resolution: a few meters (along-track and across-track)
* Correlation: optical
* Aircraft: MIRAGE III
* Incident angle: from 81° to 87°, according range (alti-
                  tude 1,000 feet).

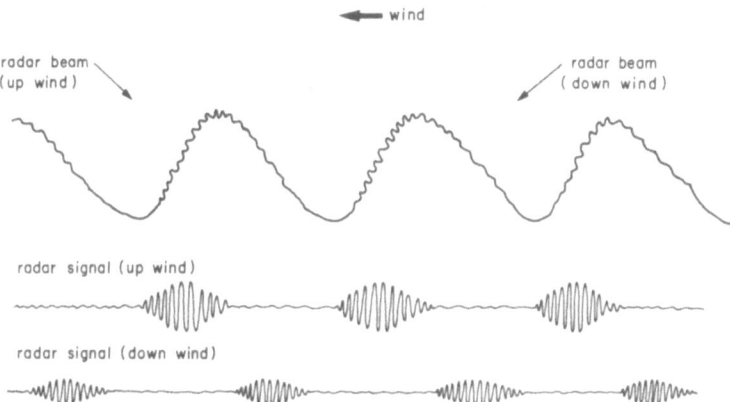

Figure 2. Sketch showing the concentration of capillary waves
          on the forward edge of long gravity waves and the
          corresponding radar signal returns.

Several physical mechanisms are accepted as responsible
for scatter from the sea, i.e. specular reflection, Bragg
scattering, Rayleigh scattering, etc. (Brown, 1976). From theoretical studies alone it is not very clear which of these three
effects dominates. From experimental results, it can be said
that, under given conditions, radars can give "images" of wave
patterns on the sea. Radars "see" gravity waves many times
longer than the wavelength of the emitted electromagnetic waves.
Furthermore an imaging radar can detect sea waves only when its
resolution is many times smaller than the sea wavelength. In
any case capillary waves must be present on the long underlying
gravity waves, and this last condition implies a wind speed
greater than 1 or 1.5 m s^{-1}.

Radar backscattering is mainly due to these capillary
waves, and the mechanism is not isotropic. It has already been
mentioned (Fontanel, 1977) that radar echo intensity depends on
the angle between the direction of the wind and the direction
of the radar beam. Echos obtained in the up-wind conditions
(Figure 2) are approximately 4 dB stronger than echos obtained
in the down-wind conditions. This may be due to the distribution of capillary waves on the underlying long waves. Because
of hydrodynamic non-linear interaction, short capillary waves
tend to be concentrated more on the leading edge of the long
gravity waves on which the radar beam impiges in the up-wind
condition.

A mechanism necessary for the formation of the image is
the slope modulation of capillary waves by the underlying long
waves. It can be assumed that the total slope to be considered

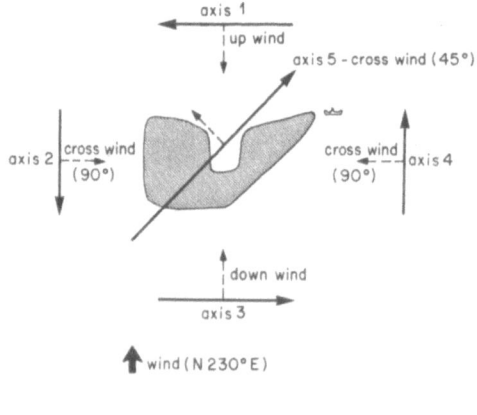

Figure 3. 1976 POLUMER campaign: Overflight diagram of slick.

is the sum of the slopes of the long and capillary waves present on the surface (Barrick, 1968). Hence, more specular (or quasi-specular) surface elements exist on the forward slope of the long waves than on the backside. The upper part of Figure 2 gives a schematic diagram of the concentration of the capillary waves on the leading edge of long waves. The lower part of the figure shows the amplitude variations of radar signals in up-wind and down-wind conditions. Furthermore owing to the usually low grazing angles of radar beams, some parts of the waves may be in the shadow of the preceding crests.

Imaging radars detect oil slicks because of a reduction in radar backscattering resulting from the calming effect of oil films on capillary waves (Brunsveld, 1976). The attenuation of backscattered signals by the oil may vary from 3 dB to 13 dB depending on experimental conditions. Consequently oil can be detected only if the energy backscattered by the water surrounding the slick is several dB above noise level. This is generally obtained when wind speed is greater than 1 or 1.5 m s^{-1}. When wind speed increases, the energy received by the radar rises until the speed is close to 15 or 20 m s^{-1}. A saturation phenomenon is then commonly observed. In some conditions radar detection of oil slicks depends not only on wind speed but also on the relative direction of the radar beam and wind vector (Figure 3), as can be observed in Figures 4 and 5.

EFFECT OF RADAR RESOLUTION ON DETECTION OF OIL

Oil films on a water surface are best detected not only when a reduction of the mean signal level is observed but also

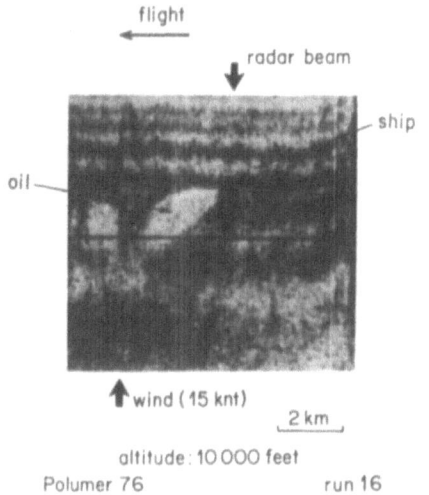

Figure 4: SLAR: Up-wind detection.

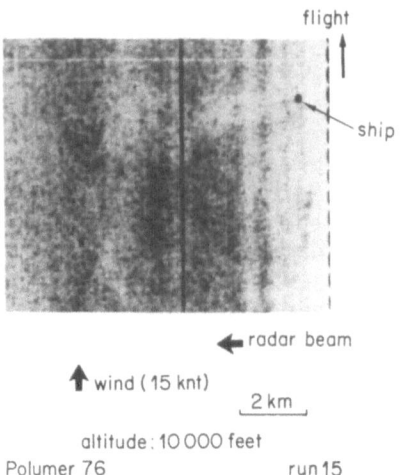

Figure 5. SLAR: Cross-wind detection (90°).

when, owing to the resolution of the imaging radar, long gravity
waves can be seen to vanish (Figures 6 and 7). This happens when

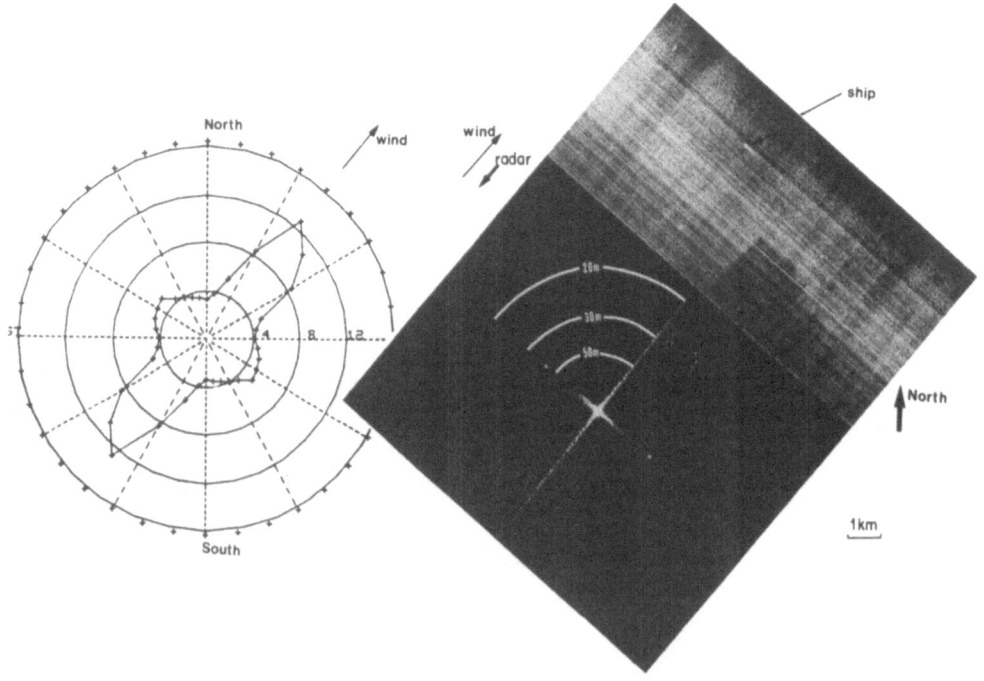

Figure 6. Synthetic aperture radar: oil slick detection.

the wavelength of the waves is several times longer than the
resolution of the radar. In this campaign, oil was detected
better with SAR because, independently of polarisation, its
resolution made it possible to "see" the waves (mean wavelength of about 27 m) and consequently their vanishing due to
the calming effect of the oil film. Such was not the case with
SLAR (30 meter resolution).

PRESENTATION OF RADAR IMAGES

Real Aperture Radar - Horizontal Polarisation (Fig. 4 and 5)

Oil appears in white on these images. Best detection was
obtained when the radar was aimed 180° from the direction of
the wind (Figure 4, up-wind detection). In most cases, it was
difficult and sometimes impossible to detect oil when the radar
beam made a 90° angle to the wind (Figure 5, cross-wind
detection). When this angle was close to zero, detection was
generally poor. Radar images were obtained at various altitudes
between 300 and 3,000 m. Oil detection does not seem to be very
sensitive to this parameter.

Synthetic Aperture Radar - Vertical Polarisation (Fig. 6 and 7)

Oil appears in black in both these figures. The first was
taken in the up-wind conditions, the second in the cross-wind
conditions. In both, the edges of the slick can be precisely
detected. With SAR, oil detection seems to be less sensitive to

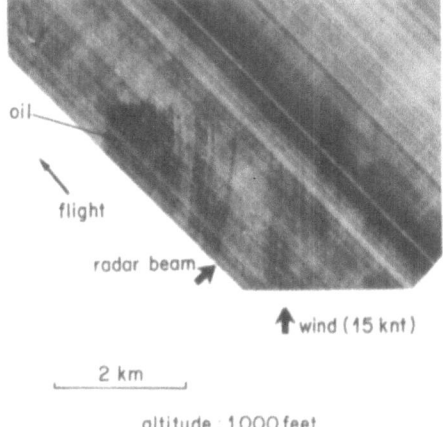

Figure 7. Synthetic aperture radar: Oil slick detection.

the angle between the radar beam and the direction of the wind, and this may be due to the better resolution of this radar.

Furthermore, the footprint of SAR (a few meters) made it possible to well detect the crests of the wave pattern except in areas covered with oil.

An oceanographic buoy located near the oil patch measured a mean wavelength of 27 m. The two dimensional spectrum of the SAR image shows wavelength spreading from 22 m to 38 m, what is in well accordance with buoy records. The wind speed was about 15 knots.

The streaks on Fig. 6 and 7 as well as the dark strip in the central part of the image are due to mechanical defects in the correlation processing.

COMPARISON OF SLAR AND SAR IMAGES: DISCUSSION

SAR flew over the oil 20 minutes after SLAR. The size of the slick was not identical, but the sea state and wind conditions were the same. Oil appears in white on SLAR images and in black on SAR images due only to film processing. The ship CAPRICORNE participating to the experiment appears on images 4, 5, and 6.

Oil is better detected on SAR images and, with this type of radar, oil detection was not found to be very sensitive to the respective directions of the wind and of the radar beam. However, only up-wind and cross-wind conditions were used in the case of SAR. No flight was made in down-wind condition. It is interesting to notice that the wave pattern disappears completely in the case of SAR on those parts of the sea covered by the oil film.

QUANTITATIVE MEASUREMENTS OF THE EFFECT OF AN OIL FILM ON THE BACKSCATTERING COEFFICIENT

At fall 1978, experiments were conducted from the Noord-wijk platform located in the North Sea off Amsterdam, using a four frequency scatterometer (RAMSES). These experiments were conducted by Institut Français du Pétrole (IFP), Centre National pour l'Exploitation des Océans (CNEXO), Rijkswaterstaat, TNO (The Hague, The Netherlands) and the Centre National d'Etudes Spatiales (CNES). The aim was to measure the backscattering coefficient of water under various sea state conditions, with and without an oil film. Due to pollution regulations, oleic alcohol (cis-octadecen) was poured at sea instead of oil. Table 1 indicates some characteristics of this alcohol and of two crude oils. Figure 8 shows the variation of the

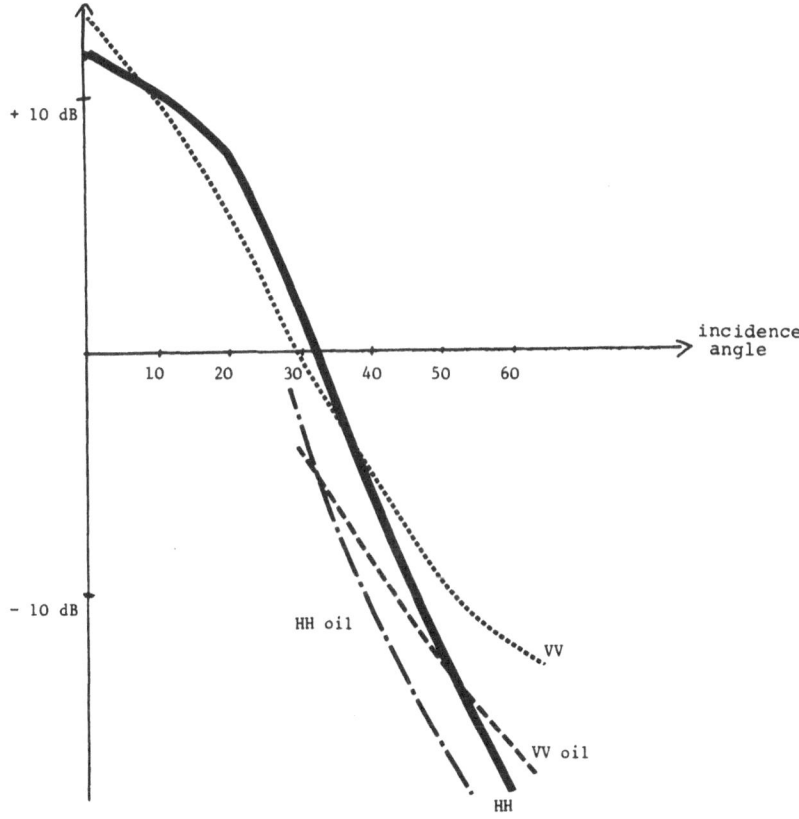

Figure 8.   Variation of the backscattering coefficient
            versus the incidence angle.

backscattered signal at 9.GHz, versus the incidence angle for
a wind speed of 28 knots. Measurements were obtained under HH
and VV polarisations with and without alcohol on the surface.

Table 1.   Characteristics of two crude oils and oleic alcohol.

Products	Surface tension	Interfacial tension with water	Viscosity at 20°C
Crude oil Irak	27 dyn/cm	8 dyn/cm	10
Crude oil Koweit	23 dyn/cm	25 dyn/cm	20
Oleic alcohol	32.9 dyn/cm	11.8 dyn/cm	47

It can be observed on Fig. 8 that the dampening of capillary waves by alcohol causes a mean attenuation of 4 dB at incidence angles going from 30° to 60°. It is difficult to explain only with this weak attenuation the disappearance of the waves on the SAR images (Fig. 6 and 7) in the parts of the sea covered by oil.

CONCLUSION

The best detection of the oil slick was obtained with a synthetic aperture radar using vertical polarisation as contrasted with real aperture radar using horizontal polarisation.

Independently of polarisation conditions, the higher resolution of synthetic aperture radar seems to favor oil detection. In contrast, with the real aperture radar, the SAR resulted, in the absence of oil, in good detection of the wave field; the slick was detected not only by a reduction in the mean signal level but also by the complete disappearance of the wave pattern.

It must be noticed that this effect cannot be explained only by the 4 dB attenuation which was measured on an oleic alcohol film, using a 9 GHz scatterometer.

REFERENCES

Barrick, D.E., 1968, Rough surface scattering based on the specular point theory, IEEE Trans. Ant. Prop., AP-16, 449.

Brown, W.E., Elachi, C., and Thompson, W., 1976, Radar imaging of ocean surface patterns J. Geoph. Res., Vol. 81, 15:2657.

Brunsveld van Hulten, H.W., 1976, Application of SLAR to coastal zone phenomena, Rijkswaterstaat Report FA 7601.

Fontanel, A., 1977, Détection d'une nappe d'huile et étude de l'état de la mer par radar latéral aéroporté, IFP Report N° 25 269

Maurer, A., and Edgerton A.T., 1976, Flight evaluation of U.S. Coast Guard Airborne Oil Surveillance System, M.T.S. Journal, Vol. 10, 4:38

OIL SPILL DETECTION

USING A MULTIPURPOSE SYNTHETIC APERTURE RADAR

F. Lacoste

Thomson-CSF
Avionics Division
Malakoff, France

ABSTRACT

The subject of this paper is to underline the ability for one single radar to provide the different functions which are necessary for an airborne detection and monitoring of sea pollution:

* Sea traffic surveillance
* Oil spill detection
* Identification and positioning of faulty ships.

This is obtained by the association of the pulse compression airborne surveillance radar "Varan", with a synthetic aperture processing unit "Anaconda" which is adding to the classical function of the radar, as sea target detection, a side looking mode. This mode delivers high resolution radar images whose exploitation can be made, either in real time on board the aircraft through a TV set, or after film or digital tape recording.

One high-light of the Varan-Anaconda radar is its ability to provide functions which usually require two distinct radars: one high performance search radar, plus one large antenna side-looking radar.

Using the small size antenna of the basic search radar to provide those both functions, the Varan-Anaconda solution permits an easy aircraft installation, in terms of weight, volume and power supply.

BASIC PRINCIPLES

Airborne radars can improve azimuth angular resolution by using synthetic aperture radar (SAR) processing. These technics generate an equivalent very large antenna by moving a small one. SAR processing can be used with a rotating antenna or with a side-looking fixed antenna.

Such a processing is often associated with a pulse compression transmitter-receiver in order to generate high resolution radar images of the ground. High resolution radars have many applications as for example:

* Navigation up-dating
* Topographic mapping
* Detection of land changes
* Monitoring oil spills
* Ship traffic, and so on.

The radar advantage is its ability to operate in all weather, night and day conditions.

The multipurpose SAR Varan-Anaconda radar consists of the association of an airborne pulse compression surveillance radar "Varan" with a SAR processing unit "Anaconda". The Anaconda unit allows real-time synthetic aperture processing by digital correlation.

Synthetic aperture radars are based upon Döppler effect due to the aircraft speed; Döppler signal processing need very fast and complex circuitry. The first SAR processors used optical correlators. Improvements of the technology made possible, now, to implement real-time digital SAR processors. So, it is possible to obtain SAR images on board the aircraft.

OPERATIONAL FEATURES

The basic Varan radar is an airborne maritime survey and mapping radar with a 360° bearing antenna. This high performance radar uses a pulse compression transmitter-receiver. The basic version has the following modes:

* Maritime target detection
* Real beam mapping.

The 360° bearing ability depends in the radar installation in the aircraft.

Pulse compression technics are used in order to obtain very short pulses and consequently a very good range resolution,

with a reasonable low peak power and a very high detection
range. The pulse compression transmitter-receiver can be easily
transformed in a fully coherent chain and so, be suited with
Döppler processing.

For example, SAR processing, which uses Döppler effect can
be associated to the basic Varan radar. This is made, using the
processing unit Anaconda. SAR technics simulate very narrow
beam antennas. This feature would be obtained, with a classical
radar, by having a very large antenna whose implantation on
board an aircraft or physical realization would be impossible.

So the Anaconda unit gives the Varan radar the function of
a high resolution side looking airborne radar (SLAR).

In the SLAR mode, the antenna is slaved perpendicularly to
the aircraft track. So the radar maps a swath parallel to the
flight line on the right or on the left. In this mode, the
azimuth theoretical resolution (or along-track resolution) does
not depend on the range. In practice, the azimuth resolution is
function of the radar processing power. Note that range (or
across-track) resolution is given by radar transmitted frequency
bandwidth.

The improvement of the radar resolution gives a better
contrast in the images and so, a good capacity to detect low
radar cross-section targets in heavy ground or sea clutter
(reconnaissance, small targets detection as life-boats, etc.).
The improved contrast also permits to distinguish two clutter
regions having different reflectivity coefficients (oil spill
detection on the sea surface).

The SLAR mode has several operational uses as for example:

* very high resolution mapping
* recognition
* oil spill monitoring
* iceberg detection and localisation
* fishing aid
* ground resources monitoring
* sea traffic surveillance, etc.

The SLAR mode has several sub-modes which are defined by
the swath width: 5, 10, 20, 40 km with different associated
resolutions, respectively 10, 20, 40 and 80 m. The resolution
limitation is the result of a trade of between processing
abilities and a low volume, reasonable cost unit design.

In each sub-mode the processed swath can be moved up to a
maximum range which depends on flight altitude and selected

sub-mode. The maximum range is 100 km and in this case the better resolution is 40 meters. This corresponds to a real antenna being 80 meters long.

SLAR exists, that have better resolution at such a range, but they need much more powerful and expensive processing.

In the SLAR Anaconda mode, the radar image is presented in real-time on a TV set. The image moves vertically according to the aircraft speed and the selected scale. The operator may stop the image at any moment and widen it to examine details around a marker. The widening scale is 2 or 4. The minimum and maximum ranges, together with other parameters (speed, aircraft coordinates, etc.) are visualized on the TV screen near the radar image.

The usual exploitation is on the TV set, but signal outputs are provided to record the radar maps on film (or paper) and on digital magnetic tape for further exploitation on ground or re-reading on the TV set.

CONCLUSION

The association of the Varan radar with the SAR processing unit Anaconda is a multipurpose radar which has the main following features:

* Real-time SAR technics give high resolution radar images which provide a better survey of the observed area;

* This radar has two main functions, 360° bearing surveillance and SLAR mode which are usually obtained with two different radars. This is a great advantage in term of aircraft installation (weight, volume, and power supply);

* The Anaconda unit has growth potential as for example aircraft ground speed precise measurement;

* The radar modular conception (line replacable units) makes easier its installation on board different carriers. The addition of special units provides supplementary operational capacities. For example the association with a data link unit makes possible real-time exploitation of the radar images in a ground station or on board ships.

INFRARED LINE SCANNER AND SIDE LOOKING AIRBORNE RADAR

USED FOR REMOTE SENSING OF OIL ON THE SEA

D.W. Nicholson

Royal Signals and Radar Establishment
Malvern, United Kingdom

ABSTRACT

The RSRE paper will outline the use of various sensors and techniques eveolved in the United Kingdom. The first attempts to use either IRLS or SLAR for oil slick detection were carried out during the TORREY CANYON incident in April 1967. The IRLS operated in the 3-5 micron band and initially a Q-band (8.5 mm) SLAR was employed; this was later replaced by X-band (3 cm).

Subsequent operations have been carried out using 8-14 micron band IRLS and Q-band SLAR.

The attributes and disadvantages of the system will be discussed along with the justification for real-time airborne displays.

THE UK EXPERIENCE FOR REMOTE SENSING OF OIL AT SEA

The first UK attempt at deliberate detection of spill crude oil was when the TORREY CANYON (120,000 tons of oil) went aground on the Seven Stones rocks in April 1967. Because of the pollution danger to our own SW coast and the French coast, we at RSRE were asked if we could do anything to assist detection of the resultant slicks. We decided to try to use a SLAR for area detection, because of its capability of operation over long range and because we could adjust the antenna to vary the main beam depression angle to optimise the clutter return signals. We were able to exercise on option on the choice of equipment; at the time, we carried both X-band (3 cm) and Q(K)

band (8 mm) equipment in a HERMES aircraft. It was decided that
initially we would use the Q(K) system on the basis that the
shorter wavelength might give more favourable sea clutter
signals; this system employed a real aperture antenna giving a
horizontal beamwidth of 0.1° and a pulse width of 0.1 µs (resolution of 50 feet by 50 feet at 5 nm).

Our first flights with this system showed us that the antenna depression angle was more critical than we had assumed;
this posed a problem because the Q(K)-band antenna was only
adjustable with the aircraft on the ground with its radome off.
We needed to be able to remotely control an antenna with the
system operating and airborne to optimise performance. It also
became apparent that very high gain levels were required and
that it would be advantageous to be able to increase the peak
power of the transmitter. Our active swath width was limited
to 2.5-3 nm.

We were able to make these assessments rapidly because
from the beginning we used a real (or near real) time display,
in the form of a rapid photo processor which gave the finished
imagery on a continuous paper strip. We were thus spared the
necessity of making a number of individual measurements, and
having to process a film to see results.

Our other option, the X-band (3 cm) system was now to be
tested; the real aperture antenna had an horizontal beamwidth
of 0.3°, the pulse width was 0.2 µs and the transmitter peak
power was approximately 100 kW. The antenna was remotely controlled with respect to beam depression angle.

Results were much improved and we were able to employ a
7 nm swath as opposed to the 2.5-3 nm of the Q(K)-band system.
The better mixer noise factor for X-band also assisted at the
high gain levels employed.

This X-band system was employed thereafter for all TORREY
CANYON operations.

For more detailed examination of the oil, it was thought
that IRLS might be employed, although it was recognized that
the field of view would be more restricted than a SLAR and its
performance in detecting oil slicks was not known. The system
employed was a 3-5 micron Indium Antimonium detector with a
seven inch aperture. The spatial resolution was approximately
1 mrad and the thermal resolution was 0.25°K. There was no
real-time display and imagery was recorded on 70 mm film.

The equipment was carried in a HASTINGS aircraft (similar

to SLAR carrying aircraft) and the film magazine could be
changed during flight.

This equipment was used for all "TORREY CANYON" operations,
8-14 micron system being available at that time.

Subsequently we at RSRE have done very little with SLAR
for oil slick detection; our preoccupation is with military
reconnaissance. In consequence, we have no experience of polarization changes for either Q(K)- or X-band, since horizontal polarization is the accepted form for the detection of
most military targets. We would have liked to investigate the
effects of sea state and oil thickness, but no further work
was done until we were brought in by Warren Spring Laboratories.

IRLS on the other hand has been employed from time to
time to overfly oil slicks and from these operations we have
come to the conclusions that operating with 8-14 micron Hg Cd
or CMT detectors is more effective than 3-5 micron. We have
therefore used 8-14 micron detectors for most of our subsequent
work. Simultaneous recordings of an oil slick using both forms
of detector from the tanker PACIFIC GLORY are shown as an
example to substantiate this conclusion.

Our present facility is a converted VISCOUNT civil air-
liner which carries both IRLS and SLAR. The SLAR is a Q(K)-
band with a remotely controlled antenna which has a horizontal
beamwidth of 0.1°. The transmitter has a peak power of 50 kW
with a pulse width of 0.1 µs. The IRLS scanner is carried in a
pod mounted on the forward end of the fuselage; the SLAR
antenna is contained in a radome pod mounted on the aft end of
the fuselage. Both equipments feed into a digital scan
converter to give TV real-time display facilities, which are
used to ensure acquisition of the target area and to optimise
the operation; changes of aircraft track and position can be
quickly determined as can the extent of the slick and its
position.

In addition to real-time display, both systems employ
films recorders and video outputs of both are tape-recorded.
The tape-recordings can be used for in-flight recall of
information using the digital scan converter to produce the
TV display facility, from which more information may be derived
about the oil or to re-organise the aircraft search track
system or operating altitude.

Our operational method, first employed when dealing with
the TORREY CANYON, has been to locate and find the extent of
the oil using the SLAR and then to use IRLS to produce more
detailed information on the build up and formation of the

individual oil streaks and patches. We have noted that in
general the SLAR will detect large areas of oil film which will
not be detected by IRLS and from that we suspect that IRLS can
only detect the denser areas.

The present Q(K)-band SLAR system in use would give a
better performance if more transmitter peak power was available.
Our early experience with the higher powered X-band system lead
us to believe that as much peak power as possible was desirable
and we would like to double our present peak output of 50 kW

We were forced to abandon our early Q-band system in favour
of X-band, purely on a basis of system facilities, our original
Q(K)-band system did not employ a remotely controlled antenna,
which we found to be essential and our X-band system had the
advantage of a higher powered transmitter; however, we
originally chose Q(K)-band because we believed that the sea
clutter signal level might be higher than that produced by
X-band as a function of wavelength and this would be a desirable
feature when the success or otherwise of detection depended on
differentiating between contrast levels. We do not know whether
this theory is born out in practice.

We also have no experience of polarization change, i.e.
horizontal to vertical; our system is permanently horizontal
with no changeover capability.

The IRLS is a modified system in as much as the scanner/
receiver is located outside the pressure cabin in a pod, the
other units are within the cabin, particularly the film
recorder which enables film magazine changes to be made during
flight, thus long duration sorties can be flown without having
to land to change magazines.

We have already mentioned that from the start of our work
we used a real-time display for the SLAR and for all our more
recent activities of both sensors have been provided with this
feature. The advantages of such a facility are:

* The SLAR system can be optimised with regard to antenna
  depression angle setting and gain setting for prevailing
  conditions;
* The IRLS can be optimised for contrast setting for
  prevailing conditions;
* The location and extent of the oil can be rapidly
  assessed;
* Rapid response is possible in consequence (particularly
  with SLAR which under adverse meteorological conditions
  may be the only sensor able to operate).

The obvious extension of the real-time facility is by data link to shore or ship base, where the information could be used at first hand by the controlling authority, giving a potential improvement in response time and overall command of limited resources.

EVALUATION OF INFRARED LINE SCAN (IRLS) AND SIDE-LOOKING AIRBORNE RADAR (SLAR) OVER CONTROLLED OIL SPILLS IN THE NORTH SEA

H.D. Parker and D. Cormack

Warren Spring Laboratory
Stevenage, Hertfordshire, United Kingdom

ABSTRACT

A series of experiments have been carried out in the North Sea jointly by Warren Spring Laboratory and the Royal Signals and Radar Establishment (RSRE) to evaluate the use of infrared line scan (IRLS) and Q-band, (H) polarized side-looking airborne radar (SLAR) for remote sensing of oil on the sea.

The preliminary results of this work are reported here and the co-ordination of the experiments and the methods used to gather sea truth data, in particular oil film thickness, are discussed. Imagery resulting from these experiments is presented for two oils, a North Sea crude and a Middle East crude, together with imagery of an accidental oil spill of heavy fuel oil.

These experiments have shown that both sensors will show the presence of oil, but the imagery from the infrared line scan is more easily interpreted and also provides some informa-tion on the quantity of oil on the sea. The SLAR technique shows considerable promise, and a SLAR is likely to be included in a future remote sensing system to extend the capability beyond the weather limitations of the infrared line scan.

INTRODUCTION

In November 1975, a collision between HMS ACHILLES and the tanker OLYMPIC ALLIANCE in the Dover Straits led to about 2,000 tons of oil coming ashore at Folkestone. The oil was at sea for 2-3 days before it came ashore but dispersants spraying

operations were hampered by fog and short daylight hours. One
of the lessons learnt from this incident was that a method
should be sought to extend the clean-up capability through the
hours of darkness and in conditions of poor visibility.

Initially, it was thought that this requirement could be
met by a shipborne sensor but it was soon realised that this
approach was inadequate because:

* the glancing angle of observation is too oblique;
* the ship's horizon is too close in terms of overall
  extent of a spill.

The work was then split into two areas of effort:

* identification of techniques for shipborne local area
  surveillance;
* identification of a system which could provide an overall view of the spill.

During the TORREY CANYON incident of March 1967, the Royal
Signals and Radar Establishment (RSRE) were able to demonstrate
the possibility of using remote sensors to detect oil on the
sea surface. An infrared line scanner and a X-band SLAR were
able to track the spill across the western approaches to the
English Channel. The advice was therefore sought of RSRE and a
preliminary trial was carried out in January 1977, with a
8-14 micron IRLS and a 50 kW (H) polarised Q-band SLAR overflying a spill of five tons of Ekofisk crude. The imagery from
this trial was compared with aerial colour photographs and
confirmed the findings of the work carried out during the
TORREY CANYON incident. In addition, it was apparent that the
infrared line scanner provided the possibility of identifying
the thicker regions in the slick. The trial also highlighted
many experimental difficulties, in particular correlation of
the sea-truth data with the imagery.

As a result of the encouraging results of this preliminary
trial, a series of experiments were set up with the following
aims:

* to demonstrate that both sensors could detect oil on
  the sea throughout the various stages of a spill
  (Detection);
* to establish the relationship between the imagery and
  the nature of the oil, in particular, the oil thickness
  (Interpretation);
* to establish the effects of changes in ambient conditions
  (Operational limits).

METHOD

The general policy adopted by Warren Spring Laboratory (WSL) on oil pollution R and D is that as much of this work as possible should be carried out under "real" conditions. This point of view was responsible for the acquisition of a research vessel SEASPRING, for oil pollution studies. R.V. SEASPRING is 56.7 m overall, 11.3 m in beam, and has a tank capacity of about 600 tons. She is fitted with an engineering workshop, a laboratory, and a conference room, and has ample deck working place.

Experiments under real conditions call for small quantities of oil to be discharged at sea. WSL holds an exemption from UK legislation with respect to oil discharges, subject to strict conditions: the quantity of oil is strictly limited and may only be discharged within an experimental area selected in consultation with the Nature Conservency Council, the Ministry of Agriculture, Fisheries and Food, and the Department of Trade (The lead Department in oil pollution matters).

In the two experiments described here, the oil was discharged at a position about 65 km off the east coast, in the southern North Sea. It was discharged as quickly as possible and allowed to develop as if from an accidental spill. Environmental conditions were recorded throughout the tests.

MEASUREMENT OF OIL THICKNESS

Once contact had been made with the surveillance aircraft (RSRE VISCOUNT) at the rendezvous position, samples were taken at various positions in the slick to measure the thickness of the oil. The technique utilised a commercial absorbent material 8.0 mm polyurethane pads manufactured by 3M UK, Ltd., cut into squares 200 x 200 mm. These were affixed to a metal plate by double-sided adhesive tape. The plate and the handle assembly is shown in Figure 1. The pad is just touched on the sea surface and any oil beneath it is absorbed on the pad. This pad is removed, placed in a sample jar and a new pad put in its place.

Pads are analysed by extracting the oil in chloroform, $CH Cl_3$. The pad is agitated with several aliquots of chloroform (depending on the degree of contamination), until the solvent becomes clear. This chloroform extract is compared photometrically with a standard calibration for the particular oil and the quantity of oil on the pad determined.

On exposure to the sea and weather, oil takes up water to form a water-in-oil emulsion, the so-called "chocolate

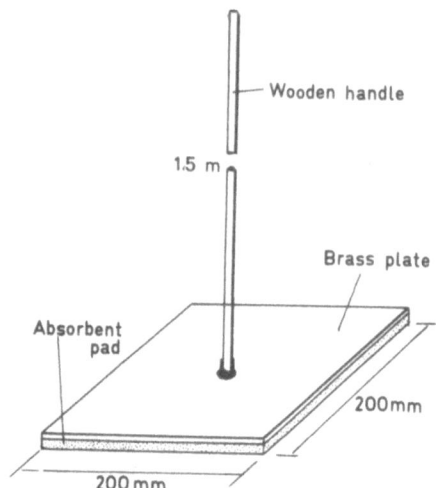

Figure 1.   Pad sampling device.

mousse". At intervals during these tests bulk samples of the
oil were taken from the sea surface and were subsequently analysed for water content. The Dean and Stark method was used in
which the emulsion is refluxed with toluene[1]. A water/toluene
azeotrope is distilled into a graduated container where the
water separates and the mount of water in a given quantity of
emulsion can be measured. Most crude oils rapidly form emulsions
and a water content of 50% can be reached in a few hours; the
maximum, about 80%, is reached within 48 hours, depending on
sea state and oil type. The quantity of oil on the pad was
corrected for the appropriate water content and thicknesses on
the sea of the actual water-in-oil emulsion was calculated.

The technique was calibrated in the laboratory giving good

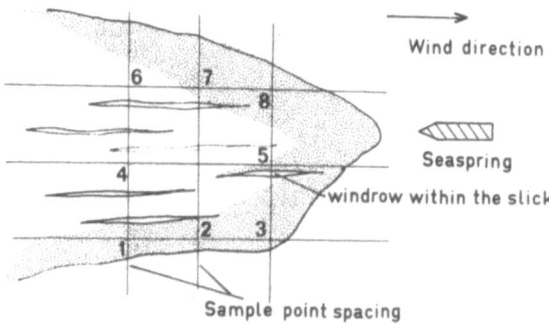

Figure 2.   Sample positioning using dynamic grid.

results for reproducibility, $\pm$ 10%, but it was found that the pad did not pick up the exact quantity of oil directly beneath it. The error was found to be related to thickness and a correction was applied to give an accuracy of $\pm$ 50%. The method did, however, provide a means of image interpretation in broad terms with respect to thickness.

SAMPLE POSITIONING

By far the greatest difficulties were encountered in establishing the sample positions in the imagery. A number of approaches were employed, the simplest being the most effective. It was found to be essential to ensure that the ship was included in all the imagery both to act as a reference point and also as a check on scaling the imagery. The ship remained in a central position while sampling was carried out from a small inflatable dinghy. For the second trial, two dinghies were lashed together, one fitted with a matt black canopy and a radar reflector (effective area 100 m^2) on a short mast. The purpose of this was to provide a target which would appear in the imagery at the sample point.

The first approach was to lay an imaginary grid over the slick, referenced from a central position marked by a buoy (Figure 2). The positions were established ($\pm$ 2½%) relative to the ship using the navigational radar (Marconi Marine Radio Locator 12). A sample was taken at the buoy (control) and then at position 1 in the grid. The dinghy then returned to Control, the position of which was re-established before moving on to position 2. The oil and the ship were moving over the ground (DECCA fix) and also relative to each other; by re-establishing the position of Control between each sample point it was thought that it would be possible to maintain a moving grid. However,

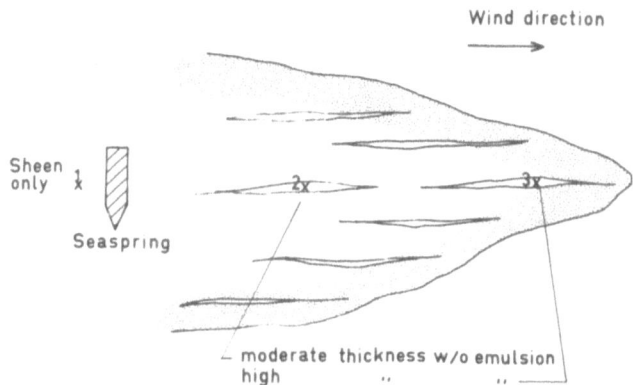

Figure 3. Sample positioning using features of the slick.

Table 1. Ambient conditions.

Test	Oil	Vol. m^3	Temperature °C Air	Sea	Oil	Wind kts	Sky conditions
1	Ekofisk w/o emulsion 40% water	16	17	14.5		10/15 WSW	Scattered cloud sunny cloud base: 5;500 feet.
2 Day 1	Kuwait Crude	7.	17	15.5	18.5	12/15 SW	Scattered cloud sunny
2 Day 2						5/10 NNE	Cloudless, sunny

in practice, the method suffered from two detects:

* the positions were arbitrary and did not relate to interesting features in the imagery, and
* the technique did not allow for manoeuvres of the ship necessary to keep station with the slick.

A simpler method was therefore adopted, utilising recogniz-able features of the slick. Three sample points only were se-lected (Figure 3) along the axis of the wind. The slick is thickest downwind and this provides a natural gradation through the length of the slick. Position 1 was in sheen only, 2 in windrows of emulsion of moderate thickness and 3 in heavy emul-sion. Each position was fixed relative to the ship by radar and for the positions 2 and 3 two sets of samples were taken, one set in sheen and the other within the windrow.

RESULTS

The experimental conditions for both trials are summarised in Table 1 and it can be seen that conditions for each were very similar.

Test 1

IRLS, 2,000 ft altitude, 1 hour after discharge (Figure 4): Since the discharge could not be made instantaneously, the slick began life about 0.8 km long, across the wind, and already wind-row formation can be seen. The darker regions (black) are thought to be relatively thin film of oil at sea temperature,

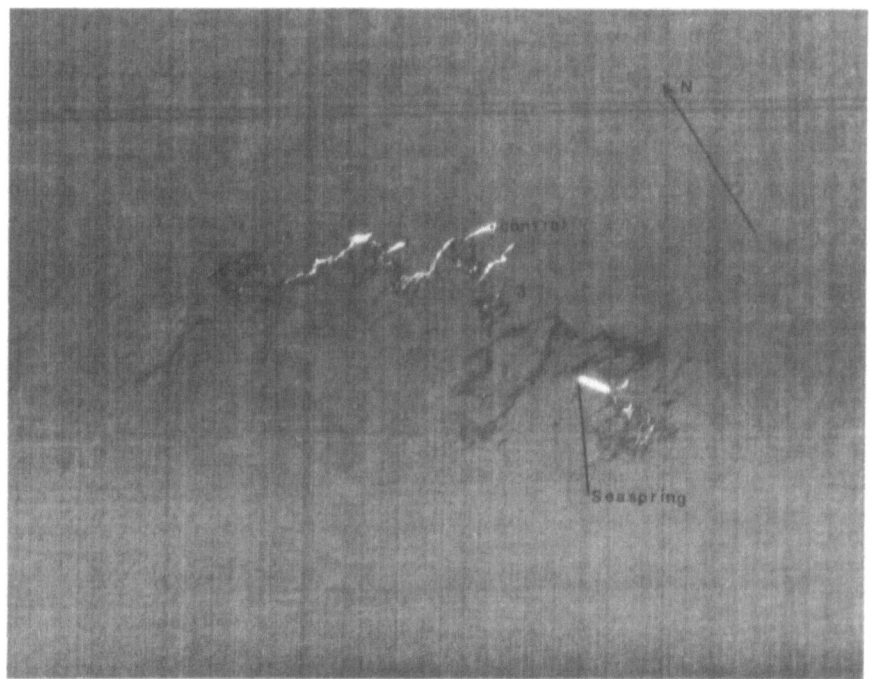

Figure 4.   10 m³ Ekofisk crude oil as 40% water-in-oil emulsion
            1 hour after discharge (2,000 ft - IRLS).
            (Reproduced by permission of the Director of WSL).

appearing cooler, because of the lower emissivity of oil. Where
the oil is sufficiently thick, it has absorbed the heat of the
sun and, because of the low thermal conductivity of oil, appears
hot. In test 2, a difference of 3°C was measured between the sea
and thick oil in a windrow.

Samples were taken shortly after this in the configuration
of the sample positions shown in Figure 2.

IRLS, 2,000 ft altitude, 3 hours after discharge (Figure 5):
Windrows were by this time well established. The size of the
windrows and spacing between them is governed by the wind speed
and is due to cross wind circulation in the air flow over the
sea. The slick had spread out downwind, measuring 0.9 km across
the wind, and 0.6 km downwind.

Samples were taken immediately after this in the corner
positions of the configuration shown in Figure 2.

Table 2. Correlations between thickness and feature in
          the IRLS imagery.

Test conditions	Position	Thickness, μm	Feature in imagery
2,000 ft alt. 1 hour after discharge	Control 1 2 3	700 windrow 10 sheen	Windrow, white nil nil feathered black nil
2,000 ft alt. 3 hours after discharge	1 3 6 8	0.5 4 125-200 windrow 4 sheen 0.5	nil nil feathered black Windrow, black Windrow, black nil

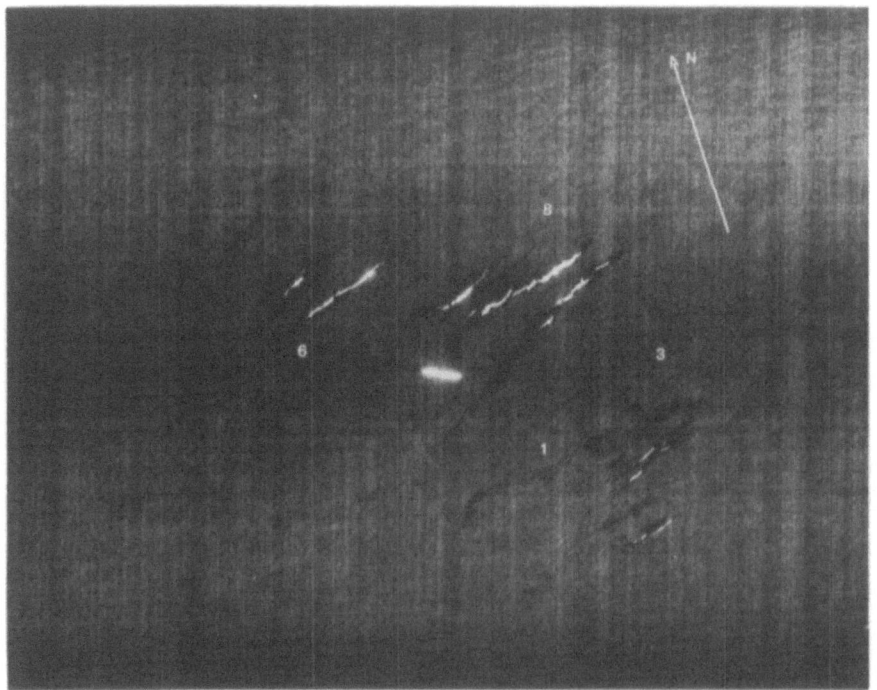

Figure 5. Ekofisk w/o emulsion, 3 hours after discharge
           (2,000 ft, IRLS). Reproduced by permission of
           the Director of Warren Spring Laboratory.

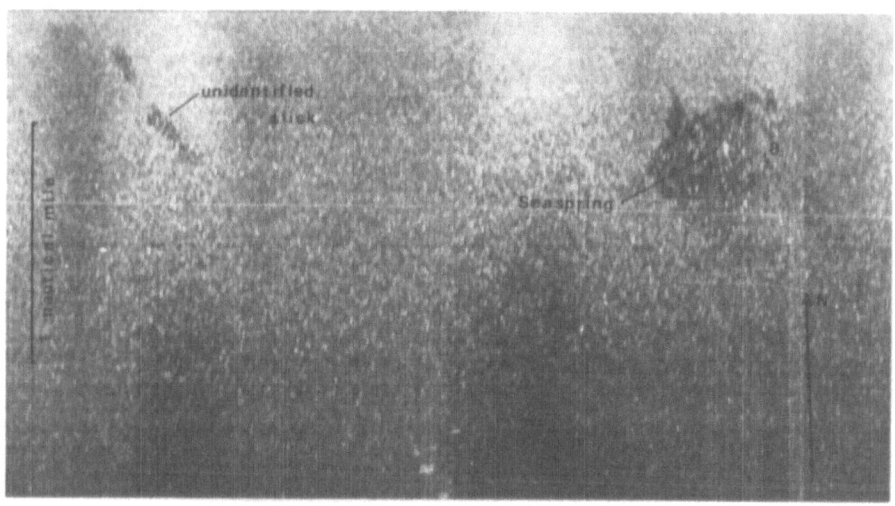

Figure 6.   Ekofisk w/o emulsion, 3 hours after discharge
             (3,000 ft, SLAR) (By permission of WSL).

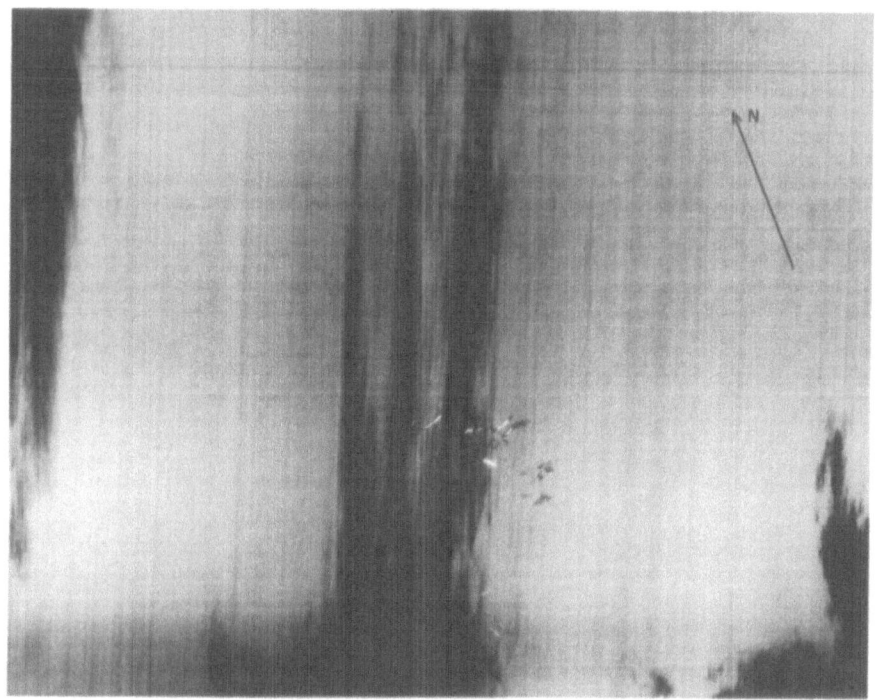

Figure 7.   Ekofisk w/o emulsion, 3½ hours after discharge
             (6,000-6,500 ft, IRLS, cloud base 5,500 ft)
             (By permission of the Director of WSL).

SLAR, 3,000 ft altitude, 3 hours after discharge (Figure 6):
The SLAR imagery shows much lower contrast between oiled and
clean water and shows oil covering a wider area of sea. The slick
measures 1.4 km across the wind, and 1.1 km along the wind. All
the sample points above were within the darker region so that
film thicknesses as low as 0.5 μm appear on SLAR imagery. Despite
this the windrows at the downwind edge of the slick appear to be
visible.

IRLS, 6,000 - 6,500 ft altitude, 3½ hours after discharge:
Figure 7 is included to show the effect of cloud on the sensor.
Cloud base was at 5,500 ft altitude. It should be noted that the
very hot features penetrate the cloud cover in this case.

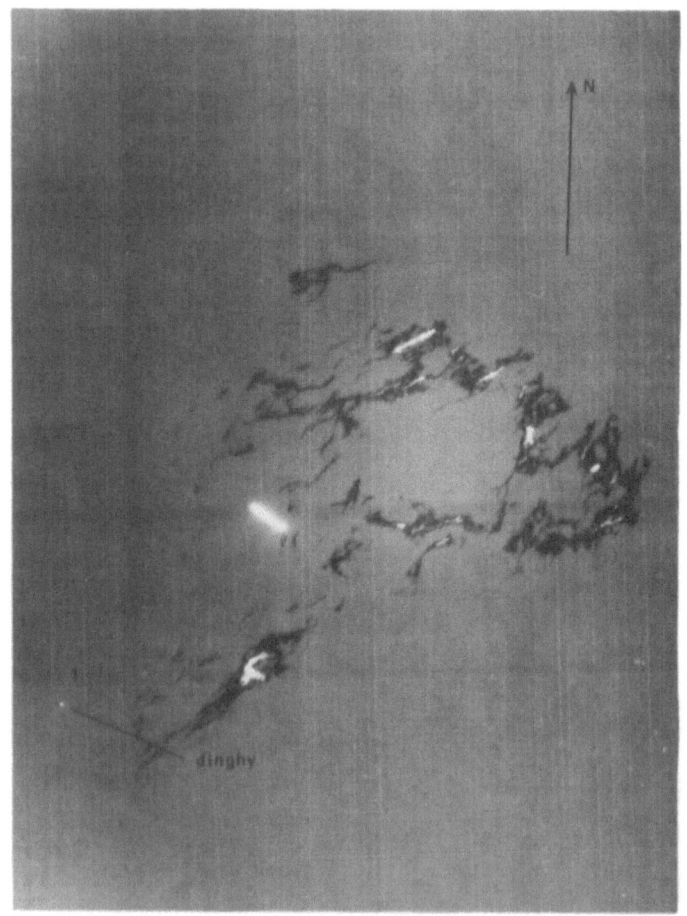

Figure 8.   7 m³ Kuwait crude oil, 2 3/4 hours after discharge
             (2,000 ft, IRLS) (By permission of the Director of
             Warren Spring Laboratory).

Test 2, Day 1

IRLS, 2,000 ft altitude, 2 3/4 hours after discharge, Figure 8: Comparing this with Figure 5, it can be seen that the oil appears in much the same format but that the windrows are less developed[2]. Kuwait oil forms a more stable and more viscous emulsion than does Ekofisk and so it might be expected that this oil would be more resistant to windrow formation. After the discharge had been completed, the tank was flodded to ensure it was gas free and another heavy portion of the slick upwind of SEASPRING resulted from this. The slick as shown here measures 0.5 km across wind and 0.7 km downwind.

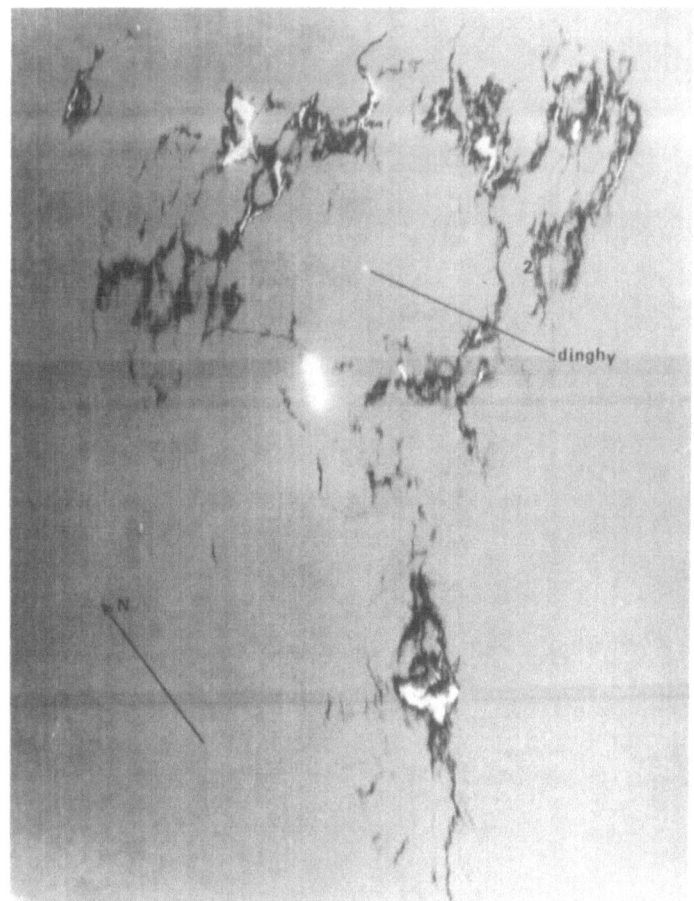

Figure 9.   Kuwait, 3 hours after discharge (1,000 ft, IRLS)
            (Reproduced by permission of the Director of
            Warren Spring Laboratory).

Table 3. Correlations between thickness and feature in
          the IRLS imagery.

Test conditions	Position	Thickness, μm	Feature in imagery
2 3/4 hours after discharge	1	0.2	nil
3 hours after discharge	2	60, windrow 5, sheen	Windrow, black nil
3 1/4 hours after discharge	3	35, sheen on extreme down-wind edge.	nil feathered black
	2		nil

Figure 10. Kuwait, 3 1/4 hours after discharge (2,000 ft, IRLS)
            (By permission of the Director of the WSL).

The sampling configuration of Figure 3 was used in this test.

IRLS, 1,000 ft altitude, 3 hours after discharge (Figure 9: Decreasing altitude does not provide any significant increase information of value to oil spill management. The slick dimensions from this imagery are 0.5 km x 0.9 km possibly elongated by the appearance of some thinner material in this imagery at this altitude.

IRLS, 2,000 ft altitude, 3 1/4 hours after discharge, (Figure 10): A visual estimate of the slick dimensions from a light aircraft measured it as 0.9 km downwind, with the main concentration of oil in the last 400 m, and only 250 m crosswind. This compares with 0.9 km x 0.6 km from this imagery. Apart this discrepancy in scaling, the visual description of the structure of the slick was very similar.

Note (Table 3) that this position 3 apparently was beyond the heaviest concentrations of oil but a sample taken in a

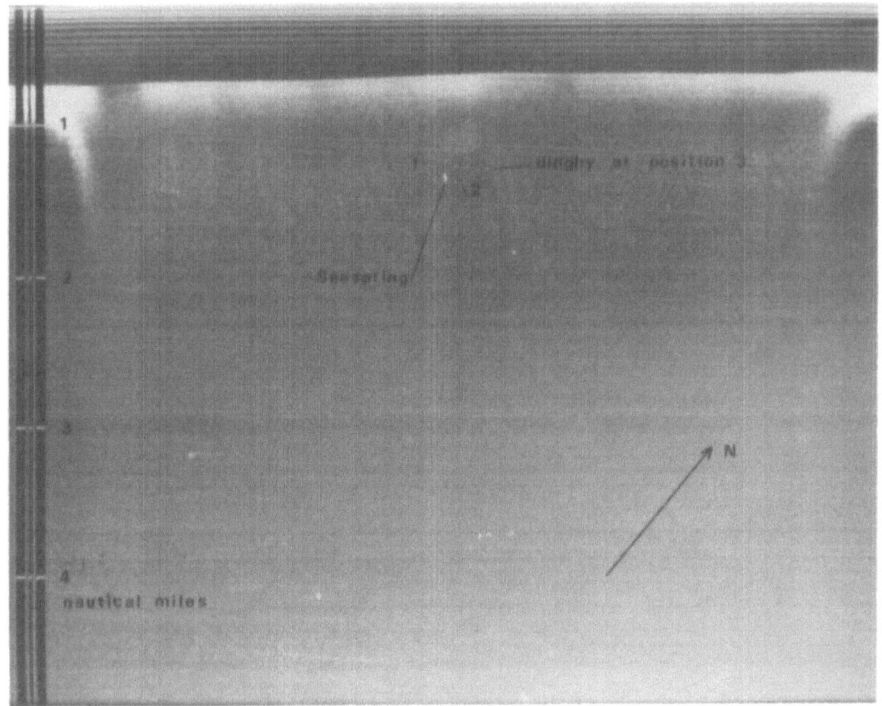

Figure 11.   Kuwait, 3 1/4 hours after discharge (4,000 ft, SLAR) (Reproduced by permission of the Director of Warren Spring Laboratory).

windrow at this downwind edge after the aircraft had left the
test area measured 600 μm.

SLAR, 4,000 ft altitude, 3 1/4 hours after discharge,
(Figure 11): Again the SLAR imagery shows rather poor contrast
when compared with IRLS and also a slightly greater area of sea
covered by oil. The slick dimensions from this imagery are
1.0 km downwind and 0.6 km across the wind. All the sample
points are within the area detected as oil, suggesting a threshold below 0.2 μm for the SLAR sensor. Indeed, the sample point
3, judged to be on the edge of the slick as observed from
SEASPRING, is put 50 m inside the slick as detected by the side
looking airborne radar.

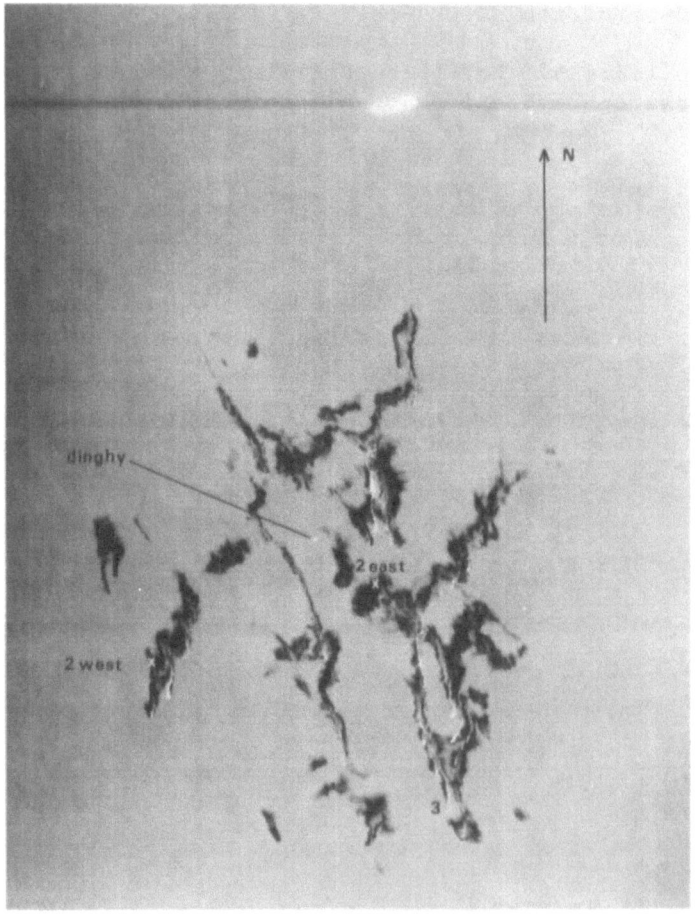

Figure 12.   Kuwait, 25 hours after discharge (2,000 ft, IRLS)
             (Reproduced by permission of the Director of WSL).

Test 2 - Day 2

   IRLS, 2,000 ft altitude, 25 hours after discharge: The
wind dropped overnight and went round to the north. The
imagery (Figure 12) shows a confused pattern of windrows as a
result of this wind change. It is of interest to note that
overnight the thicker areas are much reduced, while the overall area of the slick is much the same; it measures 0.6 km
across wind by 0.7 km downwind. This compares with a visual
estimate from the air of 0.5 km x 3.5 km with the heaviest
concentration on the downwind side covering 1.5 km. The format
of Figure 3 was followed to select sample points although two
samples were taken at position 2 and none at position 1.

DISCUSSION OF RESULTS

   The identification of sample points in the imagery was a
severe problem in the analysis of the results. The imagery was
non-rectilinear and non compensation was made for this, so
that distances and bearings of sample points could not be directly translated to the imagery. In future work a correction
is to be applied to this data so that accurate positioning can
be achieved. In cases where the dinghy was detected it may not
have been on station at the time the imagery was recorded, and
in other cases the dinghy was lost against the hot background
of the oil. There was a general difficulty in synchronising
the aircraft overflight with fixing the dinghy position and
sampling the slick. Thus the sample point was not known precisely but it was known whether or not the sample was taken
within a windrow, and its approximate position within the
slick was known. It was also known how accurately the thickness of the slick could be measured. It was thought that the
results were sufficiently reliable to make a first order approximation of the quantity of oil on the sea.

Table 4.   Correlations between thickness and feature in
          the IRLS imagery.

Test conditions	Position	Thickness, μm	Feature in imagery
25 hours after discharge	2 West	200, windrow 3, sheen	Windrow, black nil
	2 East	350, windrow 1, sheen	Windrow, black nil
	3	2400 windrow	Windrow, white

Table 5.   Thickness with respect to imagery feature (μm).

I R L S		SLAR
Black	White	
150 - 200	700	0.5
60	(600)	0.2
35		
200	2400	
350		

The results are summarised in Table 5 for two tests. From these results, it may be deduced that the IRLS threshold thickness is about 25-50 μm and for the SLAR it is of the order of 100 nm.

Order of Magnitude Calculations of Oil Quantity

If an order of magnitude thickness is taken from the results (presented in Table 5) the quantity of oil on the sea surface can be estimated from the two grey levels in the IRLS imagery. Black was therefore taken to represent a mean thickness of 100 μm and white, 1,000 μm (1 mm).

Measurements of the area covered by the two grey levels in Figure 5 (Test 1, + 3 hours) gave:

$5,580 \text{ m}^2$ @ $1,000$ μm (white) = 5.6 m^3 ⎫
and   ⎬ 14 m^3
$83,700 \text{ m}^2$ @ 100 μm (black) = 8.4 m^3 ⎭

The 14 m^3 have to be compared with the 16 m^3 discharged.

In the event of cold, or possibly even on hot overcast days, the upper grey level is lost (this was observed in the preliminary trials in January 1977, although intense black replaced the white, so that it may be possible to retrieve this level). If the entire area is taken as a single grey level, it can only be expressed as being in excess of 100 μm, and this yields a result of $>9 \text{ m}^3$.

The area analysis of Figure 10 (Test 2, Day 1, + 3 1/4 hours) yields:

$5,830 \text{ m}^2$ @ $1,000$ μm = 5.8 m^3 ⎫
                                                          ⎬ 9 m^3
$29,900 \text{ m}^2$ @ 100 μm = 3.0 m^3 ⎭

With no upper grey level, the result would have been quoted as ≥ 3.6 m^3.

7 m^3 Kuwait crude were discharged and after 3 hours it is estimated that something like 20% of this volume would have been lost by evaporation, by comparison of this situation with thoses studied for other oils. However, the oil would also have picked up about 40% water in the formation of a water-in-oil emulsion so that the anticipated volume of material on the water was actually about 9 m^3, possibly less, allowing for an unknown loss by natural dispersion into the water column.

Analysis of Figure 12 (Test 2, Day 2, + 25 hours) gave:

$$2,9800 \text{ m}^2 \quad @ \quad 1,000 \text{ }\mu\text{m} = 3.0 \text{ m}^3 \quad \left.\right\}$$
$$46,600 \text{ m}^2 \quad @ \quad 100 \text{ }\mu\text{m} = 4.5 \text{ m}^3 \quad \left.\right\} \quad 7.5 \text{ m}^3$$

With no upper grey level the estimate would have been upper than 5 m^3.

The water content was measured as 60% and about 25% of the initial discharge would have been lost by evaporation by this time. If natural dispersion and lateral dissipation (an oil film was visible for 3.5 km) are neglected, the about 13 m^3 are anticipated. The discrepancy here is probably mainly due to neglecting natural dispersion.

The results gained to date therefore enable an estimate to be made of the quantity of oil on the sea surface to within an order of magnitude. Since the grey levels correspond to thickness greater than a certain threshold thickness, it is likely that under some circumstances the quantity of oil on the sea may be underestimated, but nevertheless the technique does provide an objective means of:

* assessing the appropriate response to reported oil spill incidents, and
* identifying areas of highest contamination within a slick.

ELENI V INCIDENT, MAY 1978

Figure 13 shows IRLS imagery of part of the ELENI V incident, so demonstrating how the above technique might be used in a real spill. The upper part of the photograph shows the bow section of the wreck under tow and it is of interest to note how rapidly the oil attains a temperature above that of the sea as it flows away from the wreck.

The oil was Bunker C, residual fuel oil and formed thick platelets, 1-3 m in diameter. These appear in the imagery as individual hot spots.

CONCLUSIONS

1) Both the SLAR (Q-band, (H) polarised) and the IRLS (8-14 μm) were able to detect oil on the sea and, in addition, the IRLS provided information on the structure of the slick.

2) IRLS provided high contrast imagery that could be easily interpreted; the SLAR showed lower contrast and so interpretation was more difficult.

3) The threshold thickness for the IRLS was found to be of the order of 10 μm (probably 25-50 μm) and for the SLAR of the order of 100 nm.

Note: This very low threshold for SLAR may result in extremely large areas of sea being shown as being contaminated with oil since in the event of a real oil spill the area of sea covered by sheen is extensive. The area of interest would normally be located at the downwind edge, however.

4) The adoption of a thickness of 1,000 μm for areas appearing hotter than the sea, in the IRLS imagery, and 100 μm for those apparently colder, led to estimate the quantity of oil (as a water-in-oil emulsion) on the sea surface within an order of magnitude. From these experiments a mean thickness of 15-20 μm over the area detected by radars was found to fit the data but see note above.

5) It may be possible to distinguish broad categories of oil type from IRLS imagery because of their different behaviour at sea.

ACKNOWLEDGEMENT

Figures 5 through 13 are reproduced courtesy of Royal Signals and Radar Establishment, Malvern, United Kingdom.

NOTE

The work described in this report was sponsored by the Ship and Marine Technology Requirement Board. For further information enquiries should be addressed to the Warren Spring Laboratory, Stevenage, Herts.

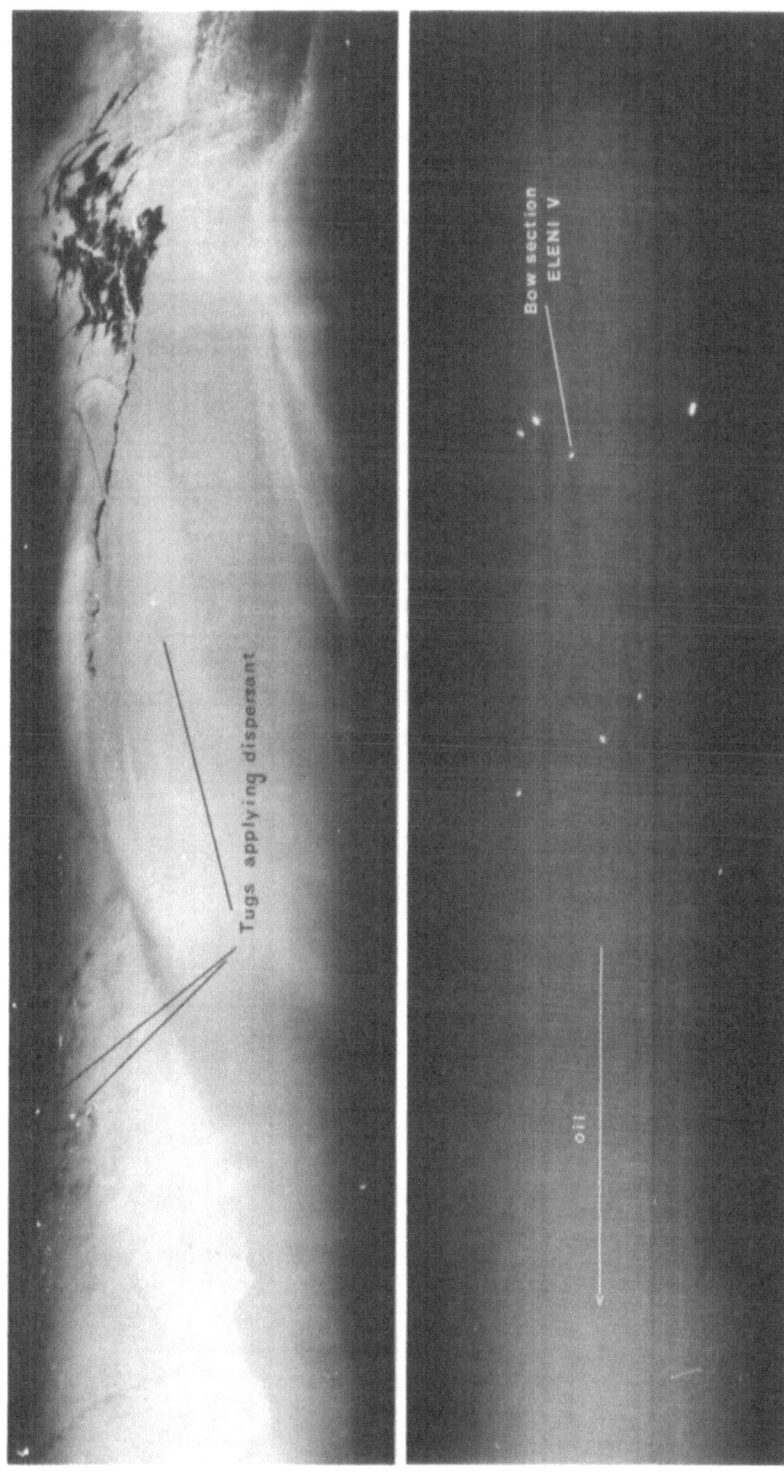

Figure 13.   IRLS imagery from the ELENI V spill of heavy fuel oil.

REFERENCES

1. P.G. Jeffery, Apparatus for the analysis of oil pollution
      Lab. Pract., 20(1) (1971).
2. D. Cormack, J.A. Nichols and B. Lynch, Investigation of
      factors affecting the fate of North Sea oils discharged
      at sea. Part 1. Ekofisk crude oil, July 1975-February
      1978, Warren Spring Laboratory Report LR273 (OP),
      Stevenage, Herts, United Kingdom, (1978).

OBSERVATION OF TWO TEST OIL SPILLS

WITH A MICROWAVE SCATTEROMETER AND A SYNTHETIC APERTURE RADAR

R.A. Neville[1], V. Thomson[1], L. Gray[2]
and R.K. Hawkins[2]

[1]Intera Environmental Consultants Ltd.
Ottawa, Canada
[2]Canada Centre For Remote Sensing
Ottawa, Canada

ABSTRACT

A microwave scatterometer and a synthetic aperture radar
(SAR) were flown over two test oil spills off the coast of New
Jersey in November, 1978. This paper examines the microwave
data obtained from this mission and draws some tentative
conclusions.

The CCRS, Ryan 13.3 Ghz microwave scatterometer was used
to quantify the depression in the microwave backscattering
coefficient, $\Delta\sigma_o$ resulting from the oil slick on the ocean
surface. This scatterometer can transmit vertically and horizontally polarized microwaves and can receive simultaneously
both like- and cross-polarized returns. Scattering cross sections were measured over a range of incidence angles from 60°
aft of the aircraft to 60° fore. Over this range of angles,
VV and HH polarizations showed approximately the same depression in radar cross section with the presence of the oil.

The depression is negligible at the incidence angles below
10° and increases to approximately 12 dB for incidence angles
between 30° and 55°. For angles of incidence greater than 30°,
the average backscattering cross section, σ_o, due to sea
clutter is greater in the VV mode than in the HH mode. This
confirms that a radar with a vertically polarized transmitter
and a vertically polarized receiver may be preferable for the
detection of oil slicks on the water surface because the sea
clutter-to-noise ratios will be higher. These results, however,

suggest that for angles of incidence likely to be found in
satellite radars (and some airborne radars) a 13.3 Ghz system
using either like-polarized mode (VV or HH) will permit
satisfactory detection of oil slicks.

The ERIM (Environmental Research Institute of Michigan)
SAR was flown over the same slick and imaged in both L- and X-
band. Horizontally polarized microwaves were transmitted and
both polarizations received while the slicks were imaged over
the narrow range of incidence angles from 33° to 49°. Initial
optical processing of this imagery shows the cross-polarized
imagery to be of very low contrast. The like-polarized X-band
imagery is superior in both resolution and contrast to the
like-polarized L-band imagery. Small signal suppression in the
vicinity of extremely bright targets such as ships (or icebergs)
has been observed and this artifact may appear to be similar
to oil. For the sea states present (sea state 1-2), little
variation in the oil-water contrast was noted when the slick
was imaged from different directions.

INTRODUCTION

The application of imaging microwave radars to the
surveillance of coastal waterways for the purpose of detection
of ocean oil pollution is now well established method. The
presence of oil manifests itself as a reduction of the ocean
radar cross section, and therefore an oil slick may be
represented as a darkened region in the sea clutter of a microwave radar image. The mechanism for this reduction, as pointed
out by van Kuilenberg[1] is first the disturbance of the air-water
interface imposed by the oil membrane and secondly, the
mechanical damping of the ocean capillary wavelets. The latter
effect is related jointly to the oil viscosity and its thickness. Because the radar and capillary wavelengths are similar,
we expect the damping effect to have an important influence on
the backscatter cross section.

There are three basic radar parameters by which a target
backscatter cross section may be characterized: the radar
frequency (or wavelength); the look or incidence angle, Θ; and
transmit and receive polarizations. Once these radar parameters
have been set, the characteristics of the target oil spill
including the physical properties of the oil itself and its
distribution, as well as the wind and sea conditions combine to
determine the contrast in the radar cross section of the non-
polluted and polluted water. The detectability of the oil spill
then depends on the ability of the radar itself to image this
contrast. We briefly review some of the work done to date.

Guinard[2], using a four frequency multipolarized SAR

(Synthetic aperture radar) working with incidence angles near
80°, studied the radar imagery of the 1970 tanker ARROW
disaster in Chedabucto Bay, NS. The results were qualitative
and due to instrumental limitations, only the VV (vertical
transmit, vertical receive) polarization combination showed
oil.

SLAR (Side-looking Airborne radar) imagery obtained by
the United States Coast Guard[3] in their AOSS (Airborne Oil
Surveillance System) programme showed very little dependence
of the oil-ocean contrast with target conditions. They observed
incidence angles $\theta > 76°$ and found that the oil-sea contrast
rose to a maximum of 12 dB near $\theta = 84°$.

Unlike imaging radars, the microwave scatterometer offers
an accurately calibrated measure of the target radar cross
section. A Ryan 13.3 GHz scatterometer (with fixed VV polarization) was used by Krishen in connection with the Chevron Oil
spill in 1973. Krishen mapped the variation of θ_0 with
incidence angle for both non-polluted and polluted ocean with
windspeeds of 18 knots. A contrast rising to 15 dB near $\theta = 60°$
was seen.

Significant gaps exist in the characterization of the
radar cross section of oil laden water. These include both the
dependence of the radar parameters themselves and those of the
environment. The scatterometer offers several advantages in
this respect since it gives quantitative information over a
wide range of incidence angles.

The Canada Centre For Remote Sensing (CCRS) scatterometer
depicted in Figure 1 is an advanced version of that used by
Krishen, transmitting a continuous fan beam in either of the
two polarizations and receiving in both polarizations. The
combination VV, VH, HH, and VH are therefore possible. The
beam is narrow across track limited to 3°, but the fan beam
extends to 60° along track fore and aft of the aircraft nadir
position. Individual incidence angles for each ground footprint can be determined unambiguously from the Doppler
frequency shift of the returning signal.

SCATTEROMETER DATA

The scatterometer[4] was flown in the CONVAIR 580 aircraft
on ten flight lines over two planned oil spills on November 2
and November 3, 1978, as part of the CCRS AMOP (Arctic Marine
Oil Spill Program) contribution to an experiment organised
by NASA at dump site 106 off the New Jersey coast.

An RC-10 mapping camera aboard serves both to define the

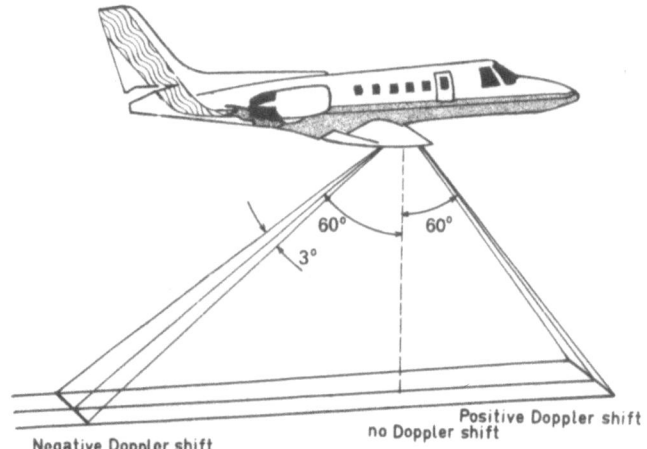

Figure 1. Fan beam geometry of Ryan 13.3 GHz scatterometer.

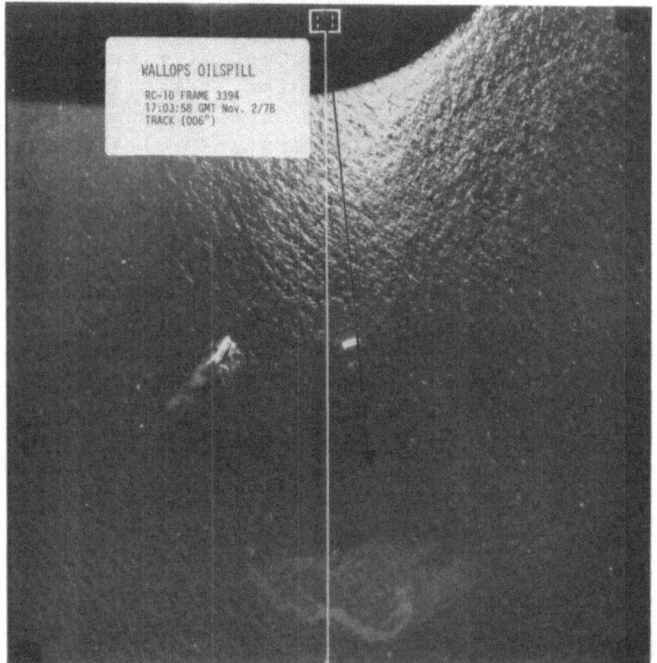

Figure 2.   Aerial photograph showing Wallops oil spill and the
            projection of scatterometer forebeam. Visible on
            the sea surface is the spill itself as well as two
            ships: the ANNANDALE, a 90 ft vessel and a 50 feet
            sport fisherman.

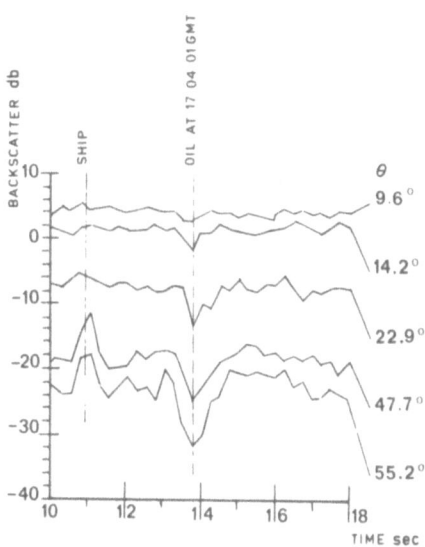

Figure 3.   Radar backscatter time history along flight line 3.
            The small ship is seen as a bright target in the sea
            background while the oil spill is seen as an increas-
            ing depression in the sea clutter background. The VV
            polarization result is shown.

target oil spill and the geometry of the scatterometer footprint itself. Figure 2 shows a frame from one of the three
flight lines in the mission which were successful in traversing
the oil. The spill shown consists of 450 gallons (2 m^3) of
Merban crude oil and the flight line occurred a few minutes
after the oil was dumped. The larger of the two ships shown is
the "ANNANDALE", a 90 (27 m) feet vessel. The line down the
centre of the figure represents the aircraft track. The skewed
arrow, seen intersecting the small sport fisherman vessel,
marks the projection of the scatterometer forebeam and coincides
with the aircraft heading. The rectangles at the top of the
figure are the footprint at the nadir and the 60° positions.

     The flight line history for the scatterometer is shown in
Figure 3 (corresponding to Figure 2) for a selected group of
incidence angles in the forebeam. The scatterometer is sensitive
to all of the slick which is visible in the RC-10 photograph as
can be concluded for the number of footprints affected. The data
has been suitably shifted so that each footprint is shown at the
time it crossed the nadir position regardless of the look angle
used. The small ship is seen as a bright target at large fore
angles.

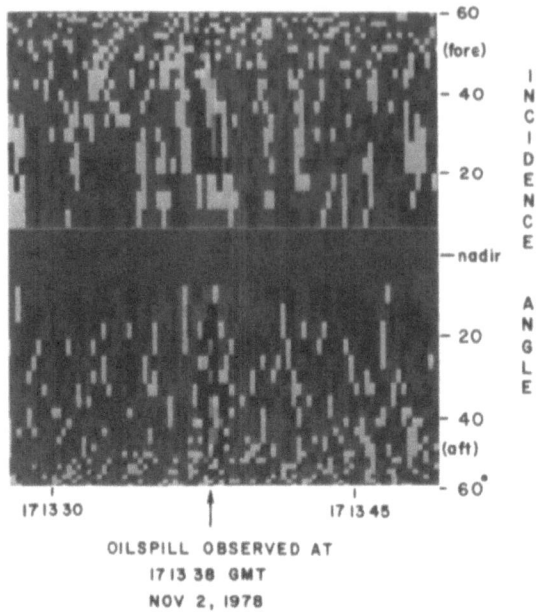

Figure 4.   Colour densigram of radar backscatter contrast from
             flight line 6, Nov. 2, 1978. In this representation
             the complete set of scatterometer incidence angles,
             fore and aft, are depicted vertically down the page
             with the time or along track history shown across
             the figure. The oil spill is seen to stand out for
             incident angles greater than 10°.

        The data shows a trend of higher oil-to-sea contrast as
the incidence angle increases, in the normal dependence.

        When the angular dependence of the sea clutter is removed,
the resultant signal-to-clutter may be represented for the complete scatterometer history as in Figure 4 for flight line 6.
The scatterometer data from flight line 6, Nov. 2, 1978, using
HH polarization are represented here with the flight track horizontally across the Figure in time steps of 0.2458 s per pixel.
The rows represent each angle selected. The departure from the
average sea return is represented as a colour or grey level.
Again the data have been shifted.

        Apparent in this Figure 4 are the low contrast at incidence
angles less than 10° and also the fall off at incidence angles
near 60°. This latter effect may be instrumental in origin.

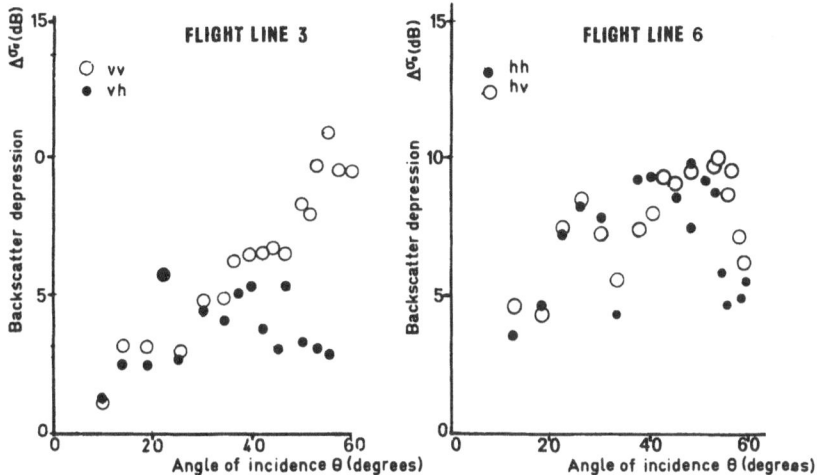

Figure 5. Backscatter depression measurements from different
           flight lines. In the data from flight line 3, the
           departure of the VH from the VV depression at
           higher incidence angles remains unexplained.

In Figures 5 and 6, we quantify the results of the backscatter depression for the three flight lines analysed during
the Wallops Island mission. The trend is similar in each set
of data with a cut off near $\Theta = 10°$ and a rise in the backscatter depression with increasing incidence angle.

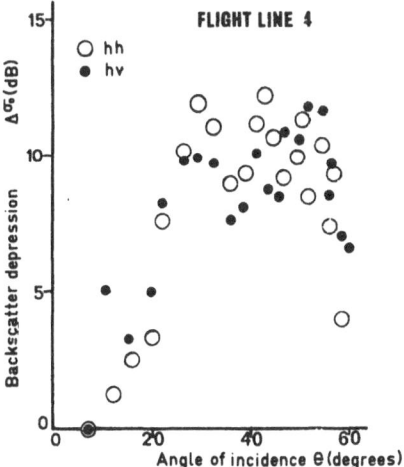

Figure 6. Backscatter depression measurements
           from flight line 4.

Polarizations does not seem to be a significant factor in these results.

Similar depressions are seen for all polarizations and it appears that the HH combination is equally effective with VV at these incidence angles and wind conditions. At higher incidence angles, where many airborne imaging radars work, there may, however, be an advantage in working with VV polarization since at these angles the sea cross section itself is higher giving better signal-to-noise ratios even though the signal depression-to-clutter caused by the oil remain equal.

Note that the departure of the like- and cross-polarized data in Figure 5, flight line 3 (VV, VH combinations) is inconsistent with the results of Figures 5, flight line 6 and 6, flight line 4 (HH, HV combinations). By reciprocity, we expect similar behaviour in HV and VH results, and therefore the VH data is suspect.

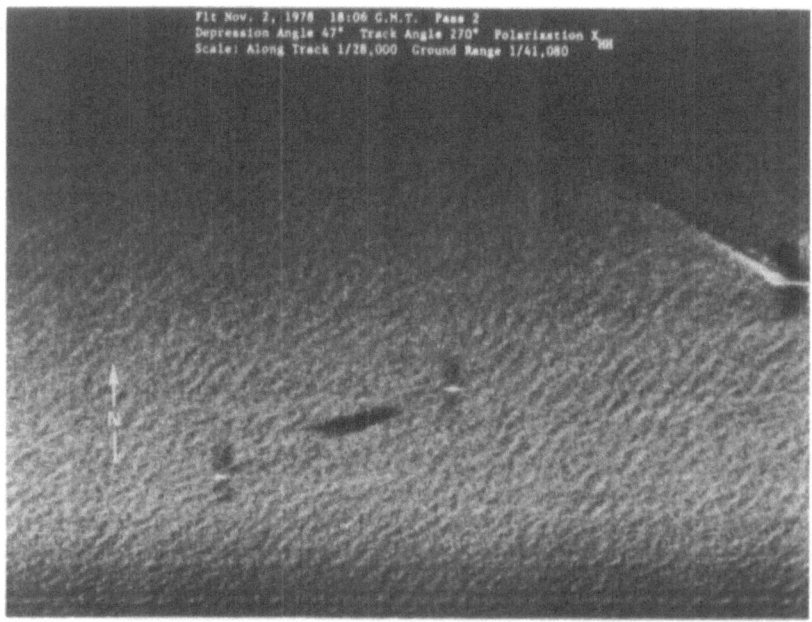

Figure 7.   L-band imagery from pass 2, November 2, 1978.
            The bright objects on the left hand side of
            the image are ships, and a ship and barge
            combination is seen on the right side. The oil
            spill is seen in the foreground as a darkened
            area between the same two ships shown in Fig. 2.

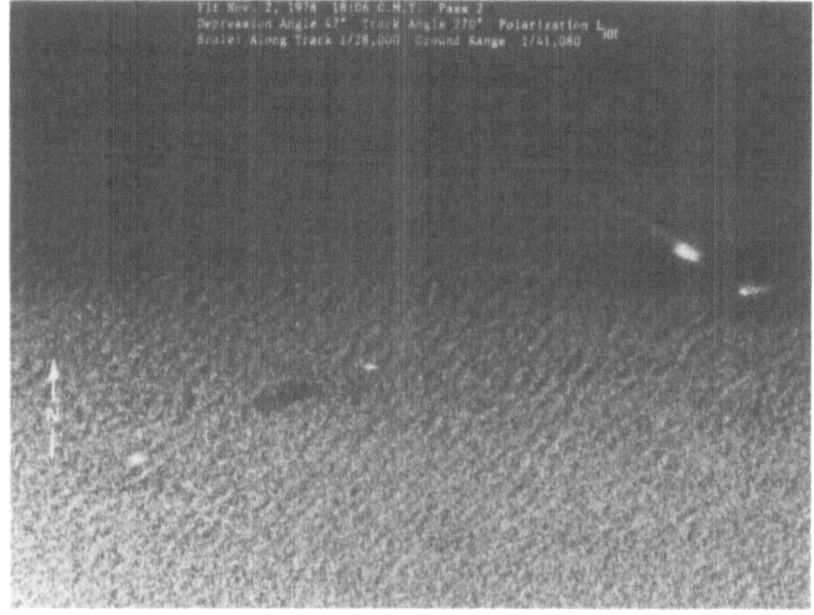

Figure 8.   X-band SAR imagery from pass 2, November 2, 1978.
            The description of the scene is the same as in
            Figure 7. The contrast and resolution are clearly
            better in the X-band imagery as seen from a com-
            parison of Figures 7 and 8.

SAR DATA

     The operation and analysis of the X- and L-band SAR aboard
the Convair 580 is contracted to ERIM and we will simply show
data (Figures 7 and 8) and comment on their results from one
flight line on November 2, 1978, the same spill imaged in
Figure 2. Figure 7 is L-band imagery while Figure 8 is from
X-band. In each case, HH polarization is used. It is clear from
the imagery that the shorter wavelength (X-band) result is
superior in both resolution and contrast especially with respect
to the oil spill. The effect of small signal suppression is also
evident in the imagery near the bright ship returns. Experienced
observer could discriminate this artifact from a real oil spill.
Wave structure is clearly defined in the imagery.

CONCLUSIONS

     The utility of a scatterometer in characterising microwave
radar cross sections and signal-to-clutter contrasts has been
illustrated by the analysis given. However, due to the extremely

limited data base of the analysis, only tentative conclusions
may be made. These include:

* The suppression of the ocean backscatter cross section
  is negligible at an incidence angle of 10° and increases
  to approximately 7 to 12 dB for incidence angles between
  30° and 55°.

* The combination of horizontally polarized transmit and
  receive radiation (HH) leads to a comparable suppression
  of the radar cross section when compared to VV data for
  incidence angles less than 60°.

* Neither cross-polarized component of the backscatter
  radiation showed larger backscatter suppression than the
  like-polarized components.

For future experiments we recommend that the following
guidelines be adopted:

* Avoid working with small slicks so that adequate
  statistics may be obtained from many sensor footprints;
* Acquire data both as a function of the age of the oil
  spill and from as many sea states and wind velocities
  (speed and direction) as possible.

REFERENCES

1. J. Van Kuilenberg, "Proc. 9th Intl. Symposium Remote
    Sensing of Environment, Ann Arbor, Michigan" (1975).
2. N. W. Guinard, "Proc. 7th Intl. Symposium Envir., Ann Arbor,
    Michigan" (1971).
3. A. T. Edgerton, J. J. Bommarito, R. S. Schwantie and D. C.
    Meeks, Dept. of Transportation report CG-D-90-75 (1975).
4. K. K. Chan and A. L. Gray, "A System Study of the CCRS 13.
    13.3 GHz Scatterometer and the Design of a Simple Digital
    Processing Scheme (1977).
5. K. Krishen, J. Geophysical Research, 76:6528 (1971).
6. W. L. Jones, L. C., Schroeder, J. L. Mitchell, "IEEE Trans.
    on Antennas and Propagation, AP-27 (1977).

MEASUREMENTS OF THE DISTRIBUTION AND VOLUME
OF SEA-SURFACE OIL SPILLS
USING MULTIFREQUENCY MICROWAVE RADIOMETRY*

J.P. Hollinger and R.A. Mennella

Department of the Navy
Naval Research Laboratory
Washington, D.C., United States

ABSTRACT

Multifrequency passive microwave measurements from aircraft have been made of eight controlled marine oil spills. It
was found that over 90 percent of the oil was generally
confined in a compact region with thicknesses in excess of
1 mm and comprising less than 10 percent of the area of the
visible slick. It is shown that microwave radiometry offers a
means to measure the distribution of oil in sea-surface slicks
and to locate the thick regions and measure their volume on
all-weather, day-or-night and real-time basis .

MEASUREMENTS OF THE DISTRIBUTION AND VOLUME OF SEA-SURFACE
OIL SPILLS

There is mounting concern by the public and governmental
agencies over the ever-increasing number of marine oil spills
and the more serious resulting pollution. Before appropriate
corrective action can be taken, a knowledge of the nature,
thickness, areal extent, direction and rate of drift of the
oil spill must be promptly established. This requires a
detection and measurement system capable of rapidly responding
to a requirement of surveying large and often remote and in-
accessible waters on a nearly all-weather and day-or-night
basis.

Reliable determination of oil-film thickness is of major
importance. It is the film thickness along with areal extent
which allows the volume of the slick to be estimated. A knowledge of the volume of oil is essential for litigation and

267

damage claims resulting from major oil spills, as well as for
assessing the impact of the spill on marine life and environment. A knowledge of the oil distribution and the location of
those regions containing the heaviest concentration of oil
would enable the most effective confinement, control, and
clean-up of the oil and perhaps is most important.

Sea-surface oil spills do not spread uniformly nor without
limit[1-2]. Thick regions which contain the majority of oil are
formed and are surrounded by very much thinner and larger
slicks. For example, in controlled oil spills of 200 to 630
gallons (760 to 2380 liters), which will be described in detail
later, the oil typically formed a region with a thickness of
1 mm or more containing more than 90 percent of the oil but
comprising less than 10 percent of the area of the slick
visible. The remaining oil formed a large slick, hundreds of
times thinner, surrounding the thick region.

Microwave radiometry offers a unique potential for determining oil-slick thicknesses greater than about 0.05 mm. The
apparent microwave brightness temperature is greater in the
region of an oil slick than in the adjacent unpolluted sea by
an amount depending on the slick thickness. In effect the oil
film acts as a matching layer between free space and the sea
enhancing the brightness temperature of the sea. The calculated
increase in microwave brightness temperature due to an oil
slick above that due to the unpolluted sea as a function of
slick thickness is shown in Figure 1 for the three microwave
frequencies at which measurements were made. As the thickness
of the oil film is increased, the apparent microwave brightness
temperature at first increases and then passes through alternating maxima and minima, due to the standing-wave pattern set
up by the sea surface. The maxima and minima occur at successive
integral multiples of a quarter wavelength in the oil. By using
two or more frequencies, thickness ambiguities introduced by
the oscillations can be removed and the film thickness determined for a wide range of thicknesses.

To verify the calculated behavior with film thickness and
to measure the dielectric properties of the three oil types
used in the controlled ocean oil spill tests, microwave radiometric measurements were made using a small test tank. The
tank was filled with fresh water, and then a known volume of
oil was added to the surface. The incremental increase in the
oil-film thickness resulting from the addition of oil was
calculated assuming uniform spreading of the oil over the
surface area of the tank. Measurements were made for No. 2 fuel
oil and No. 4 and No. 6 crude oils. Number 2 fuel oil spread
uniformly over the tank even for thicknesses as small as
0.1 mm. However No. 4 and No. 6 crude oils tended to form lenses

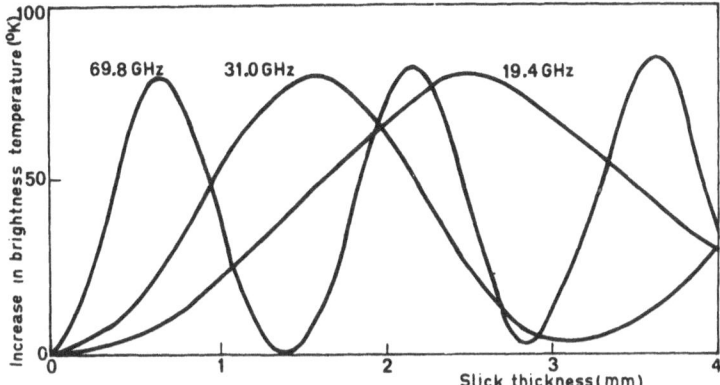

Fig. 1.   Calculated increase in microwave brightness temperature
due to an oil slick that due to the unpolluted sea as a
function of slick thickness at 19.3, 31.0, and 69.8 GHz.
(The calculations are at 0° incidence angle for a smooth
sea at a temperature of 20°C and salinity of 35 ppt covered by a uniform oil film with a complex dielectric
constant of 2.1 - 0.01j).

or blobs until the entire tank surface was covered. This required
thicknesses in excess of about 4 mm, after which the oils apparently did spread uniformly for small incremental increases.
Therefore the results for these oils are less accurate than those
obtained for No. 2 fuel oil. The complex dielectric constant of
the oil was determined by adjusting it to obtain the best fit of
the calculations to the measurements. The measurements for No. 2
fuel oil and the best-fitting calculated curves are shown in
Figure 2. The complex dielectric constant $\varepsilon_1 - j\varepsilon_2$, determined from
the measurements for No. 2, No. 4, and No. 6 oils is given in
Table 1.

The measured antenna temperature, rather than brightness
temperature, is given in Figure 2 to present the magnitudes that
were actually observed during the observations. The antenna
temperature is the average of the brightness temperature over
all directions, weighted by the antenna response pattern[4]. That
is:

$$T_A = \frac{\int_{4\pi} T_B (\theta, \varphi) f (\theta, \varphi) d\Omega}{\int_{4\pi} f (\theta, \varphi) d\Omega},\qquad (1)$$

Table 1. Measured complex dielectric constant of oil.

Oil Type	Temperature (°C)	Complex Dielectric Constant	
		f = 19.3 GHz	f = 69.8 GHz
No. 2 Fuel	19	ϵ_1: 2.10 ± 0.05 ϵ_2: 0.01 $^{+0.02}_{-0.01}$	ϵ_1: 2.10 ± 0.05 ϵ_2: 0.01 $^{+0.02}_{-0.01}$
No. 4 Crude	26	ϵ_1: 2.4 ± 0.1 ϵ_2: 0.06 ± 0.04	ϵ_1: 2.2 ± 0.1 ϵ_2: 0.07 ± 0.04
No. 6 Crude	26	ϵ_1: 2.6 ± 0.2 ϵ_2: 0.05 ± 0.05	ϵ_1: 2.6 ± 0.2 ϵ_2: 0.05 ± 0.05

where $f(\theta,\varphi)$ is the normalized antenna response pattern and $T_B(\theta,\varphi)$ is the total brightness temperature in the direction θ,φ and is composed of not only the radiation emitted by the sea surface but also the downwelling sky radiation reflected by the surface as well as the emission and attenuation of the atmosphere between the surface and the radiometer. Atmospheric effects are usually less than 10 percent for observational frequencies away from the water-vapor absorption line at 22.235 GHz and below about 40 GHz, and in general approximate corrections can be applied to partially remove them.

A series of eight controlled oil spills was conducted during the period August 1971 through August 1972 in co-operation with the NASA-Wallops Island Station, the Virginia Institute of Marine Science, and the U.S. Coast Guard to investigate the possibility of determining the thickness of an oil spill using passive microwave radiometry. The spills of from 200 to 630 gallons (760 to 2,380 liters) of either No. 2 fuel oil or No. 4 or No. 6 crude oil, were performed in accordance with the guidelines established by the Environmental Protection Agency for the discharge of oil for research purposes[5]. All of the spills were conducted in relatively calm sea conditions of less than 2-m swell and 10-m/s surface winds. The oil was transported in 50-gallon (190-liter) drums to the ocean test site about 10 mi (16 km) east of Chesapeake Light Tower off the east coast of Virginia. The drums were off-loaded, herded together, and emptied from small rubber boats in a manner so as to obtain as nearly an undistrubed point release as possible.

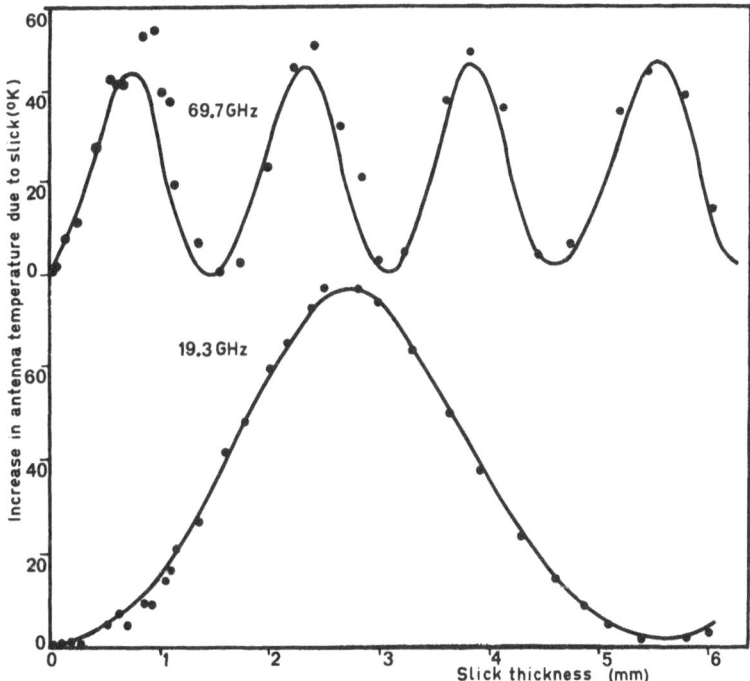

Figure 2.   The data are measurements at 19.3 and 69.8 GHz
            of the increase in antenna temperature due to
            No. 2 fuel oil spread over a smooth water sur-
            face in a test tank as a function of film thick
            thickness; The curves represent the calcula-
            tion which best fit the measurements.

     The documentation of "ground truth" gathered included the
type and volume of oil spilled, in situ measurements of oil
slick thickness, and airborne natural and color IR photography
and thermal IR imagery, as well as the environmental parameters
of sea temperature, air temperature, relative humidity, wind
speed and direction, sea state, and general weather and cloud
conditions. The oil in two spills was dyed with an oil-soluable
red dye to aid in establishing the distribution of oil over the
sea surface. The dye allowed the thick regions of oil to be
easily identified visibly. Figure 3 is a series of drawings
traced from colour photography of the July 11, 1972, oil spill.
This spill consisted of 630 gallons (2,380 liters) of No. 2
fuel oil dyed red. The sea conditions were calm, with about
1-m swell and winds of 2-4 m/s. The outer line in each drawing
represents the extreme edge of the visible slick, the next
inner line is the region of color fringing when visible in the
photograph, and the shadowed area is the region of thick

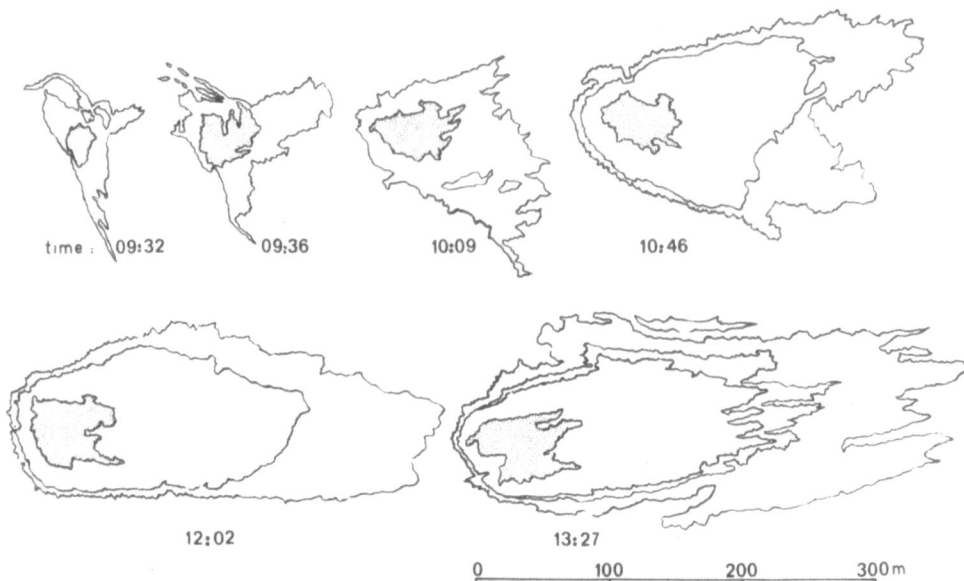

Figure 3.   Tracings of color photography of the oil slick
             resulting from a controlled oil spill of 630
             gallons of No. 2 fuel oil.

oil. The oil formed a well-defined thick region surrounded by a
very much larger and thinner region. In situ thickness measurements showed the oil to be 2.4 ± 0.3 mm thick at spots in the
shadowed region and typically 2 to 4 μm thick outside this
region. The thick inner region spread at a much slower rate than
the total slick. This is shown in Figure 4 where the area of the
inner region and the total area of the visible slick are displayed as a function of time on a log-log plot. If the dashed
lines are taken to represent the measurements, the total area
grew at a rate proportional to the time to the 0.6 power; the
thick region grew at a rate proportional to time to the 0.2
power. The spreading rate of the total area most nearly matches
the gravity-viscous spreading phase described theoretically by
Fay[6], which grows at a rate proportional to the square root of
the time. It is somewhat slower than spreading rates reported
by Guinard[7] or by Munday et al.,[8]. However, the spreading rate
is dependent on many variables, such as initial volume, age,
density and viscosity of the oil, the surface-active materials
present, interfacial surface tension, surface wind, sea state,
and surface current present, and will vary widely.

     Most significant is the dichotomous behavior of the oil,
dividing clearly into a thick, relatively compact region

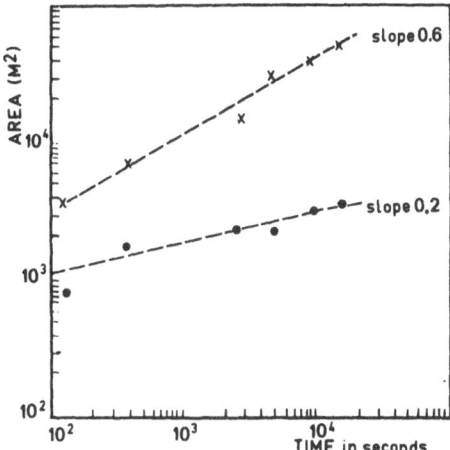

Figure 4. Area of the inner thick region (dots) and the total
area of the visible slick (crosses) measured from
the photography of the oil represented in Figure 3
is plotted versus the time the picture was taken.
The dashed lines are possible representations of
the measurements.

surrounded by a second much larger and thinner region. All of
the spills conducted of each oil type exhibited this behavior.
It may well be due to small quantities of surface-active materials in the oil which spread more rapidly than the bulk of the
oil, surrounding it, inhibiting its growth, and thus containing
and controlling the oil (from a private communication with W.
D. Garrett).

The microwave observations were taken using the NASA-
Wallops Island DC-4 aircraft. Measurements were made at 19.4
and 69.8 GHz for the initial spills and at 19.4 and 31.0 GHz
for the last three spills. The latter combination proved more
effective for the oil thicknesses of up to several millimeters
which were encountered. The half-power antenna beamwidth at all
three frequencies was 7.2 degrees, which gave a beam spot on
the surface of about 50 ft (15 m) for the aircraft altitude
used of about 400 ft (122 m). A two-dimensional antenna-temperature map of the oil slick was built up by making repeated aircraft passes over the extent of the slick. Approximately 15 to
30 minutes before and after the nominal time of the map were
required to acquire sufficient scans for the map.

Contour maps of the increase in antenna temperature above
that for the unpolluted sea at 19.4 and 31.0 GHz are shown at

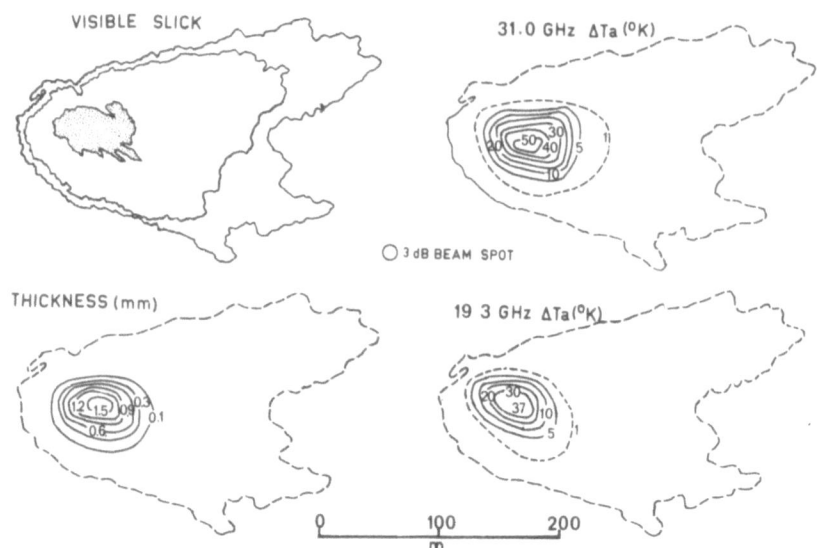

Figure 5.   The upper-left-hand drawing is a tracing of a color
            photograph of the oil slick resulting from a con-
            trolled spill of No. 2 fuel oil. The antenna temper-
            atures measured at 19.3 and 31.0 GHz are shown at
            the right superimposed on the outline of the visible
            slick. The thickness contours derived from the micro-
            wave data are shown at the bottom left.

the right in Figure 5 superimposed on the outline of the visible
slick for the spill of July 11, 1972. These antenna-temperature
distributions were used to derive the thickness contours shown
at the bottom left of the figure. The antenna temperatures and
derived thicknesses are weighted averages over the antenna beam,
as given by Eq.[1]. The half-power beam spot on the surface is
represented by the small circle. The microwave signals coincide
closely with the region of thick oil and show that average
thicknesses over the antenna beam of up to 1.5 mm are present
in good agreement with in situ spot measurements in this area
of 2.4 $\pm$ 0.3 mm. Integration of the thickness contours derived
from the microwave data gives a volume of 650 $\pm$ 65 gallons
(2,460 $\pm$ 246 liters), which taken with the volume of oil spilled
of 630 gallons (2,380 liters) indicates that nearly all of the
oil is in the thick region. This is consistent with in-situ
measurements of film thicknesses of 2-4 μm outside the thick
region, since only 15 to 30 gallons (57 to 114 liters) of oil
would be needed to cover the entire area of the visible slick
of 33 x 10^3 m^2 with a uniform film to thicknesses of 2-4 μm.
The ratio of slick thickness in the two regions of nearly 1000

also shows that nearly all the oil is located in a small region
of the slick.

The microwave measurements of all the oil spills of each
oil type showed results very similar to those just described
for the spill of July 11, 1972. The slicks always formed an
identifiable region with film thicknesses of a millimeter or
more and containing the majority of oil, which was surrounded
by a very much larger and thinner slick which contained very
little of the oil. In general the thick region contained more
than 90 percent of the oil in less than 10 percent of the area
of the visible slick. It was always possible to locate and delineate the thick region solely from the microwave observations,
and the total volume of the oil present derived from the microwave measurements was within about 25 percent of the volume of
oil spilled.

In summary, multifrequency passive microwave radiometry
offers the potential to measure the distribution of oil in sea-
surface oil slicks and to locate the thick regions and measure
their thickness and volume on an all-weather, day-or-night, and
real-time basis. As such it should prove a useful tool in the
confinement, control, and clean-up of marine oil spills.

ACKNOWLEDGMENTS

The authors thank S. R. MacLeod and J.A. Modolo for their
fine work in the design, construction, calibration and installation of the radiometers and for their help in taking the
measurements. We express our appreciation to the staff at NASA
Wallops Island Station for their invaluable assistance and aircraft support and to the Virginia Institute of Marine Science
for their aid in conducting the oil spills and obtaining ground
truth. This work was sponsored by the U.S. Coast Guard under
contract number Z-70099-2-21881.

REFERENCES

1. D. V. Stroop, "Report on Oil Pollution Experiments -
    Behavior of Fuel Oil on the Surface of the Sea",
    presented to the Committee on Rivers and Harbors, House
    of Representatives, Seventy-First Congress, Second
    Session, Hearings on H.R. 10625, U.S. Government Printing
    Office (1930).
2. W. D. Garrett, "Impact of Petroleum Spills on the Chemical
    and Physical Properties of the Air/Sea Interface," NRL
    Report 7372 (1972)
3. L. M. Brekhovskikh, "Waves in Layered Media", Academic
    Press, New York (1960)

4. R. N. Bracewell, "Radio Astronomy Techniques," Handbuch der
    Physik (1962)
5. "Discharges of Oil for Research, Development, and Demonstra-
    tion Purposes," Federal Register Document 71-5369,
    Federal Register 36, No. 75 (1971).
6. J. A. Fay, "Oil on the Sea", Plenum Press, New York (1969)
7. N. W. Guinard, Remote sensing of oil slicks, in "Proceedings
    of the Seventh International Symposium on Remote Sensing
    of Environment", Ann Arbor, Michigan, United States
    (1971).
8. J. C. Munday, Jr., W. G. MacIntyre, M. E. Penney, and J. D.
    Oberholtzer, Oil slicks studies using photographic and
    multiple scanner data, in "Proceedings of the Seventh
    International Symposium on Remote Sensing of Environment",
    Ann Arbor, Michigan, United States (1971).

Editor's note: This paper is reprinted from NRL Report 7512,
               June 5, 1973. It has been presented at the
               Washington workshop. In February 1983, J. P.
               Hollinger, called for an up-to-dating of this
               presentation, answered: "We have not done any
               further work in this area since that workshop
               and our contribution accurately describes our
               results and current views regarding airborne
               microwave radiometry of oil spills". -J.M.M.

MICROWAVE MEASUREMENTS OVER THE NORTH SEA*

G.P. De Loor[1], and H.W. Brunsveld van Hulten[2]

[1]Physics Laboratory TNO
Delft, The Netherlands
[2]Rijkswaterstaat
Directorate for Water Management and Water Research
The Hague, The Netherlands

ABSTRACT

We investigate the applicability of SLAR (Side-looking air-borne radar) to detect waves and other phenomena at sea, for future operational use. Especially, we investigate radar measurements of wave patterns, oil pollution, shipping traffic, and ice reconnaissance and relate these to sea-truth information. The SLAR program benefits from earlier clutter measurements at sea and in a wind-wave tank (carried out by the Physics Laboratory TNO and the Netherlands Interdepartmental Working Community on the Application of Remote Sensing -NIWARS-), which have been further elaborated and extended. These are summarized. Results obtained since 1974 lead to the conclusion that the SLAR is applicable for the aforementioned purposes, albeit in a somewhat restricted way. The large resolution cell used, containing a vast number of independent samples, improves the dynamic resolution to about 1 dB and so turns the SLAR into a sensor eminently suited for the detection of very slight modulations of capillary waves. For the detection of shorter gravity waves, however, especially under severe wind conditions, a finer geometrical resolution might be needed. Results emphasize the need for further fundamental studies of sensor-object interaction.

* Reprinted from Boundary Layer Meteorology 13 (1978) 119-131
  By courtesy of D. Reidel Publishing Cny., Dordrecht, NL.

INTRODUCTION

To aid in the task of controlling the Dutch continental shelf, Rijkswaterstaat (Department of Water Control) investigates remote sensing techniques as a possible surveillance system. To build up techniques for adequately handling the vast amount of data inevitably produced by such activities, a research program was started in 1971, supported by the Physics Laboratory TNO and the National Aerospace Laboratory (NLR).

Emphasis was laid on sensor-object interaction studies. The initial sensor was an infrared line scanner (IRLS), followed in 1974 by an X-band side-looking airborne radar (SLAR) of the real-aperture type with horizontal polarization, carried on a QUEEN AIR BEECHCRAFT from NLR. Because of the limited size and payload of the aircraft, both sensors cannot be flown simultaneously. Only cameras and an inertial navigation system can be flown with the SLAR or the IRLS.

As a background for the SLAR observations, earlier measurements of sea clutter, carried out in the Netherlands, were further elaborated and extended. Of primary importance was the use of SLAR to measure contrast produced by waves and oil spills.

Since the program and facilities were limited, the investigation was centered on selected objects such as waves, oil pollution, traffic density in shipping lanes, and ice reconnaissance, but in this paper only oceanographic applications are reported. From the SLAR imagery it became clear that the sensor is eminently suited for the detection of very slight modulations of waves, generally existing under light to moderate wind conditions.

In rough weather, the imagery becomes too speckled and no wave patterns can be detected. Oil patches, on the contrary, then become more distinct.

The SLAR imagery collected over the range of wind speeds from 0 to 13 ms^{-1}, provides information on the detectability of gravity waves. In addition, sand waves and internal waves have been detected through their effect on the sea surface. The latter two will be referred to as wave-like patterns because of their appearance in the imagery, although they cannot be described adequately by a wave motion. Studies of sand waves and internal waves can thus be made indirectly, and details are given elsewhere (Brunsveld, 1976). Wave rider buoys in the direct vicinity of the North Sea test site provided sea-truth measurements (i.e., wave power spectra); spectra derived from radar imagery are compared with such power spectra.

Furthermore, characteristics of the relationship between a "mechanical" spectrum generated in a wave tank and the fading spectrum are discussed. The latter provides information on the de-correlation time of samples, which is of importance in determining the number of independent samples available to build up the image in a SLAR system. The relationship among the different spectra are still under investigation because of fundamental differences in the methods of measurement (Poole, 1976), i.e., scatterometers and wave-rider buoys measure in the time domain, whereas imaging radars measure in the space domain.

RADAR BACKSCATTER MEASUREMENTS

Literature Review

The return parameter: Two common definitions of the radar backscatter coefficient are used, σ_0 and γ (Cosgriff et al., 1960; De Loor, 1976);

$$\sigma_0 = \sigma /A \text{ and } \gamma = \sigma_0 /A_i$$

with σ = the total backscatter cross-section,
  A = the area illuminated by the system,
  A_i = the area of the cross-section of the illuminating antenna beam at the position of the target.

The relationship between σ_0 and γ is then:

$$\sigma_0 = \gamma \sin \theta , \text{ or } \sigma_0 = \gamma \cos \phi$$

with θ = the grazing or depression angle, and
   ϕ = the angle of incidence (away from the vertical).

Measurements have shown that for grazing angles between 2 and 45°, many surfaces behave as diffuse scatterers and that γ is reasonably constant over this range of angles (De Loor, 1974). This provides the possibility of averaging data from different sources over different grazing angles. In this way, we obtain a picture having a reasonable coherence from other-wise seemingly disjointed observations, so that ranges for the value of γ (for grazing angles from 5 to 45°) can then be given (Table 1). As may be seen from this table, ranges can be fairly narrow for specific applications. The complete dynamic range as indicated here for sea clutter seems to be fairly large, but this is only because γ is given for a large range of wind speeds (from 1 to 20 ms^{-1}). For the observation of specific patterns on the sea surface (gravity waves, swell, oil patches, etc.) under a narrow wind-speed range, the dynamic range of γ becomes much smaller. Our SLAR observations suggest

Table 1.   Ranges of the return parameter γ for the X-band
for grazing angles from 5 to 45°.

	γ	(dB)
Sea clutter	-40	to -15
Bare soils	-35	to -15
Vegetation	-15	to -3
"Natural" surfaces	-10	to 0
Man-made structures		> 0

a range of the order of 10 dB. An important case is the oil-
water contrast for the detection of oil pollution at sea.

Dynamic range and de-correlation time: Sea scatterers are
Rayleigh distributed, and independent samples vary considerably
in value. The number of samples collected depends on the de-
correlation time. For vegetation observed from a fixed platform, this value lies between 0.1 and 1 s, depending on wind
speed (De Loor at al., 1974). For the sea, the de-correlation
time is smaller by more than a factor of 10, of the order of
milliseconds (Long, 1975).

To obtain accurate values for γ, measured as grey tones
in a SLAR image, integration over a fair number of independent
uncorrelated samples is required, but processing of a SAR
(synthetic aperture radar) signal is only possible for correlated samples. This dilemna can be solved by sub-aperture processing and/or by using a higher resolution than finally needed
and integrating spatially over a group of samples, trading-off
in this way spatial resolution for dynamic resolution (Moore,
1976).

The de-correlation time determines the length of the synthetic aperture in a SAR. For vegetation, the de-correlation
time is fairly large, and reasonable imagery can be obtained
with a SAR system. The de-correlation time for the sea, however,
is very short. Still, some good-looking imagery has been obtained with SAR systems (Brown et al., 1975).

Measurements

The return parameter γ: Two instruments for sea-return
measurements are used in the Netherlands. The first, adopted
by the Physics Laboratory TNO, is a non-coherent pulse system
mounted on a platform high on the Norwegian coast overlooking
the Atlantic. Measurements (extending over about 8 yr) have
been reported recently (Sittrop 1974, 1976). The second,

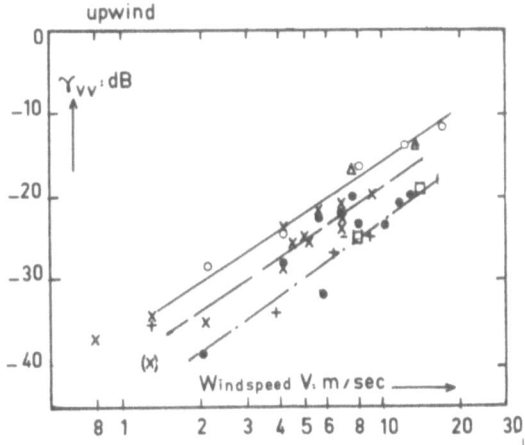

Figure 1.   Measurements of γ_{vv} as a function of wind speed.

adopted by NIWARS, employs a short range X-band FM/CW system in a wind-wave tank at the Delft Hydraulics Laboratory. This tank is 100 m long, 8 m wide and 0.5 m deep. Wave generation is by means of hydraulic plungers and wind (wind speed up to 18 m s^{-1}).

Figure 1 shows some measurements of γ for X-band in relation to the wind speed, together with some reported data (Daley et al., 1971; Myers and Fuller, 1969; Grant and Yaplee, 1957). In this Figure the signification of the different signs used is as follows:

● NRL, JOS I, θ = 5-45°
✗ NRL Puerto Rico, θ = 5-45°
○ NIWARS, tank, θ = 5-30°
▲ Sittrop, θ = 1-10°
✛ Grant and Yaplee, θ = 10-45°
◻ Wiltse, θ = 15-45°

The inclination of the curves found by different authors are the same, but the figure clearly suggests differences in calibration. Measurements at frequencies other than X-band are only available from the Naval Research Laboratory (NRL). Analysis of these data (Daley et al., 1971), after reducing σ_0 to γ and averaging over the grazing angles from 5 to 45°, shows a variation of inclination with frequency. The inclination decreases to lower frequencies, suggesting that the dependency of γ on wind speed practically disappears at L- and P-band (De Loor, 1976).

Table 2. Radar contrasts of oil slicks with surrounding sea echoes (after Van Kuilenburg, 1975).

Oil type	Layer thickness (μm)	Wind speed 5 ms^{-1} Polarization		
		HH	VH	VV
Crude	5	8 dB	8 dB	11 dB
Crude	9	7	5	11
Motor-oil-SC30	4	5	7	11
Motor-oil-SC2OW	4	6	6	11
Average	5	7	6	11
		Wind speed 10 ms^{-1} Polarization		
		HH	VH	VV
Crude	9	4 dB	7 dB	9 dB
Crude	144	9	6	9
Motor-oil-SC30	10	7	5	7
Motor-oil-SC2OW	12	6	6	6
Average	11	6	6	8

Variations in γ caused by other effects such as gravity waves, swell, oil film, etc., are much smaller. But because radar study of waves is very interesting in oceanography, knowledge of the magnitude of the contrasts involved (i.e. variations in γ) is important. In this respect, the decrease in γ due to an oil film has been studied using measurements in a wind-wave tank by NIWARS (van Kuilenburg, 1975). At about the same time, the Department of Water Control (Rijkswaterstaat) made observations with a X-band SLAR over the North Sea, of controlled oil spills.

Table 2 lists the oil-water contrasts measured in the wave-tank. It must be remarked that the values for $v = 5$ ms^{-1} are minimum figures; the signal received with oil present was so low that it approached the noise level of the measuring system. The variation in γ due to an oil layer (i.e. the total contrast) is between 5 and 10 dB, dependent on polarization and wind speed. The SLAR observations carried out with HH-polarization confirm these results (see section Results and Conclusions). They also indicate that contrasts due to other surface phenomena such as gravity waves and swell are of the same order of magnitude. Hence, it can be concluded that for radar

Table 3.   Mechanically generated spectrum and fading spectrum, as measured in a wave tank.

Wind speed (ms^{-1})	Peak of mech. spectrum (Hz)	Peak of fading spectrum (Hz)
0	1.26	1.25
4	1.12	1.12
8	1.04	0.97
12	0.91	0.89
17.5	0.77	0.72

observations in oceanography, the dynamic range of γ is limited, with the window varying with wind speed.

The fading component: Due to movements of the sea surface, γ will vary and the measured echo shows fading. Spectra of this fading have been measured by Sittrop (1974, 1976) and by NIWARS (van Kuilenburg, 1975). Sittrop's measurements with HH-polarization indicate the sea clutter contains a slow and a rapid component. Through the composite scattered theory, these two components are attributed, respectively, to the facets on the wave slopes and to the small ripples. Long (1975) found the same effect and observed that the power of the slow component compared to that of the rapid one usually is substantially greater for HH- than for VV-polarization. The slow component also has a fairly large de-correlation time. This effect could perhaps explain the build-up of SAR images of the sea.

The fading spectra measured in the wind-wave tank clearly show the mechanically-generated wave spectrum. With a fixed setting of the plunger, the wind modulates the mechanically generated wave spectrum to a certain degree. The relationship between this wave spectrum and the fading spectrum is good, as shown in Table 3, leading to the conclusion that radar back-scatter also provides information on the power spectrum of waves.

SLAR OBSERVATION OF THE SEA

SLAR Characteristics

The radar used for the present observations is an EMI X-band SLAR of the real aperture type. The antennas are rigidly attached to the aircraft, so movements of the platform can distort the imagery, especially under severe weather conditions. Scales can be chosen at 1:50, 100, 250 and 500 K (K = 1000) of

Table 4. EMI-X-band SLAR characteristics.

RF frequency	9.6 GHz
Transmitted pulse	
Peak power	25 kW (situation A)
	80 kW (situation B)
Pulse length	0.2 μs
P.R.F.	2520 Hz (A)
	1260 Hz (B)
Antenna	
Beamwidth, horiz.	54'
vert.	23°
Polar diagram	$cosec^2$ (approximately)
Resolution	
Range (across track)	30 m
Azimuth (along track)	16 mrad
IF amplifier	
IF	45 MHz
IF bandwidth	6 MHz
Windows	2 x 3.25 km (1:50 K)* (A)
	2 x 6.50 km (1:100 K) (A)
	1 x 24.0 km (1:200 K) (B)
	1 x 63.0 km (1:500 K) (B)
Range markers	0.5, 1 and 4 nm
Map, film, size	5"

* K = 1000

which the latter two are one-sided. Further characteristics are included in Table 4. For detailed information on SLAR, Skolnik (1970) and De Loor (1976) are recommended references.

The images are recorded on film because magnetic tape recording was not available. The distortion in the imagery, due to the combined effect of the line-by-line composition of the scene and wave propagation, results in apparent wave-crest directions different from those of the actual waves (Pierson, 1976). Assuming a more or less constant factor, no corrections have been made for the results presented in the following; but it is realized that the crest direction might be affected slightly.

Waves and Wave-Like Patterns

The speed and form of waves in shallow seas are dominated

by the bottom stress. The North Sea, being shallow, exhibits
this effect strongly off the coast of the Low Countries. In
this particular area, a balance exists between the abrasive
action of the sea on the sandy coast and bottom and the formation of the coast. On the sea bottom, sand dunes and sand waves
are continuously formed and abraded, while strong tidal currents
and storms continuously redistribute the sand masses along the
coast. The presence of sand waves on the sea bottom is seen at
the sea surface by the interference of the tidal current with
the bottom topography, producing a very weak wave-like pattern
which modulates the capillary waves in combination with the
velocity fluctuations at the sea surface. Detection of such
weak phenomena seems possible only with sophisticated techniques such as SLAR. Refraction and diffraction of waves are
common phenomena in shallow seas. Bottom features, such as
slopes and shoals, can be studied by means of radar under favourable conditions.

In the ocean, internal waves travel at a speed which is a
fraction (relative mass difference) of that of long gravity
waves (gh; g acceleration due to gravity, h depth of the
thermocline). The associated wave-like pattern at the surface
travels at the same speed (Brunsveld van Hulten, 1976), and it
can be considered as stationary over the scan period.

Internal waves are associated with slicks, which are numerous long, narrow, quasi-parallel smooth streaks at the
surface when the wind is light (Neuman and Pierson, 1966,
Ch. 12). During experiments carried out in the Gulf of Biscay,
it was thought that the smooth streaks produced by the spilled
oil from the neighbouring shipping lane existed during very
light winds (< 2 m s^{-1}, Bocard et al., 1976). However, no proof
is available of the existence of a thermocline which could
support internal waves. Analysis of the applicable weather
charts excluded the presence of swell. The modulation of the
capillary waves by the surface-projected wave pattern combined
with the slick appearance made detection of the wave-like
pattern by radar possible under very calm weather conditions.

Oil Detection

The detection of oil spills at the sea surface by radar is
made possible by the suppression of the capillary waves by the
oil. Experiments to detect oil spills with SLAR have been carried out on the North Sea. These oil spills were made deliberately with 2 m^3 of light topped crude oil, during times when
the wind speed ranged from 1 to 5 m s^{-1}, and at different tidal
phases. The spills have been traced over a period of 24 h,
after which they were lost. Further oil-spill detections have
been carried out on the Atlantic Ocean in a co-operative

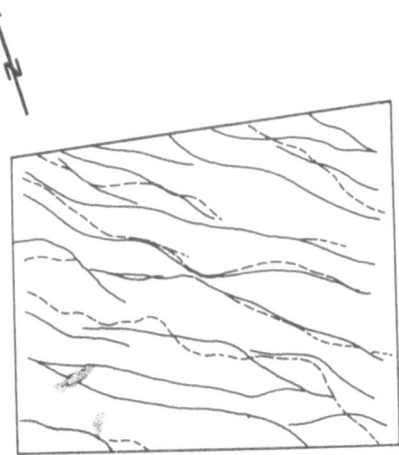

Figure 2. Matching of linear features on SLAR images
           over the same area, separated in time by a
           couple of minutes. Drawn and dashed lines
           represent two matched patterns as obtained
           from different images. The successive posi-
           tions of the assisting ship are indicated
           by the shaded areas. Deviations due to sec-
           ondary effects are recognized.

venture with the Institut Français du Pétrole of France, the
results of which have been reported elsewhere (Bocard et al.,
1976).

RESULTS AND CONCLUSIONS

SLAR Results

     The SLAR imagery obtained covers the wind-speed range from
0 to 15 m s^{-1}. In some experiments, the optimal flight direction for the highest contrast was established on the basis of
quick-look imagery during the mission. In other experiments, a
flight pattern in differently oriented squares was adopted to
observe the same area from different view-points. In the case
of stationary phenomena, such as sand waves, this latter procedure was of proven advantage. Analysis of the imagery combined
with weather information resulted in the determination of wave-
propagating directions. In some exercises, different wave-propagating directions could be discerned in the same series of
images. Sometimes the wave-propagating direction coincides with
the wind direction ($\pm$ 180°), but deviations occur as a consequence of the lag of the wave field to changes in the wind.

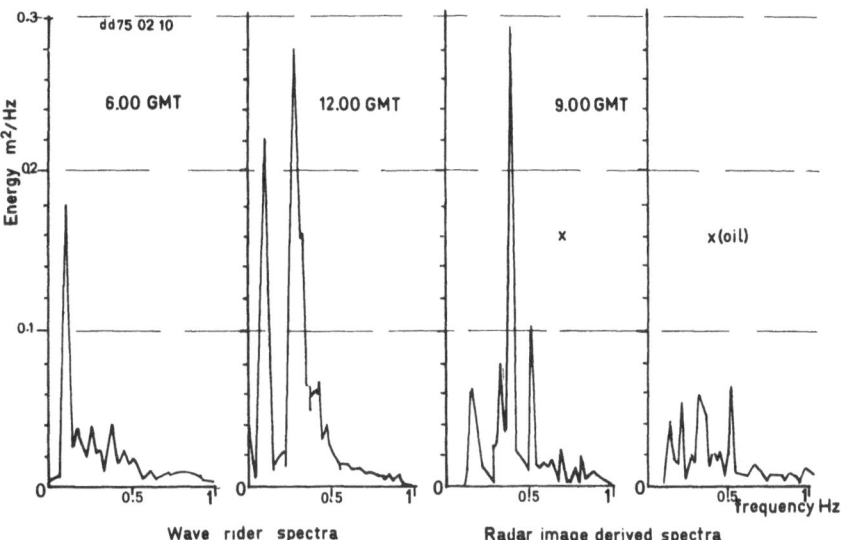

Figure 3. Two power spectra obtained by a wave rider buoy near
the test site six hours apart, and two spectra derived from a radar image. Unit of the image-derived
spectrum is film density. Notice the damping of the
gravity waves by oil.

The derived wave-lengths are those of gravity waves (0.6-600 m).
Other imagery obtained over the North Sea also exhibits a wave
pattern, but with a longer wavelength (800 m) and having a persistent crest direction of 120°. This pattern has been related
to the sand waves in that particular area which have exactly
the same crest direction and wavelength. The variance in the
directions deduced is larger for gravity waves (20°) than for
sand waves (10°). No sand wave patterns have been detected with
the turning of the tide but rather only when diurnal current is
present.

Imagery obtained over the Atlantic also exhibits a longer
wavelength (650 m) with a smaller variance (10°) in crest direction than that of gravity waves. Detection was made possible by
the appearance of slicks which were photographed during the
exercise.

The variance of ± 20° in crest direction for gravity waves
corresponds to that found by Pierson (1976), implying an accuracy of ± 20° for the detection of the direction of wave propagation with SLAR. The low variance of ± 10° for the crest direction in the case of internal waves and sand waves is due to

their being more or less stationary or slowly changing.

For a series of images in which sand waves are clearly
seen, a matching within the same region, as seen from different
view-points, has been carried out (Figure 2). Aside from secondary effects (such as fluctuating irregularities at the
surface, the movements of the aircraft and the differences in
the wave-propagating direction in relation to flight direction),
features in the imagery match strikingly. Again, this provides
evidence that the stationary character of a long wave-like patterns at the surface are caused by sand waves at the bottom of
the sea.

Directional spectra, derived from some imagery including
oil slicks, have been compared to wave power spectra measured
in the same region and at the same time by a Datawell wave-
rider buoy, but we note that a wave-rider buoy measures in the
time domain whereas SLAR-derived information concerns the space
domain. Although the measuring techniques are completely different, the shapes of the curves are strinkingly similar (Fig. 3).
A transformation of the apparent spectrum derived from the
imagery into a real spectrum can be derived from linear wave
theory (Poole, 1976). The transformation, however, is essentially non-linear and still under investigation. Two spectra
have been derived from radar images parallel to one another,
one for the oil spill and one outside. The damping of energy
due to the oil layer is evident (80%, Figure 3). Spectral studies of the damping of the energy might indicate the thickness
of the oil layer. The imagery was obtained for different wind
speeds ranging from 0 to 15 m s^{-1}. For calm conditions, the
wave crest are long, whereas for heavy winds they become much
shorter, and randomness of the wave distribution over the sea
surface increases. In the imagery, one notices that the speckled
character is very pronounced at high wind speed, so that wave
patterns can no longer be discerned (Figure 4). This results
from the large resolution cell, which fails to resolve the
short-crested waves produced under heavy weather conditions.
It is also reasonable to suppose that the wave distribution at
the sea surface changes faster under heavy wind conditions and
so contributes to an increased randomness in the imagery. For
wave detection under severe wind conditions, a small resolution
cell is needed.

From the quality of the imagery showing oil spills, we may
conclude that detection of oil under light wind conditions is
marginal whereas detection under stronger wind conditions improves. Boundaries can be established quite clearly for strong
wind conditions, whereas for light winds, the oil spills are
faintly discerned.

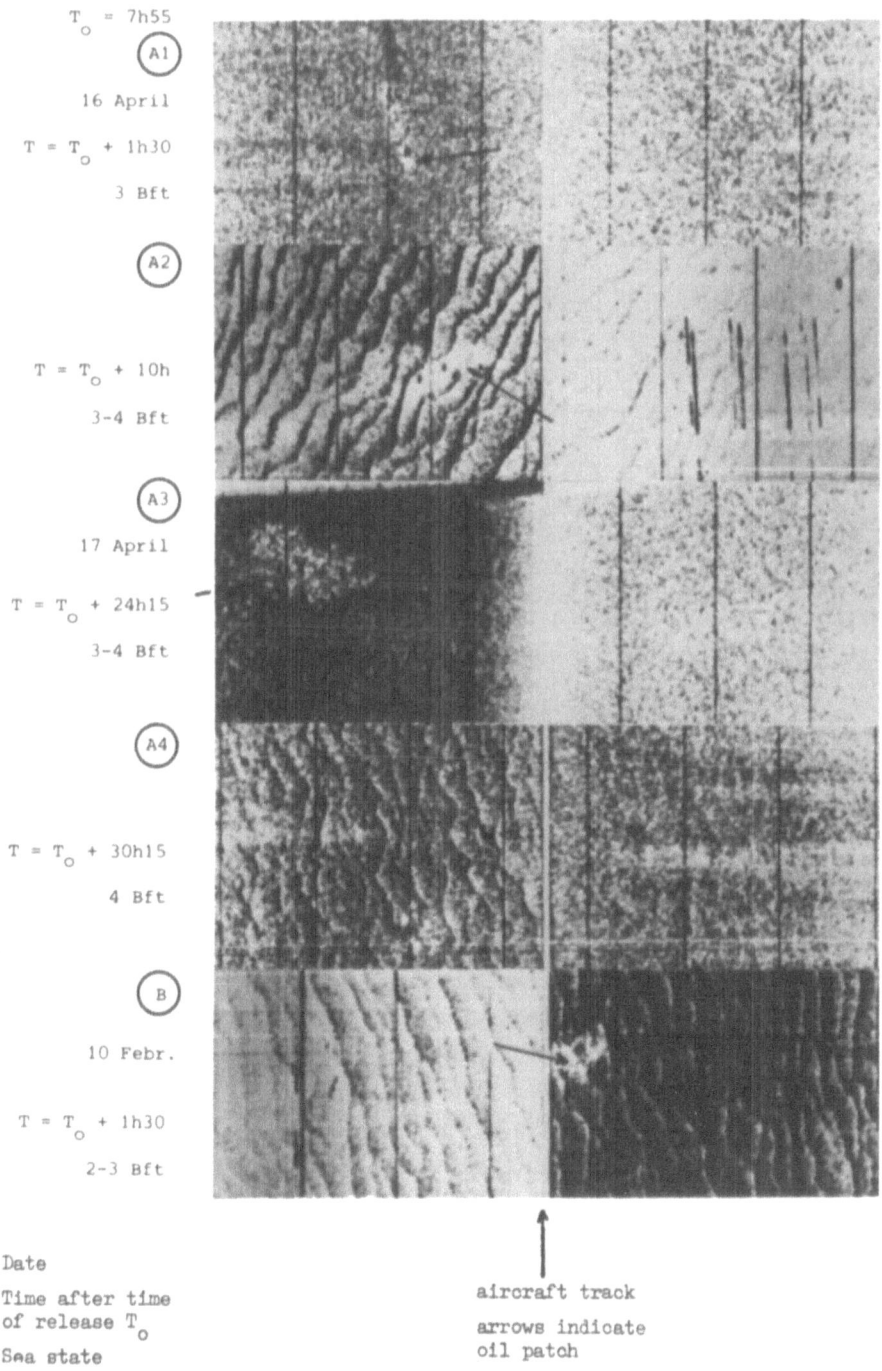

T_o = 7h55

(A1)

16 April

$T = T_o + 1h30$

3 Bft

(A2)

$T = T_o + 10h$

3-4 Bft

(A3)

17 April

$T = T_o + 24h15$

3-4 Bft

(A4)

$T = T_o + 30h15$

4 Bft

(B)

10 Febr.

$T = T_o + 1h30$

2-3 Bft

Date

Time after time
of release T_o

Sea state

aircraft track

arrows indicate
oil patch

Figure 4.   Oil spills resulting from 25 tons of light-topped
             crude oil

In Figure 4, series A has been observed in successive stages under moderate wind conditions (4-8 m s^{-1}). B has been observed under light wind conditions (2-5 m s^{-1}). Sand wave patterns show up distinctly in the presence of tidal currents. Photographs give double images (left and right side of aircraft; gap in centre is about 2 nm). Distance between range markers (vertical lines) is 1 nm. Alongside photos: data and time after spilling To. The speckled character is more pronounced in the absence of a sand-wave pattern.

As mentioned previously, the total dynamic range for sea clutter is about 25 dB. However, only a small range is used for the detection of waves for a certain sea state. The dynamic range covered by an image is 20 dB. By measuring the density variation in an image, we obtain a quantified impression of the contrasts for an oil patch and for surface reflected bottom relief. In one case, wind speed was 2.5- 3.5 m s^{-1}, and the direction of the tidal current was perpendicular to that of the wind. The difference between the average gray scale levels of the left- and right-half image is partly due to the difference in gain between the respective antennas. The contrasts in the dark-half image are about 8 dB and those in the light-half 4-7 dB. The difference in the average gray-tone level between both halves is 11-13 dB. The variation in the average gray-tone level over one scan line is due to the polar diagrams of the antennas and their positions relative to the horizon. Measurements (See Table 2) indicate figures of 6 to 7 dB for contrasts of oil spills for 5 and 10 m s^{-1} wind speed, respectively, which is in fair agreement with the above observations. Hence, the supposition that only a small part of the total dynamic range, available in a radar, is actually used for the detection of waves and oil spills is demonstrated.

DISCUSSION AND CONCLUSIONS

The backscatter coefficient γ, although not exactly constant over the range of grazing angles from 2 to 45°, provides a means for comparing different measurements. The complete dynamic range of γ for wind speeds from 1 to 20 m s^{-1} is about 25 dB, but it only is partially used for the detection of waves and the observation of details at the sea surface. The modulation of the sea returns by waves and oil spills has a much limited dynamic range for a specific sea state and is of the order of 5 to 15 dB. This variation of γ is dependent on polarization and wind speed. The fading spectrum, observed with measuring radars, provides information on wave-slope facets and capillary waves. The wave-slope facets, related to a slowly moving component, show larger power magnitudes for HH- than for VV-polarization and also exhibit a fairly large de-correlation time. Eventually, this could explain the

formation of SAR images and the excellent results obtained with
the EMI SLAR with HH-polarization. Furthermore, spectra derived
from SLAR imagery can provide information on the wave power
spectra, although this relationship is not yet completely established.

From the imagery, it may be concluded that the SLAR is a
sensor eminently suited to the detection of very slight modulations of capillary waves and to the detection of oil spills
under moderate wind conditions (up to 15 m s^{-1}); however, oil
detection under calm weather conditions can sometimes be disturbed by specific oceanographic phenomena, such as sand waves
and internal waves. Wave detection under heavy wind conditions
becomes impossible because of the speckled character of the
imagery. A higher geometrical resolution appears to be needed
for such situations.

Finally while horizontal polarization gives satisfactory
results, vertical polarization should also be investigated.
Research is continuing along these lines.

REFERENCES

Bocard, Ch., Fontanel, A., Rivereau, J. C., Loubersac, L., and
     Brunsveld van Hulten, H., 1976, "Detection et échantil-
     lonnage de nappes d'hydrocarbures traitées par produits
     dispersants", Institut Français du Pétrole, Report I.F.P.
     24013.
Brown, W. F., Elachi, C., and Thompson, T. W. 1975, Radar
     imaging of ocean surface patterns, J. Geoph. Res. 81:
     81:2657-2667.
Brunsveld van Hulten, H. W., 1976, Application of SLAR to
     coastal zone phenomena, Lecture held at the 1976 Alpbach
     Summer school, Austria. Rijkswaterstaat, Nota FA 7601,
     The Hague, The Netherlands.
Cosgriff, R. L., Peake, W. H., and Taylor, R. C., 1960, Terrain
     scattering properties for sensor system design, in :
     Terrain Handbook II, , Engineering Experimental Station
     Bull. 181, Ohio State University, Columbus, United
     States.
Daley, J. C., Ranson,Jr., J., and Burkett, J.A., 1971, "Radar
     Sea Return-JOSS-I", Naval Research Laboratory, Report
     7268, Washington, D.C.
De Loor, G. P., 1974, Measurements of radar ground returns, in:
     "Proceedings URSI Specialists Meeting on Microwave Scatterin
     Scattering and Emission from the Earth", Berne, Sept.
     23-26.
De Loor, G. P., 1976, Radar methods, Ch. V, in: "Remote Sensing
     for Environmental Sciences", Ecological Series No. 18,
     Springer-verlag, Berlin, Heidelberg, New-York
     (E. Schanda (ed.)).

De Loor, G. P., Jurriëns, A. A., and Gravesteijn, H., 1974, The radar backscatter from selected agricultural crops, IEEE Trans. Geosc. Electronics GE-12:70-77

Grant, C. R., and Yaplee, B. S., 1957, Backscattering from water and land at cm and mm wavelengths, in; "Proceedings IRE", 45:976-982.

Long, M. W., 1975, "Radar Reflectivity of Land and Sea", Lexington Books, Lexington, Toronto, London.

Moore, R. K., 1976, "SLAR Interpretability Trade-offs Between Picture Element Dimensions and Non-coherent Averaging", Kansas University, Remote Sensing Laboratory, Technical Report 287.

Myers, G. F., and Fuller, I. W., 1969, "Nanosecond Observations of Sea Clutter Cross-section vs. Grazing Angle", Naval Research Laboratory, Report 6933, Washington, D.C.,

Neuman, G. and Pierson, W. J., 1966, "Principles of Physical Oceanography", Prentice-Hall, Englewood Cliffs, N.J.

Pierson, W. J., 1976, "The Theory an Applications of Ocean Wave Measuring Systems at and Below the Sea Surface, on the Land, from Aircraft and from Spacecraft", NASA, CR-2646, Washington, D.C., (NTIS N 76-17775).

Poole, L. R., 1976, "Transformation of Apparent Ocean Wave Spectra Observed from an Aircraft Sensor Platform", NASA TND-8246, Washington, D.C.

Sittrop, H., 1976, Characteristics of clutter and targets at X-band and K -band, Paper 27 in "Proceedings AGARD Symposium: New Devices, Techniques and Systems in Radar", The Hague, June 14-17.

Sittrop, H., 1974, X-band and K -band radar backscatter characteristics of sea clutter, in: "Proceedings URSI Special-stists Meeting on Microwave Scattering and Emission from Earth", Berne, Sept. 23-26, pp. 25-37.

Skolnik, M., 1970, "Radar Handbook", McGraw-Hill, New York.

Van Kuilenburg, J., 1975, Radar observations of controlled oil spills, in: "Proceedings 10th International Symposium on Remote Sensing of Environment", Ann Arbor, Oct. 6-10, pp. 243-250.

DETECTION OF OIL POLLUTANTS

USING REMOTE SENSING TECHNIQUES

C. Smorenburg and H. Visser

Technisch Physische Dienst, TNO-TH
Delft, The Netherlands

INTRODUCTION

During the last decade, the oil pollution of the sea has become an increasing problem. For an effective solution of that problem, good detection methods will be necessary. Therefore in a number of countries, investigations have been started to see which techniques can be used for detection. It is thought that in a latter stage the sea surface is inspected with various sensors by an aircraft, in order to have a rapid information about the presence of oil spills and, if possible, about the kind and the amount of oil spilled.

Technisch Physische Dienst TNO-TH is studying a technique that is based on the fluorescence properties of oil. With this technique it is possible to perform a rough classification of the oil and to get an indication of the thickness of the oil layer. With this active method powerful lasers must be used to "illuminate" from an airplane the oil spill, in order to create an acceptable measuring distance. Experience in Canada has already shown that oil detection is possible at a distance of a few hundred meters.

THE INVESTIGATION

Until now the investigation at the TPD has been limited to the laboratory. Starting point were the user requirements that were fixed after consultation of the State Public Works Department (RWS). They are elaborated separately.

Detection of the Presence of Oil Spill by Day and Night

This can be done by a lot of techniques such as: multi-spectral, side-looking radar, infrared scanning, and also fluorescence techniques. Surveying the waters by an airplane, these sensors can be used simultaneously, each sensor giving supplementary information about a possible oil spill. Fluorescence techniques seem to be promising for unambiguous detection, especially at night.

Identification of the Oil

Therefore the fluorescence properties have to be considered into more detail. In our laboratory test-equipment has been manufactured, with which the fluorescence spectra of 42 different crude oils, fuels and lubricating oils have been measured. In the Figure 1, typical fluorescence spectrum of each of the three types of oil is given for an excitation wavelength of 337 nm (N_2-laser).

After these measurements, it is tried to find a correlation between the spectrum and the type of oil. A characteristic number is found to discern from the spectra in a simple way between the three types of oil.

Moreover it appeared to be possible to get information from the fluorescence spectrum about the specific gravity and the absorption of visible light of the oil.

Estimation of the Amount of the Oil Spill

The total amount of an oil spill only can be estimated when the thickness of the oil film is determined. Existing measures to measure the thickness from the thermal infrared radiation of the water-oil surface give a coarse impression

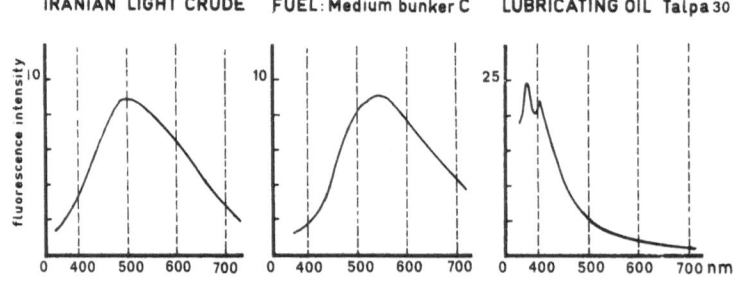

Figure 1.   Fluorescence spectra of three types of oil.

over a very limited range of thicknesses. It appeared possible
to use the fluorescence technique hereto. Because of the penetration depth in the oil, radiation with longer wavelength must
be used. In the laboratory, experiments are performed using a
He-Ne laser with a wavelength of 633 nm. Here it appeared that
a thickness varying from a few micrometers to just over 20
micrometers for strongly absorbing oils and to several hundreds
of micrometers for the more transparent oils could be determined.
Hereto the fluorescence intensity in the near infrared (about
730 nm) is measured, while at the same time the properties,
that are determined by the kind of oil, are found from the UV
excited fluorescence spectrum.

CONCLUSION

    From this investigation it appeared that within laboratory
conditions, all user requirements could be met. For the future,
verification of the results outside the laboratory is planned.
When that phase can be completed successfully, an operational
system for use in an aircraft can be designed and manufactured.

OPTICAL DIAGNOSTIC OF OIL POLLUTION

P. Camagni, N. Omenetto[1], A. Pedrini,
G. Rossi[1] and G. Tassone

Joint Research Centre
Establishment of Ispra
Electronic and Chemistry[1] Divisions
Ispra,(Varese) Italy

Some three years ago, the JRC-Ispra started a research activity aimed at defining and testing fluorescence methods, to serve as a basis for future actions in remote prospection of hydrocarbons at sea. In addition to the detection capability of existing airborne fluorosensors, thickness measurements and more comprehensive oil finger-printing are envisaged; the latter goal includes a refinement of known techniques of spectral correlation as well as a new development of time-decay signatures and the cross-linking of these two approaches.

Applied research is paralleled with suitable actions at the laboratory scale, aimed at implementing basic diagnostic and simulation experiments required as a guideline for any project of a Lidar fluorosensor. Hence the project, as a whole, is composed of two research themes, namely:

* Spectral and temporal characterisation of oils by means of laser-excited fluorescence; and
* Development of field methods for fluorescent monitoring of oils.

The experimental facilities include:

* A tunable dye laser assembly, pumped by an excimer laser operated with Xe Cl (308 nm), with frequency doubling crystals; tunability from 217 to 700 nm;

* A fast-pulsed nitrogen laser (0.3 ns) coupled to a

297

goniometric sample/detector arrangement in various ex-
citation/emission geometries and to a fast detection system including a transient digitiser, a TV display and
photographic recording, for the study of the fluorescence
time response;

* An indoor Lidar assembly consisting of:

a) twin laser source;
b) associated Newtonian mirror as optical receiver,
mounted colinearly with the two laser beams;
c) parallel detection by a fast photomultiplier and a
fast-gated multi-channel analyzer.

This apparatus operates on an outdoor target structure
for simulation of remote fluorescence monitoring.

So far various measurements have been undertaken for the
spectral and temporal characterisation of representative commercial oils, as well as of their fractions obtained by distillation and of mixtures of these fractions.

The spectral investigations indicate a peculiar signature
associated with the various oil fractions, consisting in a
shift towards the visible and a progressive spectral broadening,
which are a function of the oil-fraction volatility. On the
other hand, the spectrum of the original crude appears to be
relatively insensitive to the lightest fractions and their mixtures.

Fluorescence lifetimes (measured at fixed spectral channels between 400 nm and 560 nm) have been found to increase
with the decrease of oil fraction volatility. For any given
crude, the lifetime of the distillation residue has been found
to be similar to that of the original oil, both lifetimes being
shorter than those associated with the lighter oil fractions
distilled from 175 to 370°C. However, it was found that different crudes are characterized by different trends in the wavelength dependence of their lifetime. This might well constitute
a new element of identification, to be exploited in practical
remote operation.

As a conclusion, while the spectral studies alone confirm
the gross features of crude, heavy refined and light refined
oils reported in the literature, considerable advantages can be
expected from the combination of decay time and spectral measurements in properly selected channels.

OIL SPILL DETECTION AND IDENTIFICATION

USING A LASER FLUOROSENSOR

L. Buja Bijunas[1], R.A. O'Neil[1], V. Thomson[2],
R.A. Neville[2], K. Dagg[3] and D. Rayner[4]

[1]Canada Centre for Remote Sensing,
Ottawa, Canada
[2]Intera Environmental Consultants Ltd.
Ottawa, Canada
[3]The Genesys Group
Ottawa, Canada
[4]National Research Council of Canada
Ottawa, Canada

ABSTRACT

The Mk III airborne laser fluorosensor (built for the
Canada Centre for Remote Sensing by Barringer Research Ltd.)
was designed to detect and identify oil spills by means of the
characteristic fluorescence emission of the oil. Fluorescence
is induced in the water surface by means of a very short pulse
of ultraviolet light (337 nm wavelength) from a nitrogen laser.
A receiver consisting of a telescope, a spectrometer, a gated
image intensifier and a photo-diode array collects the induced
fluorescence return. The spectrometer dispurses the fluorescence emission spectrum into 16 channels covering the spectral
range from 380 to 680 nm, each nominally 20 nm wide. The spectrum is digitized and presented on a real-time display as well
as being written to magnetic tape for subsequent more detailed
analysis.

In a joint experiment with NASA, the U.S. Coast Guard and
the American Petroleum Institute, the laser fluorosensor was
flown over spills of rhodamine WT dye, Merban crude oil and
La Rosa crude oil. The results of these experiments are discussed.

Observation of the real-time display during these flights

showed clear differences between the dye, the two types of crude
oil, the ocean water and a ship. In particular, the oil slicks
suppress the return in the 380 nm channel normally seen due to
the laser light being Raman scattered by the water. This suppression is characteristic of substances floating on the water
and is a very useful initial indicator of oil slicks.

A correlation technique has been developed in which the
spectral finger-print of the oil, in all 16 fluorosensor channels is compared to a standard spectrum obtained in the laboratory or from other fluorosensor data. By this means it has
been shown the Merban crude oil, La Rosa crude oil, rhodamine
dye and ships can all be distinguished one from another and
from the general fluorescence background of the ocean water.

Because the algorithm is quite simple, it is possible to
implement a sophisticated real-time oil detection alarm as part
of the existing fluorosensor data acquisition system.

CHARACTERISTICS OF EMISSION SPECTRA

There has been considerable interest and development
during the last few years in the application of fluorescence
spectroscopy for the detection and identification of remotely
sensed targets. By comparing their respective wavelengths of
maximum emission λ_m, oils can be generally divided into three
main groups as illustrated in Figure 1 (Rayner and Szabo, 1976):

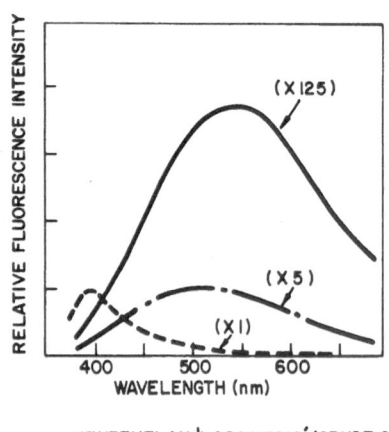

Figure 1.   Emission spectra of example oils
            from the three main oil groups.

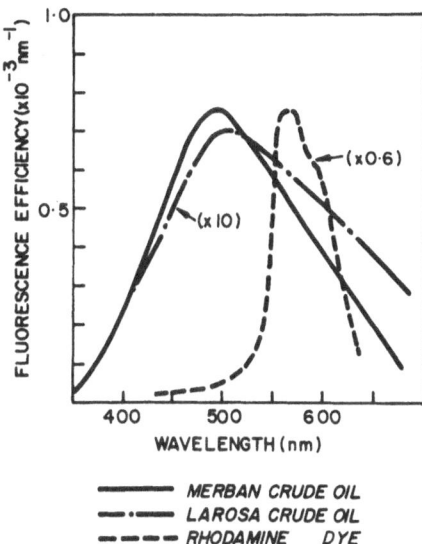

Figure 2.   Emission spectra of Merban and La Rosa crude oils
and rhodamine WT dye used in Wallops test flights.

Group 1: Light refined petroleum products with $\lambda_m = 400$ nm.
Emission spectra are narrow, sometimes structured.

Group 2: Crude oils with $\lambda_m = 500$ nm. Emission spectra are
low and broad.

Group 3: Heavy residual petroleum products with $\lambda_m = 550$ nm.
Emission spectra are weak and broad.

Figure 2 shows the emission spectra of the two crude oils
and the rhodamine WT dye used as spills in experiments held at
Dump site 106, off the New Jersey coast, in November 1978. The
spectral signatures (emission spectra) for the two crude oils
were measured in the laboratory, whereas the rhodamine dye
spectral signature was extracted from actual flight data taken
over the dye spill. Except for the lower fluorescence efficiency for the case of the heavier La Rosa crude oil, the signatures of the two crude oils can be seen to be quite similar,
their respective wavelengths of maximum emission differing by
only 20 nm. The rhodamine WT dye, on the other hand, is characterized by a narrow, structured emission spectrum, centred at
570 nm.

EXPERIMENTAL RESULTS

Figure 3 (flight line No. 6) gives an example of data

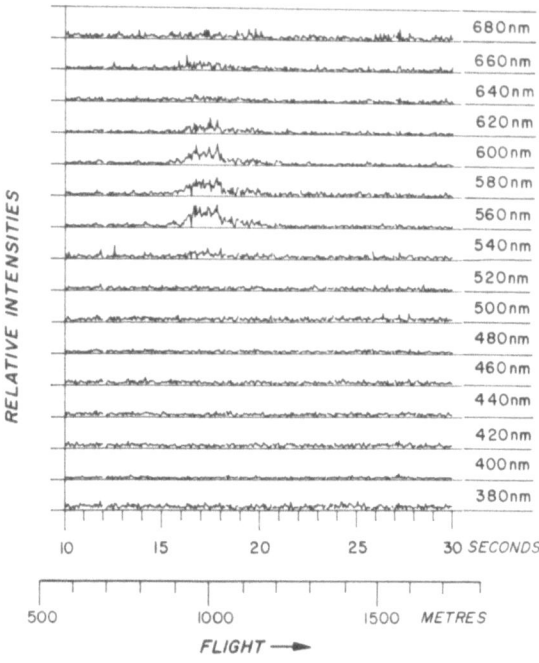

Figure 3.   Fluorescence data gathered over rhodamine dye spill.

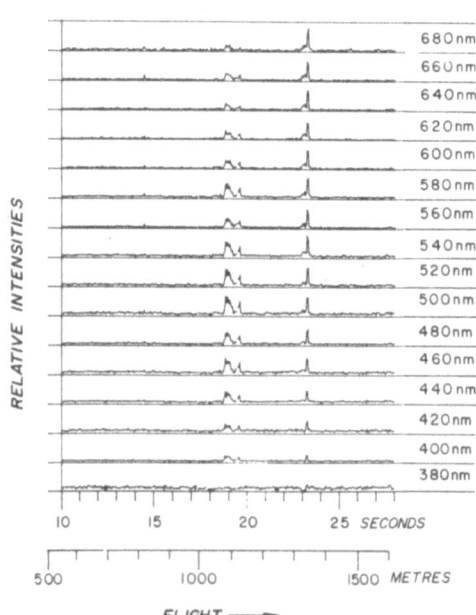

Figure 4.   Fluorescence data gathered over Merban crude oil
            spill (Flight line 12).

gathered on a test flight over the dye spill, while Figures 4
and 5 (Flight lines 12 and 17, respectively) show data from
flight over Merban crude oil and Figures 6 and 7 (Flight lines
6 and 16, respectively) over La Rosa crude oil. The abscissa
represents the time or the distance travelled along a flight
line, with the fluorosensor return divided into the 16 wavelength bands acquired simultaneously by the sensor.

In the cases of the dye and Merban crude oil spills, the
presence of an anomalous substance on the water surface can be
readily seen by the sharp increase in fluorescence within
certain wavelength bands. This is not the case with La Rosa
crude oil, which, as shown in Fig. 2, fluoresces 10 times less
intensely than the Merban crude oil. In this case, the flight
data does not show significant fluorescence. The presence of
the oil is indicated, rather, by a suppression of the water
Raman scattering of the laser light. The layer of oil absorbs
the laser pulse and no photons penetrate the oil film to be
scattered by the underlying water. The Raman scattered photons
have a wavelength of 380 nm and so the suppression of this
signal is clearly visible in the 380 nm channel shown in Fig. 7.

CORRELATION ANALYSIS

Viewing the increase in fluorescence gives a definite indication of the presence of some anomalous substance, however,

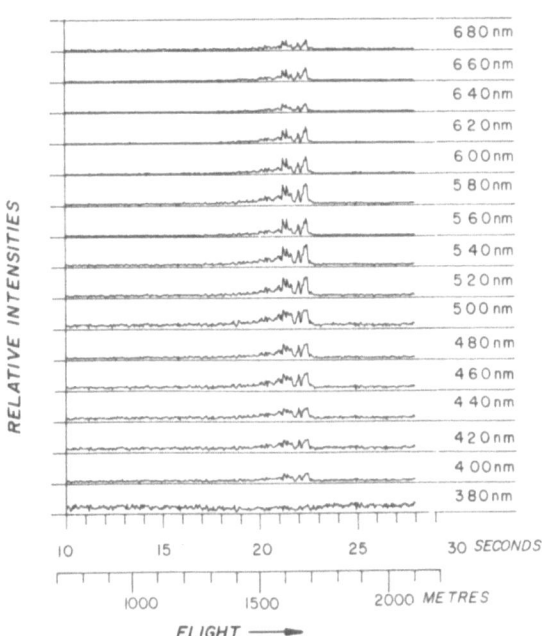

Figure 5.   Fluorescence data gathered over Merban oil spill.

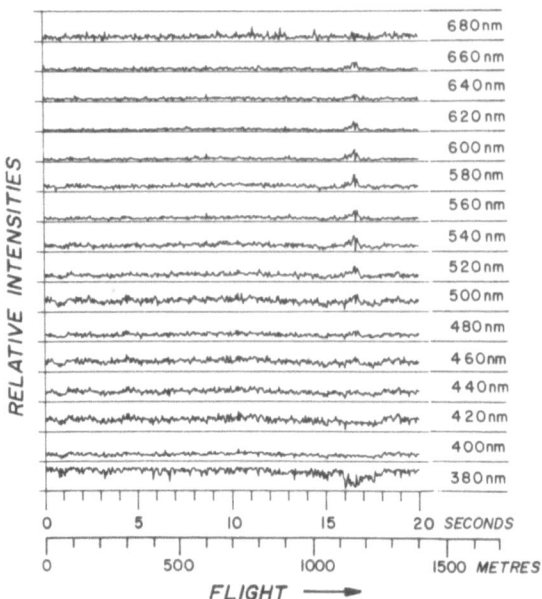

Figure 6.   Fluorescence data gathered over La Rosa
            crude oil spill (Flight line 6).

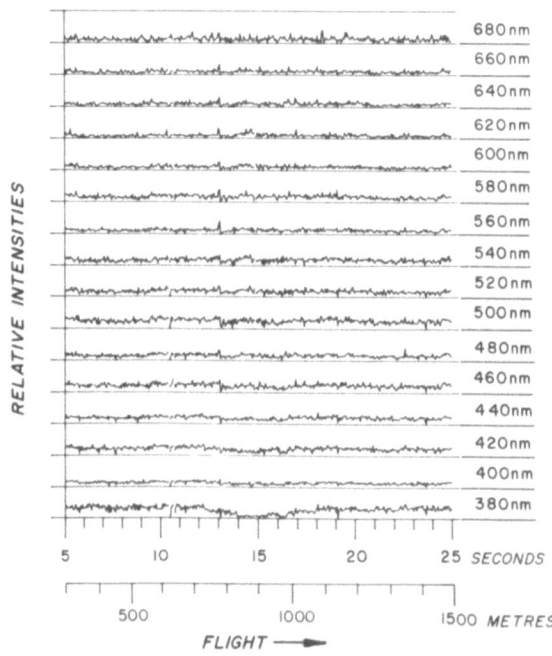

Figure 7.   Fluorescence data gathered over La Rosa oil spill.

by itself, it does not identify that substance. To do this,
one must look at the relative intensity of fluorescence as a
function of wavelength (emission spectrum) as one crosses the
anomaly and compare it to known fluorescence signatures. A
mathematical correlation procedure was used thereby a known 16
channel signature is passed along a flight line and compared
with the 16 channel in-flight fluorescence data. A complete
"match" or correlation of the two signatures yields a value of
ρ equal to +1.0, no correlation yields a value of ρ equal to 0
and perfect anti-correlation a value of -1.0. All values between the extremes represent varying degrees of correlation.
Figures 8, 9, 10, 11, 12, 13 and 14 show the results of these
correlation calculations.

In each case, one of the 16 wavelength bands is presented
as a reference, as well as a plot of the values of the correlation coefficient, ρ, along the flight line, the correlation
coefficient, ρ, being defined for each laser pulse as follows:

$$\rho = \frac{N\Sigma_i\ X_i\ Y_i\ -\ \Sigma_i\ X_i\ \Sigma_i\ Y_i}{\sqrt{N\Sigma_i\ X_i^2\ -\ (\Sigma_i\ X_i)^2}\ \sqrt{N\Sigma_i\ Y_i^2\ -\ (\Sigma_i\ Y_i)^2}}$$

where X_i and Y_i are the intensity values of the ith of sixteen
spectral channel as measured in flight and in the laboratory.

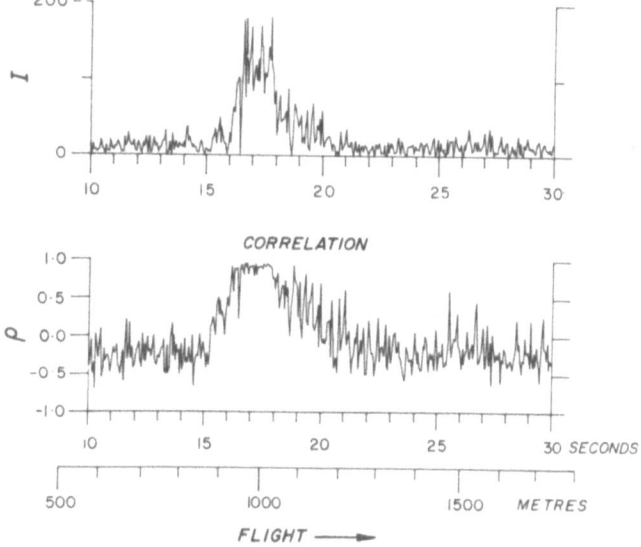

Figure 8. Correlation of in-flight fluorescence data gathered
over rhodamine WT dye spill (flight line 6) with
rhodamine WT dye spectral signature.

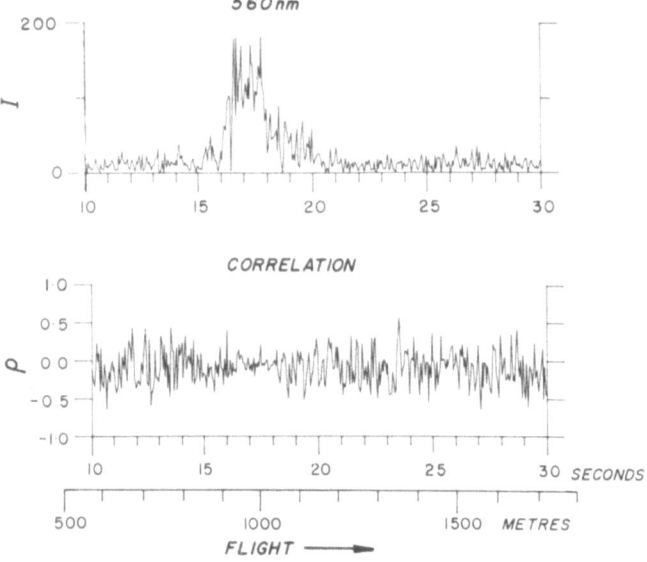

Figure 9.   Correlation of in-flight fluorescence data gathered
            over rhodamine WT dye spill (flight line 6) with
            Merban crude oil spectral signature.

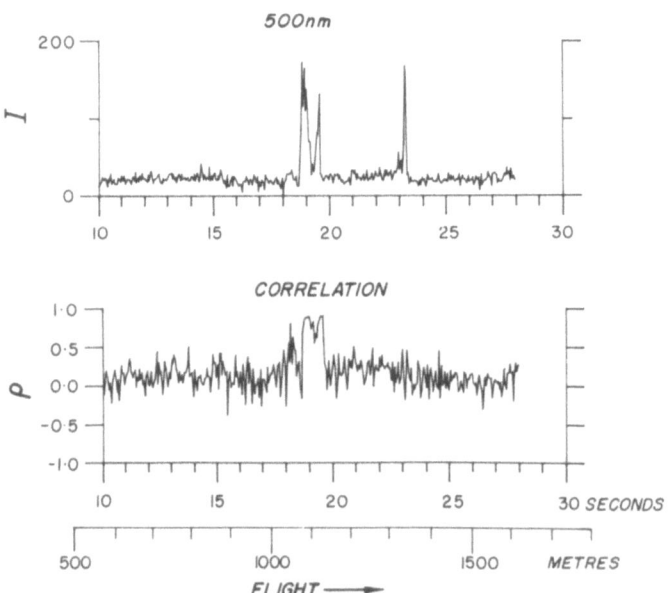

Figure 10. Correlation of in-flight fluorescence data gathered
            over Merban crude oil spill (flight line 12) with
            Merban crude oil spectral signature.

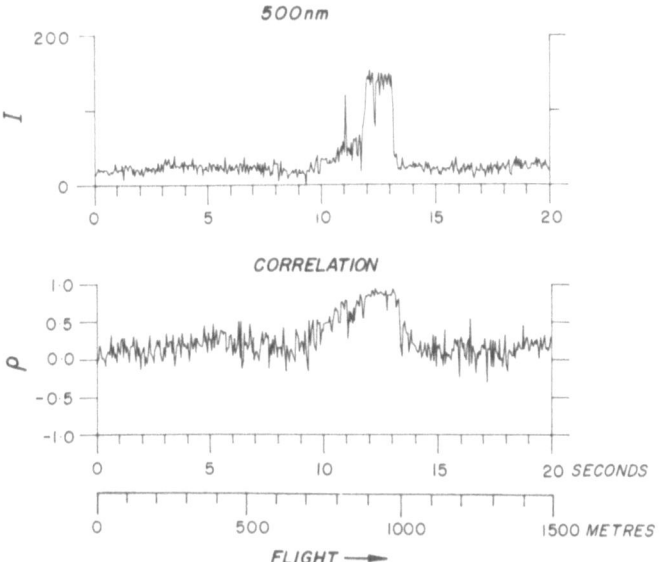

Figure 11. Correlation of in-flight fluorescence data gathered over Merban crude oil spill with Merban crude oil spectral signature.

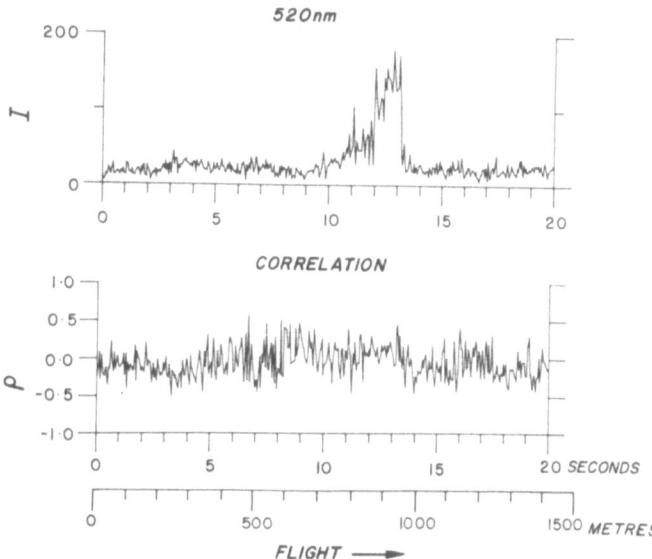

Figure 12. Correlation of in-flight fluorescence data gathered over Merban crude oil spill with rhodamine WT dye spectral signature.

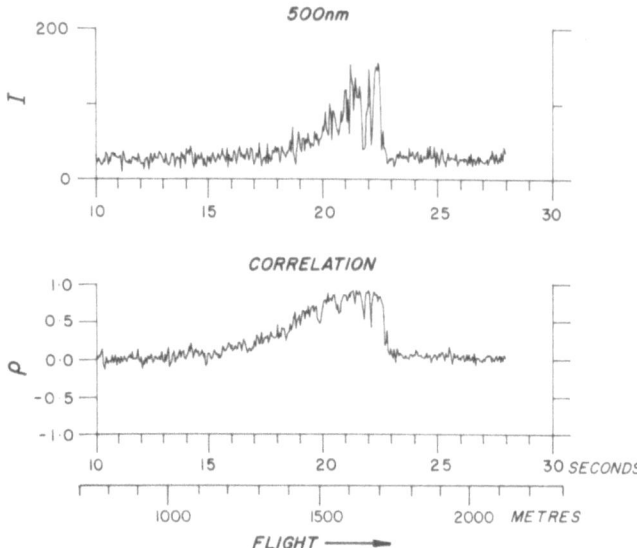

Figure 13. Correlation of in-flight fluorescence data gathered
over Merban crude oil spill with Merban crude oil
spectral signature.

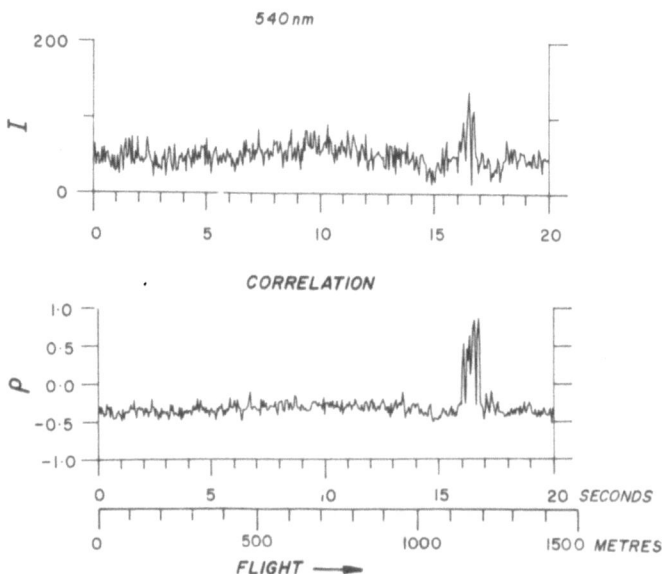

Figure 14. Correlation of in-flight fluorescence data gathered
over La Rosa crude oil spill with La Rosa crude oil
spectral signature.

   The fluorescence return in Figure 3 shows an anomaly 16 to
18 seconds into the flight line. By using the correlation technique, the results of which are presented in Figures 8 and 9,
in conjunction with Figure 2, the anomaly can be clearly identified as a rhodamine dye spill.

   The usefulness of this method is further illustrated when
considering the fluorescence return of Figure 4. Two anomalies
are present along the flight line. One might, at first, mistak-
enly conclude both indicate the presence of oil until a correlation calculation is carried out using the spectral signature
of Merban crude oil. Figure 10 shows the second anomaly not to
be Merban oil and, in fact, 70 mm photographs taken during the
flight line showed it to be a ship close to the oil spill.

   To be noted also is the extent of correlation compared to
the fluorescence peak itself. Figures 15 and 11 show the correlation procedure accentuates the portions of the slick with
weak fluorescence returns, in other words, thin portions of a
slick thereby making a larger area of the slick discernible.

   This ability to accentuate is found to be particularly
crucial in the case of a heavier oil, such as La Rosa crude oil.

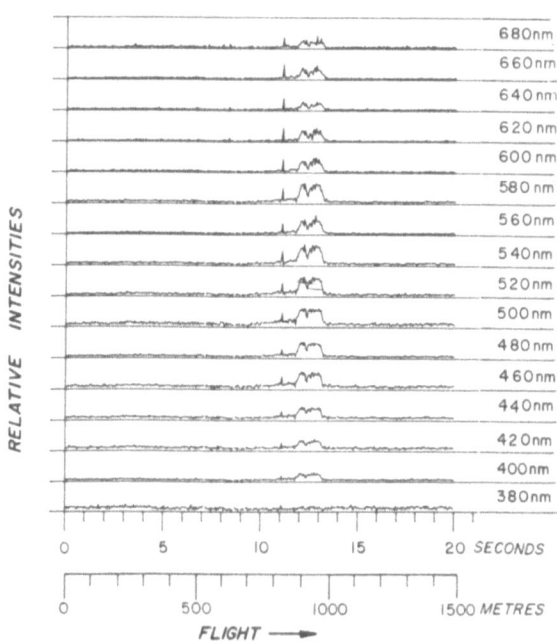

Figure 15.   Fluorescence data gathered over Merban
             crude oil spill.

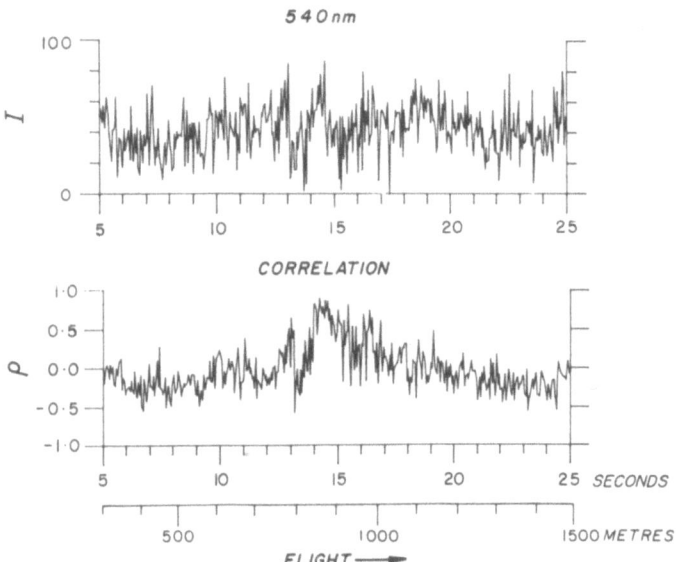

Figure 16.   Correlation of in-flight fluorescence data gathered
             over La Rosa crude oil spill with La Rosa crude oil
             spectral signature.

Figures 6 and 7 show very little significant fluorescence,
however, even with this very low fluorescence return, one can
extract valuable information about the presence of oil through
the correlation analysis procedure as shown in Figures 14 and
16.

DISCUSSION

     Although there can be no doubt as to the value of fluor-
escence spectroscopy analysis for oil identification, the
present implementation of the laser fluorosensor is limited in
its ability to differentiate between closely related spectral
signatures.

     The degree of correlation is comparable for the oil spill
test flight data presented whether one used the La Rosa or the
Merban crude oil signature. This is a result of the similarity
between the spectral signatures of these two oils which, except
for a difference in absolute fluorescence conversion efficiency,
appear almost the same when used in the correlation analysis.
There is no difficulty in the case of a significantly different signature, as in the case of the rhodamine WT dye, or in
comparison with different types of oil, such as light and
heavy petroleum products which showed no correlation.

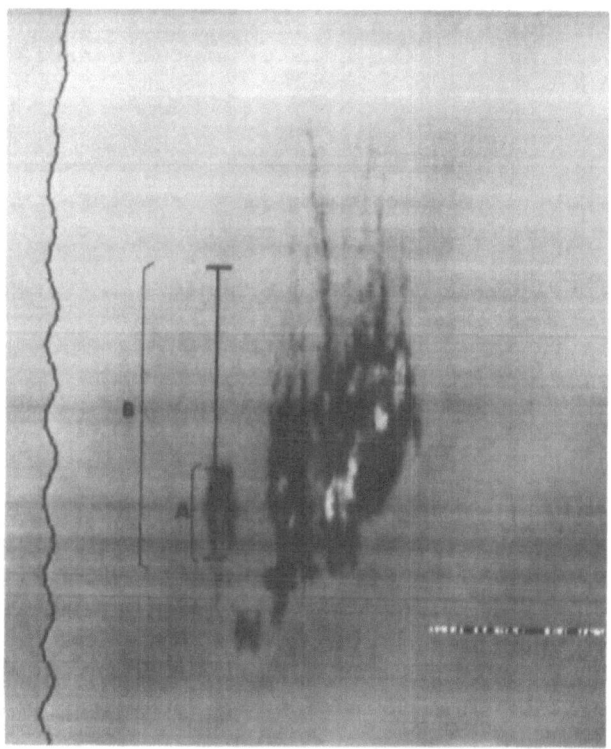

Figure 17.   A comparison of the responses of any IR scanner and
             the laser fluorosensor for test flight line 13 over
             Merban crude oil. The solid line represents the
             flight path of the fluorosensor. "A" marks the
             region over which the raw data showed discernable
             oil spill return and "B" shows the extended region
             accentuated by the correlation analysis of the raw
             data.

        Another point of interest already mentioned is the ability
of the correlation procedure to accentuate the thinner regions
of an oil slick whose fluorescence may not be easily discernible otherwise. The question then arises of how thick oil must
be to be "seen" by the fluorosensor. Figure 17 is the IR image
corresponding to the fluorosensor flight line shown in Fig. 15.
It can be seen that the fluorosensor picks out the thicker
portions of the slick at least as well as the IR and can even
extend into the thinner regions by applying correlation analysis. The minimum thickness of oil which can be seen in the IR
imagery is about 10 μm, and so the fluorosensor is sensitive
to thicknesses less than 10 μm, but the exact figure is

difficult to gauge at present. There have been some improvements
to the fluorosensor since the data were taken which should
reduce noise and further the ability of the sensor to see thin
oil patches.

CONCLUSIONS AND FUTURE PROPOSALS

The analysis of fluorosensor data from three test spills
has shown the effectiveness of this technique to:

1 - Identify oil (that is to discriminate between widely
    varying objects such as ships, oil and dye);
2 - Classify oil into different groups (crude, light
    refined and heavy refined);
3 - Relate the spectra obtained from the sensor flown over
    the target spills to spectra obtained in the labora-
    tory.

The data presented are from spills which had weathered no
more than a few hours. At that point, there was no difficulty
in matching laboratory measured emission spectra with actual
flight data.

Because the correlation algorithm is straightforward, it
is possible to implement it as part of the real-time data acquisition system. Two approaches may be taken. Representative
signatures of the different oil groups may be stored and then
used as a comparison with incoming fluorescence data. Even more
simply, a real-time comparison may be carried out using the
signature of the water background, leaving those areas to be
considered where the "match" no longer holds. A combination of
the two approaches would provide an effective alarm for the
in-flight observer who would then also have the capability to
classify the oil in the spill.

REFERENCES

Rayner, D.M., and Szabo, A.G., 1976, A laboratory study of the
        potential of time-resolved laser fluorosensors, Report
        No. By-76-2(RC) prepared for the CCRS.

THE DUTCH EXPERIENCE

WITH REMOTE SENSING OF OIL POLLUTION

R. Spanhoff[1] and W. Wijmans[2]

[1]Rijkswaterstaat
Directorate for Water Management and
Hydraulic Research
The Hague, The Netherlands
[2]Rijkswaterstaat
North Sea Directorate
Rijswijk, The Netherlands

INTRODUCTION

According to the Agreement for Co-operation in Dealing
with Pollution of the North Sea by Oil, done at Bonn on the
9th of June 1969, The Netherlands are responsible for a relative
large area of the North Sea (Figure 1), where an intensive
shipping traffic exists[1]. Since 1975, the Dutch government is
carrying out visual inspection of this area with a light aircraft. This inspection is necessarily limited to day-time operations. The inspections have shown that the majority of the
illegal oil dumps is done at night, presumably as a consequence
of the day-time surveillance. Furthermore, it is impossible in
this manner, within reasonable constraints of flying hours, to
cover the whole area regularly. Thus, the use of remote sensing
techniques is appropriate.

Since 1971, the Rijkswaterstaat of the Dutch government
investigates remote sensing techniques as a tool to aid in its
tasks on the North Sea. A research programme is performed in
which on the one hand basic studies of sensor-object interactions are carried out[2] and on the other hand experience is
gained with the operational use of remote sensing techniques.
The latter activities concern up to now mainly oil detection.

Since the workshop on the NATO/CCMS pilot study on remote
sensing for the control of marine pollution held in Washington

313

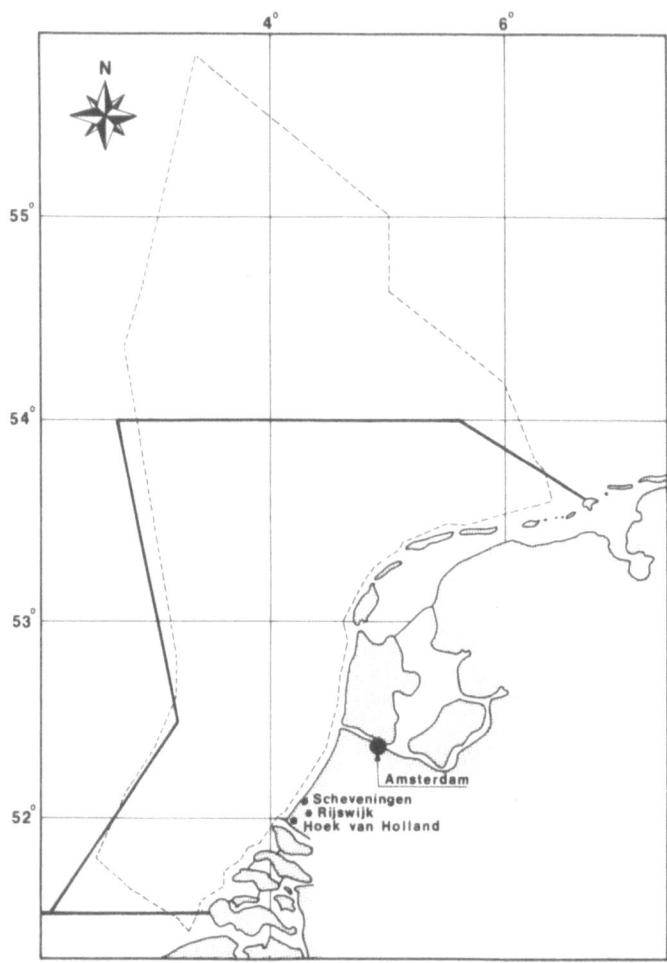

Figure 1. The Dutch control area of the Continental Shelf.

in 1979[3] several more exercises have been performed by the
Netherlands on accidental as well as on intentional oil spills,
some of which are briefly mentioned in the following.

EXPERIENCES WITH OIL SPILLS

The British-Swedish-Dutch Oil-slick Experiment of June 1980

        As a result of the contacts within working group I of the
NATO/CCMS pilot study on remote sensing for the control of
marine pollution, a tripartite experiment under the auspices
of the Warren Spring Laboratory (UK) was performed in June 1980,

involving a British, a Swedish and a Dutch airplane. Ten tons
of crude oil were released in the morning of June 17th 1980 in
the North Sea, 35 miles east of Benacre Ness (52° 24' N, 02°
42' E) under good weather conditions, with a southerly wind of
about 20 knots. The spill was overflown during two runs, one
starting just after the release and the other some four hours
later. Each run consisted of three consecutive sorties by the
British, Swedish and Dutch aircraft, respectively. The British
aircraft of the Royal Signals and Radar Establishment was
equipped with a Q-band SLAR (HH) and an IR line scanner, the
aircraft of the Swedish Coast Guard with a prototype X-band
SLAR (VV) built by LM Ericsson on specifications by the Swedish
Space Corporation, and the Metro II laboratory aircraft of the
Dutch National Aerospace Laboratory NLR with the Dutch experimental X-band SLAR (HH), which has a digital recording.

A comparison of the results of the various sensors has not
been performed thus far. For the Dutch SLAR, it can be concluded that the oil slick was observed moderately well, but
that the sensitivity of this experimental system is too low
for practical purposes. Figure 2 shows a digitally processed
image of the slick.

In order to achieve the final specifications for a Dutch
operational airborne oil-detection system, a semi-operational

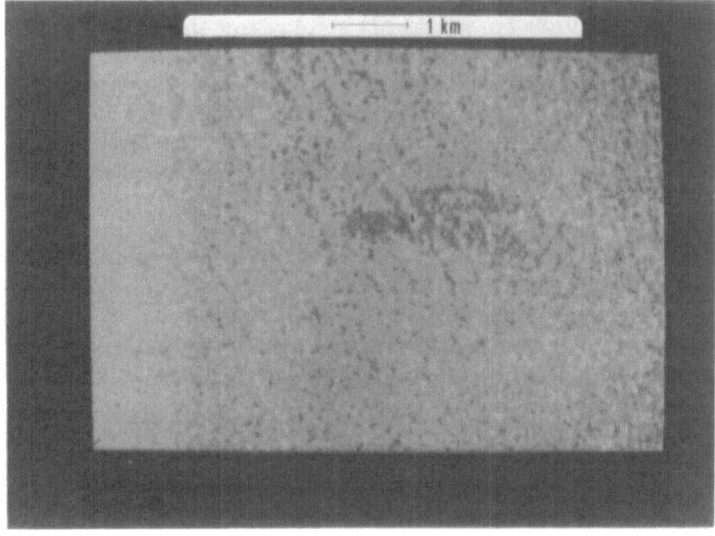

Figure 2. The oil slick of the British-Swedish-Dutch experiment
          of 1980, observed by the Dutch experimental SLAR.

flight programme was performed in November 1980.

The Dutch Semi-operational Exercise of November 1980 Involving the Dutch Experimental SLAR and the Swedish Operational SLAR System

This semi-operational flight programme was performed in November 1980, involving the Dutch SLAR system as above and a CESSNA 402C of the Swedish Coast Guard, equipped with a fully operational surveillance system based on the Ericsson SLAR. The IR/UV scanner system which forms also part of this surveillance package was not installed yet at that time.

The performances of both remote sensing systems were compared with the results of visual observations of the standard maritime surveillance, which was performed simultaneously by the Dutch airplane dedicated to this task. During five days, with varying weather conditions, flights were made with both Dutch airplanes, following the routine operational flight patterns of the visual surveillance, which cover the major part of the Dutch control area of the Continental Shelf. On one of these days also the Swedish aircraft took part. During one more day all three aircraft participated in an Isowake experiment, which is presented in the specific session.

From the simultaneous flights, it is concluded that the SLAR systems are able to detect all oil spills which are observed during visual surveillance. The reverse is not true, that is, the range of weather conditions under which oil slicks can be observed is significantly extended with SLAR systems compared with visual observations. Moreover, the area which can be covered during one track of the airplane is drasticly enlarged with the Swedish SLAR, which covers a total swath of about 40 km. The Dutch experimental SLAR has a one-sided antenna and a limited sensitivity, which restricts its total effective range to about 5 km.

The five-day exercise covered also rather rough weather conditions. It turned out that the main limitations of the use of the SLAR systems are imposed by the turbulent movements of the airplane under these conditions. Moreover, during wind speeds greater than 25 knots no oil at all is observed at the sea surface.

From the flights with the Dutch SLAR performed during the night, it was concluded that spills, which consisted very likely of oil, were observed, but no means were available to check this nature. For this purpose an IR/UV scanner is indispensable.

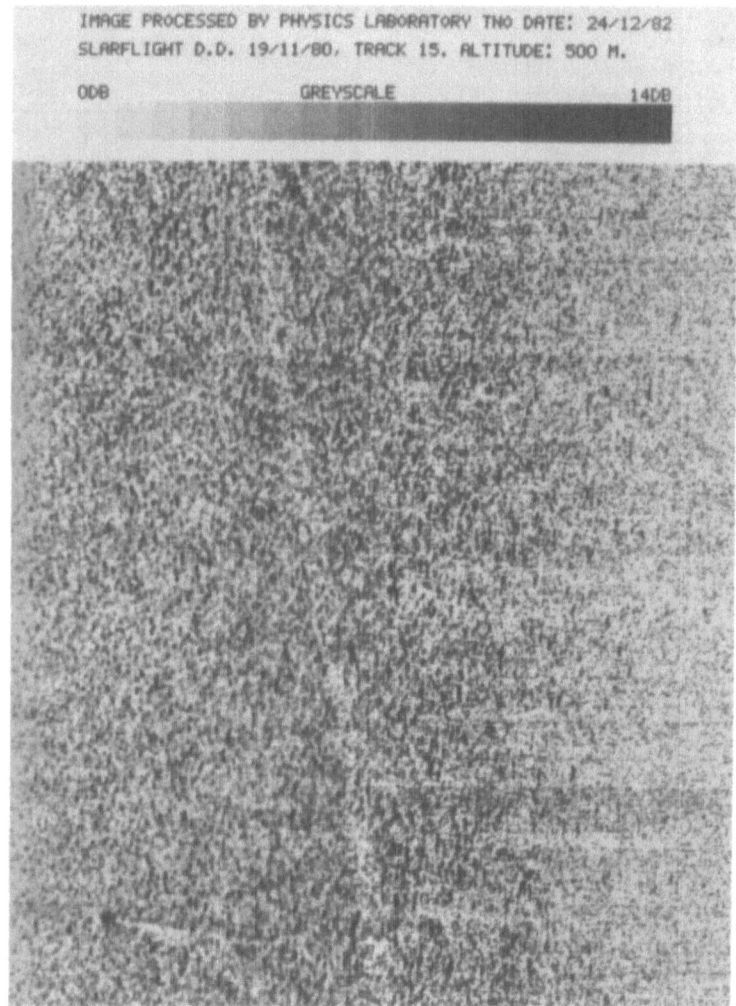

Figure 3. Computer processed image of an illegal oil spill,
observed by the Dutch experimental SLAR during the
semi-operational exercise of November 1980.

Figure 3 shows an example of an illegal oil spill detected
with the Dutch SLAR system during these exercises. The above-
mentioned limited sensitivity is reflected by the speckle in
the image, notably in the far range. It should be noted that
the Dutch SLAR is not designed as an operational oil-detection
system, but rather as a research tool, for which quite different requirements apply, such as a high dynamic resolution and
an absolute calibration. An analysis of the specifications of
the Swedish and Dutch SLAR systems learns that the higher

sensitivity of the latter arises mainly from the choice of the
polarisation (VV).

The Isowake Experiment of July 1981 Performed by the JRC in the Frame of the SAR-580 Campaign

Rijkswaterstaat cooperated with the Joint Research Center
(JRC) of Ispra (European Communities) in the Isowake experiment
on the North Sea during the SAR-580 campaign. This cooperation
was requested by the JRC since it failed to obtain permission
for an oil spill in the Mediterranean. The experiment involved
the Convair-580 with an X-band/C-band SAR, the oil-detection
system of the Swedish Coast Guard, an aircraft of Intradan
(DK) equipped with an X-band SLAR (HH), and the Dutch patrol
aircraft for visual surveillance. Preliminary results of the
Swedish and Danish SLAR systems showed interesting differences
between the several oil concentrations at the water surface
and between the two kinds of oil used. The complete results
of this exercise are elaborated by the Joint Research Center
of Ispra.

Figure 4. The vessel COSMOS in action, following the oil-
spilling, oil recovery vessel SMAL AGT.

The Oil Discharge of July 1981

In July 1981 took place a 138 m^3 oil discharge to test the mechanical oil-recovery capacities of the multipurpose oil-combatting vessel COSMOS. It was a happy coincidence that the planned acceptance test of the vessel COSMOS fell on the day after the Isowake experiment in the SAR-580 campaign.

The Swedish and the Danish participants of this previous experiment took the opportunity to overfly the large oil spill involved in the acceptance test.

A "chocolate mousse" (138 m^3) was produced from a mixture of 12 per cent diesel oil, 18 per cent heavy fuel oil and 70 per cent water. The emulsion was discharged at the sea surface and deliberately spread out by turbulent movements induced by ships. The spill was subsequently removed by the vessel COSMOS, which is a hopper-dredger which can be transformed within 8 hours into an oil-combatting ship with amongst other facilities for mechanical oil recovery (Figure 4). In Figure 5 images of the IR/UV line scanner of the Swedish aircraft are given, showing the COSMOS in action.

The Accident with the Oil Tanker KATINA of June 1982

On June 7th 1982 the Greek tanker KATINA lost within 2 hours, as a consequence of a collision, an estimated amount of about 1,100 m^3 heavy fuel oil at a position near the entrance to the Rotterdam harbour, about 15 km off the Dutch coast. The subsequent extensive operations, involving amongst others the three Dutch oil-combatting vessels, were co-ordinated from the air by the Dutch airplane which usually performs the visual surveillance of the control area in the North Sea. The characteristics of the oil spill, like its position, evaporation, the oil-water mixing etc., were continuously monitored.

The computer program OILPROG, based on a mathematical model, was able to calculate such features. The amount of oil recovered mechanically by the three Dutch combatting vessels during the days after the accident amounted up to 88 per cent of the total spill, according to these calculations. The remaining fraction, about 200 m^3 oil-water mixture, ultimately reached the beaches, where a serious contamination occurred.

Two important conclusions concerning the use of remote sensing techniques were drawn from this accident. Firstly, the lack of an operational remote sensing system was explicitly felt the morning after the accident when fog prevented aerial surveillance and, thus, the abatement of the floating oil.

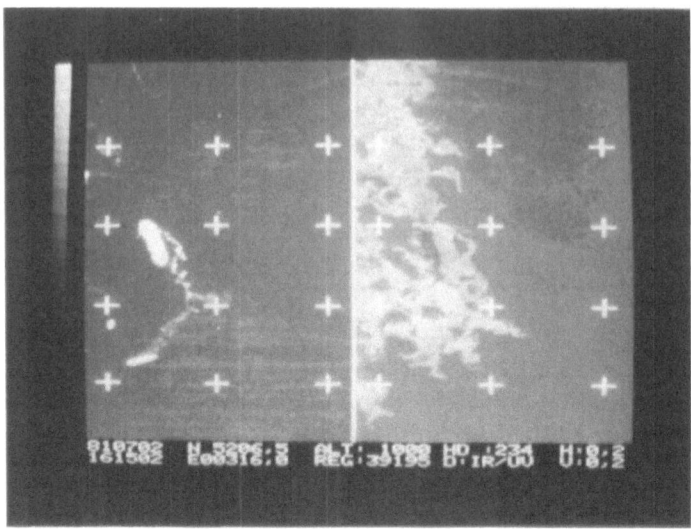

Figure 5.   IR/UV images obtained by the aircraft of the Swedish
Coast Guard of the vessel COSMOS in action.
IR image at the left, UV image at the right.

Secondly, the remaining 12 per cent of spilled oil reached the
beaches without being observed in advance. Presumably, the oil
floated below the water surface, due to mixing with the sea
water and to the increase of its specific gravity after evaporation of the lighter oil fractions. So it remained unobserved
despite intensive visual surveillance flights. Also SLAR systems
and UV/IR line scanners would not have been able to detect this
oil. A laser-fluorescence system with a sufficient penetration
depth in turbid waters is likely able to detect such floating
oil masses, for which in this case the position could be
reasonably well predicted. The experience with this oil catas-
trophe clearly provokes the need of developing such operational
systems with laser-fluorescence techniques.

As a further experience gained from this actual accident
with a significant oil spill, it should be mentioned that the
communications with the other North-Sea countries, as agreed
within the Bonn Agreement, worked out quite satisfactorily.
In fact, several countries sent observers and/or equipment.
Figure 6 shows an IR image, produced by Eurosense bv (NL),
featuring the KATINA on her way to Rotterdam harbour the
morning after the accident and some remaining leaking oil.

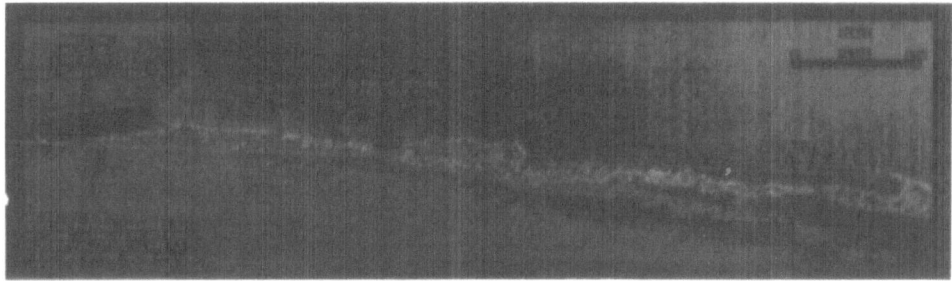

Figure 6.   IR image of the tanker KATINA heading for Rotterdam
            harbour. Leaking oil is clearly visible.
            (By courtesy of Eurosense bv.)

THE DUTCH PLANS FOR THE NEAR FUTURE

     Based on the extensive experience gained by Rijkswater-
staat thus far with visual observations of the Dutch control
area of the continental shelf as well as with remote sensing
of deliberate and accidental oil spills, specifications for an
operational surveillance system have been defined. Starting
1983, a CESSNA 404 will be flown for Rijkswaterstaat on behalf of its surveillance tasks (Figure 7).

     The primary sensor will be the Ericsson SLAR, combined
with its available data-handling and presentation-system
centered around a video monitor. For verification purposes, a

Figure 7.   The Dutch aircraft for remote sensing surveillance.

Daedalus IR/UV line scanner will be added. Room has been
reserved in this particular type of aircraft for additional
future sensors. Provisions have been made to enable the
release from the aircraft of relocatable oil-sampling devices.

     Apart from the contamination with oil, Rijkswaterstaat
monitors on a routine basis the water quality in general of the
North Sea. A monitoring programme along the Dutch coast, extending up to at most 150 km off shore, is performed every
fortnight, involving a laboratory vessel. It is investigated
presently if and how far satellites and airborne data can op-
timize, complement or partly replace these sea-truth measurements. As an example, a satellite image of the Coastal Zone
Colour Scanner (CZCS) is shown in Figure 8. This image was
processed at Joint Research Center of Ispra (European Communities) by Jørgensen and Larsen[4] in the frame of their work for
Rijkswaterstaat. The image, which is corrected for atmospheric
influences, is believed to reflect phytoplankton pigment concentrations of the North Sea, but definitely more work,
involving more images and sea-truth data, is needed.

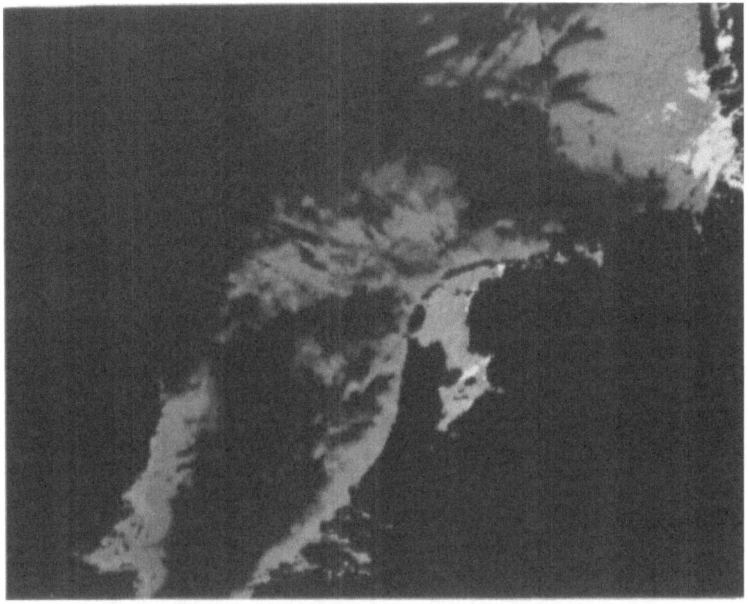

Figure 8.   Computer processed CZCS image of the North Sea,
             representing phytoplankton pigment concentrations
             (Larsen and Jørgensen, 1981).
             The darkest areas represent land and cloud-covered
             areas.

SUMMARY

From experiences gained with visual observations during day-time surveillance of the Dutch control area of the continental shelf, and from (semi-operational) experiments with remote sensing systems, specifications for a Dutch operational remote sensing surveillance system were obtained recently. Of especial importance in this respect are the more effective coverage of the whole area considered as well as the possibility of surveillance during the night and during poor-visibility conditions. Starting 1983 routine surveillance flights will be made with the remote sensing system.

From an accident with a tanker, which resulted in a serious contamination of the Dutch shore, it was learned that significant amounts of oil can reach the shore without being observed in advance, since the oil floats below the water surface. The development of an operational laser-fluorescence system with a sufficient penetration depth in turbid waters is recommended.

REFERENCES

1. R. Spanhoff, National requirements in the Netherlands, in: "Airborne Remote Sensing of Oil Spills in Coastal Waters", Proc. 1st workshop sponsored by Working Group I of the French pilot study for the NATO Committee on the Challenges of Modern Society, held at Washington, D.C., 18-20 April 1979 (1980).
2. G. P. De Loor and H. W. Brunsveld van Hulten, Microwave measurements over the North Sea, Boundary-layer Meteorology 13:119 (1978).
3. "Airborne Remote Sensing of Oil Spills in Coastal Waters", Proc. 1st workshop sponsored by Working Group I of the French pilot study for the NATO Committee on the Challenges of Modern Society, held at Washington, D.C., 18-20 April 1979 (1980)
4. P. L. Larsen and P. C. Jørgensen, "Coastal Zone Color Scanner Data Applied for Mapping of the Phytoplankton Pigment Concentration in the North Sea", Rijkswaterstaat, North Sea Directorate (1981).

PRESENTATION OF MAPS SHOWING THE DRIFT OF THE OIL

FROM THE AMOCO CADIZ*

A. Fontanel

Institut Français du Pétrole
Rueil Malmaison, France

CHRONOLOGY AND MEANS OF ACTION

For the purpose of detecting illegal deballasting operations at sea, the merchant navy has recently developed a system
consisting of a light aircraft equipped with a SAT (Société
Anonyme de Télécommunications) scanning radiometer working in
the thermal infrared and a Hasselblad camera.

At the time of the AMOCO CADIZ disaster, it was proven
necessary to have an aircraft equipped with similar sensors but
capable of flying over the polluted area for many hours and of
carrying several scientific specialists in addition to the
regular crew.

In this way, the day after the disaster, the Centre National pour l'Exploitation des Océans (CNEXO), the Institut Français du Pétrole (IFP) and the Institut Géographique National
(IGN) became associated for the first remote sensing mission to
follow the movements of pollutants on the shore and at sea.

Knowledge of these movements is indispensable for
effectively directing operations to fight against the "black
tide", to treat offshore slicks, to protect fishing zones and
tourist sites and to clean-up the coast. Afterwards, in the
light of the findings, the French Navy, as part of the POLMAR
Plan, asked for the services of the equipped aircraft and
specialists from the three organisations.

---

* Excerpts from "Remote sensing of hydrocarbon pollution"
  (Preliminary report CNEXO-IFP-IGN).

The airplane was equipped with a Daedalus scanning sensor recording radiations in the thermal infrared and two aerial cameras with a 24 x 24 cm format and a focal length of 152 mm, one using infrared film and the other color film.

From the 18th of March to the 7th of April, the IGN aircraft operated various missions during ten days and produced 1,352 km of thermal infrared data, 1598 infrared photos and 422 color photos. Beginning on the 8th of April, the aircraft remained on standby for any further action should to prove necessary. A new flight took place on May 26, to evaluate the evolution of the phenomenon as a whole.

The daily organization of flights fulfilled a three-fold mission:

(1) Visual reconnaissance, usually performed at an altitude of 1,500 to 2,000 feet because of unfavorable weather conditions and generally run along N-S axes. The results of this observation were radioed to the POLMAR center and to the vessels treating the slicks;

(2) After visual detection of oil, a thermal infrared sensor and a camera were put in operation to gather more detailed information. The thermal data were looked over on ultraviolet paper aboard the aircraft, then interpreted, and the conclusions of this survey radioed back.

These two missions were performed to serve the immediate needs of the staff in charge of the POLMAR PLAN. Upon the return of the aircraft, the data obtained during the flight were interpreted and synthetized, and the results presented in the form of a daily report and cartographic themes.

(3) During the flights, depending on the data obtained by sight and by remote sensing, the crew took the initiative of adding more flight axes to the predetermined flight plan so as to make a more in-depth survey of particularly sensitive zones at sea and on the coast.

A preliminary report gave a few examples of characteristics zones showing specific phenomena and the interpretation that can be made of them thanks to the complementary nature of the data gathered. In addition, this type of interpretation completed by simultaneous visual observations was extended to the area as a whole for the different flights. A lot of maps (Figure 1) attempt to make an exhaustive and chronological description of the pollution evolution until the 31st of March.

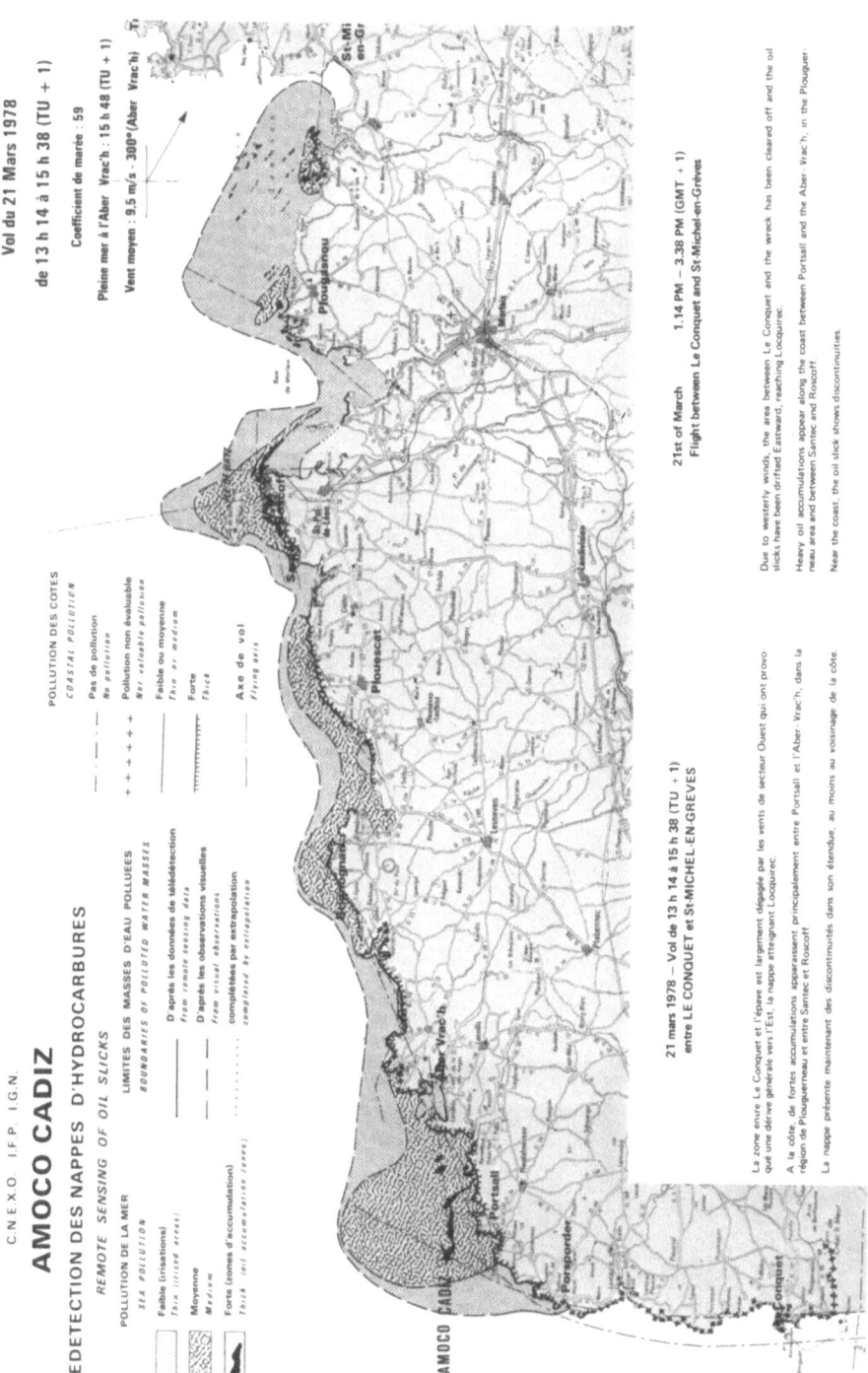

Figure 1. A synthetic map of the AMOCO CADIZ pollution on the 21th of March 1978. from remote sensing coverage

THE USE OF SATELLITES

FOR OIL SPILL DETECTION

AN ASSESSMENT OF THE USE OF SPACE TECHNOLOGY

IN MONITORING OIL SPILLS AND OCEAN POLLUTION

W.F. Croswell[1] and U.A. Alvarado[2]

[1]National Aeronautics and Space Administration
[2]General Electric, Space Division
Philadelphia, United States

ABSTRACT

Satellite sensors are selected for monitoring of oil spills
and measuring the parameters required by oil trajectory (fate)
models. The methodology of the ongoing study sponsored by the
National Aeronautics and Space Administration (NASA) considers
total system concepts encompassing operationally interactive
elements which include aircraft, ships, buoys and harbor facilities.

INTRODUCTION

Recognizing the importance of conservation of the oceanic
environment, the United States Congress has directed NASA to
assess how spacecraft may help in the monitoring of oil spills
and other types of ocean pollution. As part of the NASA sponsored effort of laboratory and aircraft experiments, analysis
of spacecraft data, and systems studies, General Electric Cny.
has been performing a study to evaluate current sensing techniques, projected technology, spacecraft and operational aspects
related to ocean pollution. This paper summarizes those interim
results of the study that are relevant to remote sensing of oil
spills. Effort to date relates to sensors and operations.

METHODOLOGY OF THE STUDY

A brief exposition of the methodology may be useful in
placing the interim results within the proper perspectives. In
order to assess the role of space within the context of the
whole operational system, the scope of the study was broadened

331

to include considerations of a total system for U.S. coastal
pollution monitoring. This is visualized as encompassing aircraft, ships, buoys, harbor installations, and data processing
centers, as well as satellites. Space activities, therefore,
are analyzed considering the complementary roles of these
diverse system elements.

The study starts with an analysis of the needs in ocean
pollution monitoring of users such as the U.S. Coast Guard, the
the Department of the Interior, the National Oceanic and

Table 1.   Assessment of sensor suitability
for measurements from space.

The methodology consists of state-of-the-art assessments,
parametric trades, and design-point evaluations relative
to four basic filters.

FILTER (A) DEVELOPMENT AND OPERATIONAL STATUS

Is the relationship (signature) between the observable
and the required measurements sufficiently understood
or developed ?

Is there sufficient operational experience with the
measurement technique ?

FILTER (B) DETECTIVITY / SENSITIVITY

Can adequate signal-to-noise ratios be attained ?

Is there sufficient sensitivity, considering the range
and atmospheric effects from orbital altitudes ?

FILTER (C) SENSOR PERFORMANCE SUITABILITY

Can the sensor meet the user requirements and corresponding measurements specifications for oil spill
and ocean pollution monitoring ?

Primary parameters are accuracy, resolution, coverage,
and range.

FILTER (D) COMPATIBILITY WITH SPACECRAFT

Are the required size, weight, power and other support
requirements of the sensor compatible with near-term
spacecraft platforms ?

Atmospheric Administration, and the Environmental Protection
Agency. Mission plans and operational scenarios are postulated
for the fulfillment of the knowledge objectives of these agencies.

From these objectives and in consultation with these agencies, we determine the measurement requirements for the system.
A key aspect of the implementation approach is the selection of
sensors and measurement techniques. Table 1 shows an example
in the selection process, wherein various space sensors are
analyzed for suitability to a specific measurement requirement.
Current and planned spacecraft programs are surveyed to evaluate possible utilization of present or proposed instrumentation,
and to determine new sensor capabilities that are to be incorporated in these satellites

Throughout the implementation analysis, the overall
mission operations are factored in, thus ensuring that the approach is viable in the light of proper sequencing of events,
orbit trajectories, and data management.

MEASUREMENTS

Three types of monitoring are considered in this study,

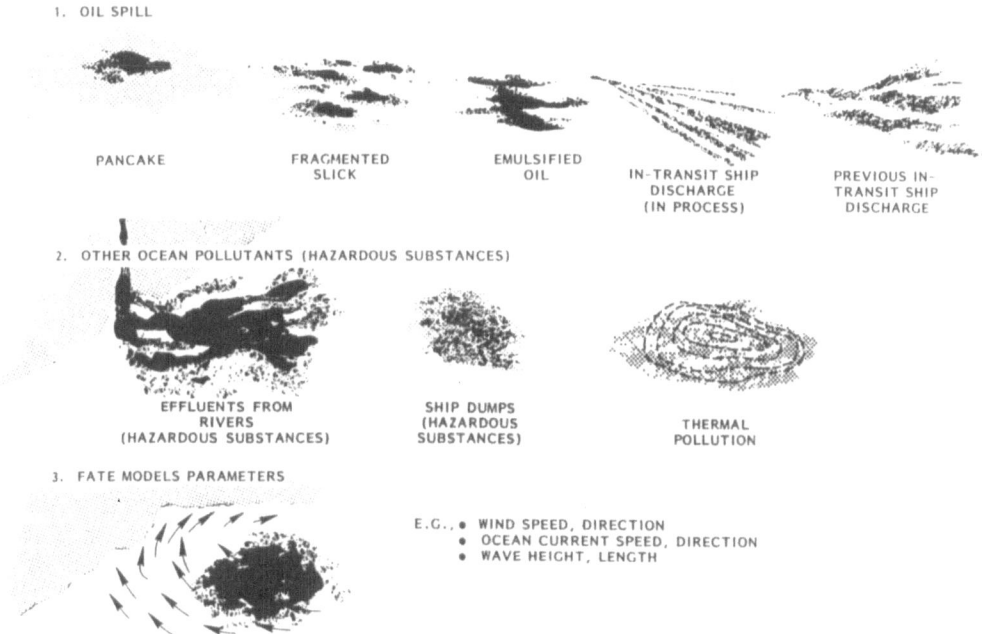

Figure 1. Types of monitoring by MOPS.

as depicted in Figure 1. The first of these is the monitoring
of the spill itself, which exhibits different types of geometries, from the somewhat symmetrical "pancake" spill to the
long and thin in-transit ship discharge. In type No. 3, we deal
with parameters such as wind, ocean currents, and wave phenomena; these are important inputs to the oil fate models which
determine the trajectory of the oil slick.

Table 2 shows a summary of the measurements specifications
for the missions that the parameters will support, and the
measurement characteristics such as accuracy, frequency and
spatial resolution. Note that items 1 through 8 deal with the
oil spill monitoring parameters, whereas items 9 through 22
deal with the oil fate model parameters. The "frequency of
measurements" tabulated on the sixth column is a very important
parameter due to its impact on the orbit and number of satellites required. The "delay" represents the period in hours between the acquisition of the data and the actual availability
of the data to the user. Most of the potential improvements in
system capabilities from space observations today are in the
area of decreasing the delay since imaging data is tradition-
ally slow in reaching the user in present experimental systems.
Perhaps the greatest impact on the sensors and their demands on
the satellite system is due to the spectral and spatial resolution. You will notice the stringent spatial resolution requirements associated with oil spill size and shape as compared to
those for wind, currents, etc., for the oil spill fate modeling.

The resolution requirement on oil spill is related in great
measure to the threshold of detectability of a particular slick
of a given size, and will be the subject of continuing analysis.

The selection of sensors and sensing techniques to satisfy
the measurements requirements discussed above has produced the
results summarized on Tables 3 and 4. For each of the measurement requirements, the tables show the sensors selected for
three time frames: a near-term period from 1984 to 1987, a mid-
term period from 1988 to 1991, and a far-term period from 1992
to the year 2000.

The selection shows a predominance of microwave sensors as
compared to the optical sensors; this is due to the all-weather,
day and night capability requirements of the system. However,
several optical sensors are deemed necessary to operate in conjunction with the microwave sensors to corroborate some of the
microwave observations such as those obtained through the synthetic aperture radar and the passive microwave radiometer.

The most important measurement in the detection and mapping
of oil spills is considered to be the synthetic aperture radar,

Table 2. Summary of measurement requirements.

PARAMETER	MISSION TYPE	RANGE OR SCOPE	PRECISION ±	ACCURACY ±	SPATIAL RESOLUTION OR GRID SIZE	FREQUENCY Every N h.	DATA DELAY Hrs
1 OIL SPILL AREAL DISTRIBUTION	SURV. & MONITOR MODELING	10 m 15 m	5 % 10 %	5 % 10 %	10 m 15 m	12 12	3 3
2 OIL SPILL COORDINATES	SURV. & MONITOR MODELING	US COASTAL AREA US COASTAL AREA	0.5 km -	1 km 250 m	N/A N/A	12 12	6 6
3 OIL SPILL THICKNESS	SURV. & MONITOR MODELING	0.1 μm - 2 mm 0.1 μm - 2 mm	- -	5 % -	- -	12 12	6 3
4 OIL CLASSIFICATION	SURV. & MONITOR MODELING	MAJOR TYPES GROSS CLASSIFICATION	N/A N/A	N/A N/A	- -	12 12	6 6
5 POLLUTANT DUMP AREAL DISTRIBUTION	SURV. & MONITOR	30 m	-	-	30 m	24	1/4 to 3
6 POLLUTANT DUMP CLASSIFICATION	SURV. & MONITOR	ACID/INDUSTRIAL OR SEWAGE	GENERIC CLASS	GENERIC CLASS	N/A	12 for acids	3
7 POLLUTANT DUMP COORDINATES	SURV. & MONITOR	US COASTAL ZONE	200 m	200 m	N/A	12 for acids	3
8 POLLUTANT SOURCE IDENTIFICATION	SURV. & MONITOR	TANKERS/BARGES/ RIVER EFFLUENT/ NATURAL SOURCE/	SUFFICIENT FOR LEGAL EVIDENCE		N/A	12	3
9 WIND SPEED	MODEL	0-50 m/sec	0.5 m/sec	2 m/sec	10 km	3	3
10 WIND DIRECTION	MODEL	0-360 °	5°	10°	10 km	6	3
11 OCEAN CURRENT SPEED	MODEL	0-300 cm/sec	5 cm	5 cm	10 km	6	3
12 OCEAN CURRENT DIRECTION	MODEL	0-360 °	10 °	10 °	10 km	6	3
13 ICE COVER AREAL EXTENT	MODEL	0-100 %	-	2 %	10 m	24	6
14 ICE THICKNESS	MODEL	0-50 m	0.2 m	0.5 m	10 m	24	6
15 SIG. WAVE HEIGHT	MODEL	0.3-25 m	0.3 m	0.3 m	10 km	3	3
16 WAVE LENGTH	MODEL	0.3-1000 m	10 %	10 %	10 km	3	3
17 WAVE DIRECTION	MODEL	0-360 °	10 %	10 %	10 km	3	3
18 AIR TEMPERATURE	MODEL	-30° to +40 °C	1°C	1.5 °C	10 km	12	3
19 WATER TEMP (SURFACE)	MODEL	- 2° to +30 °C	0.25°C	1°C	10 km	24	6
20 WEATHER FRONTS	MODEL	-	-	-	10 km	12	6
21 PRECIPITATION	MODEL	-	-	-	10 km	12	6
22 SUSPENDED SEDIMENT	MODEL	-	-	-	10 km	12	6

Table 3. Sensor selection for various time-frames.

MEASUREMENT	SENSORS FOR 1984-87 SYSTEM	SENSORS FOR 1988-91 SYSTEM	SENSORS FOR 1992-2000 SYSTEM
MARINE WASTE DUMP AREAL DISTRIBUTION AND CLASSIFICATION (5,7)	POINTABLE OPTICAL LINEAR ARRAY COASTAL ZONE COLOR SCANNER/AVHRR THEMATIC MAPPER	POINTABLE OPTICAL LINEAR ARRAY COLORIMETER (ADVANCED CZCS) MULTISPECTRAL LINEAR ARRAY (OERS)	HIGH RESOLUTION OPTICAL/IR GEO-SYNCHRONOUS SENSOR
MARINE WASTE DUMP COORDINATES (6)	GLOBAL POSITIONING SYSTEM	GLOBAL POSITIONING SYSTEM PRECISION ALTITUDE DETERMINATION SYSTEM	GLOBAL SERVICES SYSTEM FOR POSITIONING AND POINTING
POLLUTION SOURCE (8)	SYNTHETIC APERTURE RADAR POINTABLE OPTICAL LINEAR ARRAY (to aid high resolution observations by airborne instruments)	SAME AS PREVIOUS TIME FRAME*	HIGH RESOLUTION SAR SYNTHETIC APERTURE PASSIVE MICRO-WAVE RADIOMETER
WIND SPEED (9)	MICROWAVE SCATTEROMETER PASSIVE M-WAVE RADIOMETER	MICROWAVE SCATTEROMETER PASSIVE M-WAVE RADIOMETER	MICROWAVE SCATTEROMETER WIND LIDAR PASSIVE M-WAVE RADIOMETER
WIND DIRECTION (10)	MICROWAVE SCATTEROMETER	MICROWAVE SCATTEROMETER	MICROWAVE SCATTEROMETER WIND LIDAR
OCEAN CURRENT SPEED AND DIRECTION (11,12)	MICROWAVE ALTIMETER COASTAL ZONE COLOR SCANNER	MICROWAVE ALTIMETER* COLORIMETER	WIDE-SWATH ALTIMETER ADVANCED OCEAN CURRENT SENSOR
ICE COVER AREAL EXTENT (13)	SYNTHETIC APERTURE RADAR PASSIVE M-WAVE RADIOMETER	SYNTHETIC APERTURE RADAR PASSIVE M-WAVE RADIOMETER*	SYNTHETIC APERTURE RADAR SYNTH.APERT. PASSIVE M-WAVE RADIOM.
ICE THICKNESS (14)	NO SUITABLE SENSOR AVAILABLE	-	-
SIGNIFICANT WAVE HEIGHT (15)	MICROWAVE ALTIMETER	MICROWAVE ALTIMETER	WIDE-SWATH ALTIMETER SWEPT FREQUENCY M-WAVE RADIOMETER
WAVE LENGTH (16)	SYNTHETIC APERTURE RADAR	SYNTHETIC APERTURE RADAR	SYNTHETIC APERTURE RADAR
WAVE DIRECTION (17)	SYNTHETIC APERTURE RADAR	SYNTHETIC APERTURE RADAR	SYNTHETIC APERTURE RADAR
AIR TEMP. (NEAR SURFACE)	VERT. TEMP. PROFILE RADIOMETER	VERT. TEMP. PROFILE RADIOMETER	ADVANCED VTPR
WATER TEMPERATURE (19)	PASSIVE M-WAVE RADIOMETER	PASSIVE M-WAVE RADIOMETER	HIGH RESOLUTION PASSIVE M-WAVE RAD.
WEATHER FRONTS (20)			
PRECIPITATION (21)	PASSIVE M-WAVE RADIOMETER	PASSIVE M-WAVE RADIOMETER	HIGH RESOLUT. PASSIVE M-WAVE RADIOM.
SUSPENDED SEDIMENT (22)	COASTAL ZONE COLOR SCANNER/ADV. VERY HIGH RESOLUTION RADIOMETER	COLORIMETER (ADVANCED CZCS)	COLORIMETER (ADVANCED CZCS)

Table 4. Selected space sensors vs. time-frame.

MEASUREMENT	SENSORS FOR 1984-1987 SYSTEM	SENSORS FOR 1988-1991 SYSTEM	SENSORS FOR 1992-2000 SYSTEM
OIL SPILL DETECTION & AREAL DISTRIBUTION	SYNTHETIC APERTURE RADAR	SYNTHETIC APERTURE RADAR	GEOSYNCHRONOUS SAR
		POINTABLE OPTICAL LINEAR ARRAY	ADVANCED MULTISPECTRAL LINEAR ARRAY
	THEMATIC MAPPER	MULTISPECTRAL LINEAR ARRAY	HIGH RESOLUTION PASSIVE M-WAVE RADIOMETER
OIL SPILL COORDINATES	SAME AS #1, AIDED BY GPS	SAME AS ABOVE AIDED BY GPS & PRECISION SENSOR POINTING SYSTEM	GEOSYNCHRONOUS SAR AIDED BY GLOBAL SCES. SYSTEM POSITIONING AND POINTING
OIL SPILL THICKNESS	PASSIVE MICROWAVE RADIOMETER (LARGE SPILLS ONLY)	PASSIVE MICROWAVE RADIOMETER (LARGE SPILLS ONLY)	LASER STIMULATED FLUORESCENCE (RAMAN SCATTEROMETER)
			HIGH RESOLUTION PASSIVE M-WAVE RADIOMETER
OIL CLASSIFICATION	NO AVAILABLE SENSOR	NO AVAILABLE SENSOR (WILL USE AIRBORNE LASER FLUOROSENSOR)	LASER STIMULATED FLUOROSENSOR

although the Thematic Mapper and pointable optical linear array sensors will also play an important part in the detection and monitoring of such spills.

Table 5 shows the synthetic aperture radar parameters required in the ocean pollution missions in comparison with the Ice Processes and Climate Missions and the SEASAT missions.

It is significant to note that the power requirements increase rapidly with higher resolutions and wider swaths, so that at 30 meters resolution and a swath width of 400 kilometers, the power requirements of the SAR are quite appreciable. The measurement of wind speed and direction will be attained through the microwave scatterometer for the early and mid-term periods. However, the use of lidar is envisioned in the far-term area. The wave height measurements using microwave altimeter present a problem due to the fact that the instrument looks at the nadir and thus a reasonably good grid size (e.g. 10-25 km) is difficult to attain daily with a single satellite. It is in an area like this that multiple sensors and sensing techniques come into play; for instance, the significant wave height measurement over the ocean will depend not on the altimeter measurements but also wave modelling based on wind and/or atmospheric pressure measurements and altimeter measurements from aircraft in selected sites.

Table 5. MOPS-SAR parameters and comparison with
          IPAC and SEASAT-SAR.

PARAMETER	MOPS VERY NARROW SWATH	MOPS NARROW SWATH	MOPS MEDIUM SWATH	MOPS WIDE SWATH	SEASAT SAR	IPAC SAR
FREQUENCY (MHz)	9375	9375	9375	9375	1275	12000
ALTITUDE (km)	567	567	893	567	808	777
NOISE FIGURE (dB)	4	4	4	4	-	4.5
SYSTEM LOSS (dB)	6	6	6	6	-	6
NO OF LOOKS	1	1	1	1	4	2
GROUND RESOLUTION (m)	30	30	30	30	25	100
SLANT RESOLUTION (m)	15	15	15	15	8.6	50
AZIMUTH RESOLUTION (m)	30	30	30	30	25	100
δ_o NRCS (dB)	-25	-25	-20	-20	-	-20
SNR, PATCH EDGE (dB)	3	3	3	3	-	3
INCIDENCE ANGLE (deg.max.)	45	45	38.4	38.4	28.4	37
GROUND RANGE (Km) MAX	500	500	600	600	376	506
SWATH WIDTH (Km)	110	180	280	392	100	264
AVE. RF POWER (WATTS)	300	1000	1000	2000	55	128
PK RF POWER (WATTS)	-	-	3000	6000	-	-
BUS POWER ESTIMATE (WATTS)	1500	3800	3800	6800	500	650
ANTENNA AREA (M^2)	15	8.2	7.7	5.5	24.8	7.2

MISSION OPERATIONS

The operational analysis showed two basic missions in which
space observations have significant impact: the surveillance and
monitoring and modelling missions, as shown in Table 6. Several
scenarios have been exercised in the operational analysis in
order to determine the role of space observations in the overall operational scene. Figure 2 shows a typical scenario for a
surveillance and monitoring mission. Spacecraft carrying the
primary sensors for oil spill detection will transmit data
through the "Tracking and data relay satellites" to White Sands,
where the data will be relayed via DOMSAT to the "Data processing center" envisioned for the ocean pollution.

Direct transmissions will be possible to the data processing center, while the ground station will be within the access
cone of the satellite. The recipients of data will be the
National Response Center and the corresponding National Response
Team, as well as the on-scene co-ordinator, who will direct the
entire operation. Coastal patrols will send data directly to
the on-scene co-ordinator, and in long-range missions they will

Table 6. Target missions and sub-missions.

Missions	Sub missions
Surveillance and Monitoring	Detection
	Mapping and tracking
	Quantification
	Pollutant classification
	Polluter identification
	Synoptic U.S. coastal pollution monitoring and data base building
	Synoptic global pollution monitoring and data base building
Modelling	Fate modelling
	Impact/risk modelling
	Synoptic oceanographic/meteorological ecosystem monitoring and data base building.

send data through DOMSAT. The data collection platform will send data to the data processing system for correlation with the

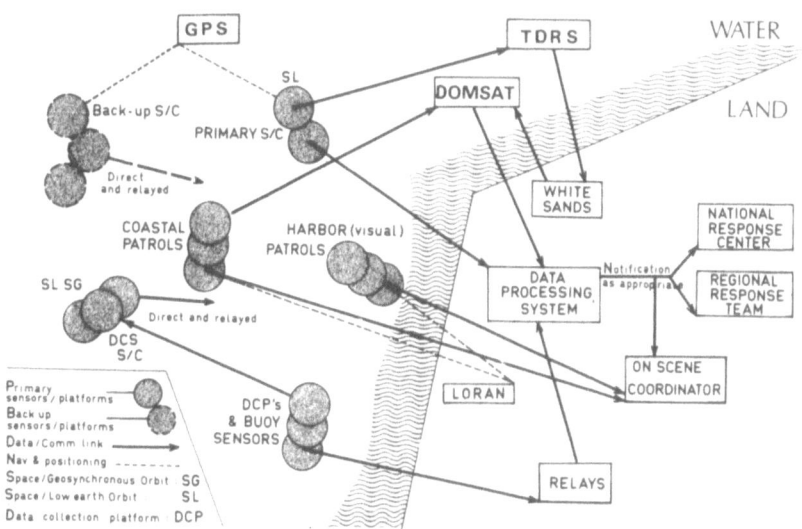

Figure 2.   Pollution surveillance and monitoring: scenario I.

remote sensing observations. Two important system concepts are
highlighted here; the multi-platform co-ordinated and co-operative observations focused on the data processing system, and
the near real-time capability that the system will be able to
afford through the use of relay satellites.

In assessing the duty cycle of the satellites in the performance of the ocean pollution mission, it is clear that the
capabilities of the space elements of the system will exceed
in great measure the requirements based solely on the coastal
areas of the United States. This suggest the desiderability of
eventually establishing a co-operative set of marine environment monitoring missions with participating countries to perform those measurements as required by the user organizations
in those countries.

RESULTS TO DATE

The following preliminary results are based on the status
of the study which is approximately 50% complete:

* Based on user requirements, an initial measurement specification has been synthetized and is in the process of
  being reviewed by the various user agencies;

* The combined space, air, and surface elements of the
  system are able to provide a large portion of the daily
  coverage within the 200-mile coastal zone;

* The emerging space elements of the system are predominantly an all-weather system employing microwave sensors,
  aided by selected optical sensors;

* The implementation of the space system elements will re-
  require significant development effort, particularly in
  the area of orbital testing of new sensor concepts;

* Technology gaps exist in the following areas: measurements of oceanic currents, the classification of oils in
  oil spill observations, the quantification of oil spills
  which requires the measurement of the thicknesses.

CAPABILITIES OF SEASAT SYNTHETIC APERTURE RADAR

FOR IMAGING THE OCEAN SURFACE

J.R. Apel[1] and W.E. Brown Jr.[2]

[1]National Oceanic and Atmospheric Administration
Pacific Marine Environmental Laboratory
Seattle, United States
[2]Jet Propulsion Laboratory
Pasadena, California, United States

The synthetic aperture radar (SAR) on the ocean-looking spacecraft, SEASAT, gathered approximately 1.5 x 10 km of 100 km-wide imagery prior to the satellite's demise on 10 Oct. 1978. The imagery processed to date has shown a wide variety of features on the ocean's surface, some of it of interest in mapping oil spills.

The radar characteristics are:

* Wavelength	25 cm
* Polarization	Horizontal
* Mean look-angle off nadir	20.5°
* Swath width	100 km
* Resolution	25 m
* Integration time	0.5 μs
* Peak power	1.13 kW
* Ocean Bragg wavelength	30 cm

The unusual choice of look-angle and wavelength appear to be near-optimum for imaging roughness variations of the ocean surface and ice on all-weather, day/night basis. The reason for this may be found in the probability distribution function for the slopes of small-scale roughness variations of the sea surface. This function, which is approximately gaussian, has an rms slope of $\langle S^2 \rangle^{1/2} = 15°$ (for wind speeds of about 5 m s^{-1}); which is near to the mean radar look-angle.

Smooth areas on the sea surface are specular reflectors

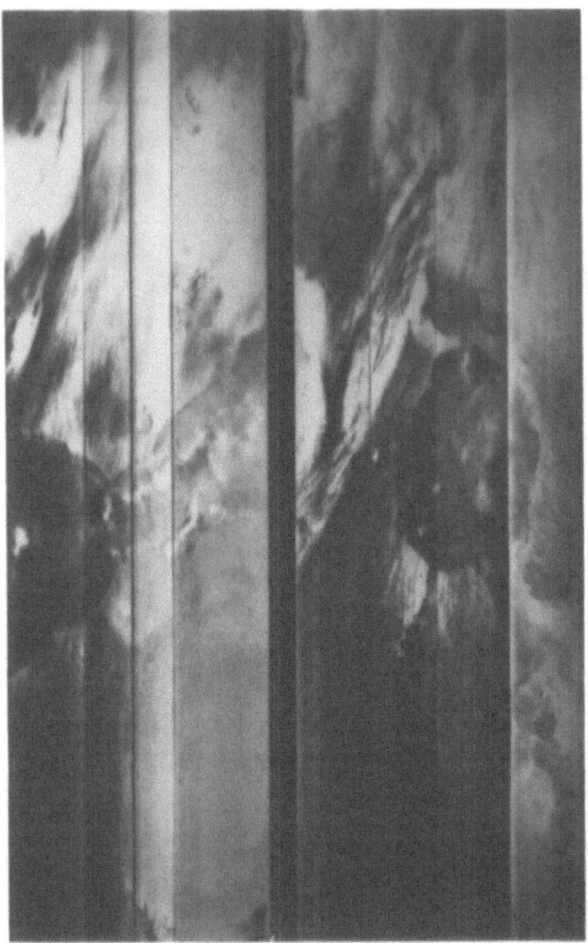

Figure 1. Two geographically adjacent **SEASAT SAR** images east
            of Palm Beach, Florida. The edge of the Gulf Stream
            is visible as long linear features tending NNE-SSW
            in the upper left-hand side. In the center of each
            pass are dough-nut-shaped squall regions showing
            rough regions (light) with small dark centers
            thought to be due to rainfall smoothing the sea
            surface. Quasi-periodic wave packets are due to
            internal waves. Each image is 100 km wide.

and scatter little energy back to the radar, while normally
rough seas backscatter considerable power because the rms slope
distribution is such that an appreciable portion of the surface
returns power to the radar. The result is that the depth of
backscatter modulation (or image contrast) is great, allowing

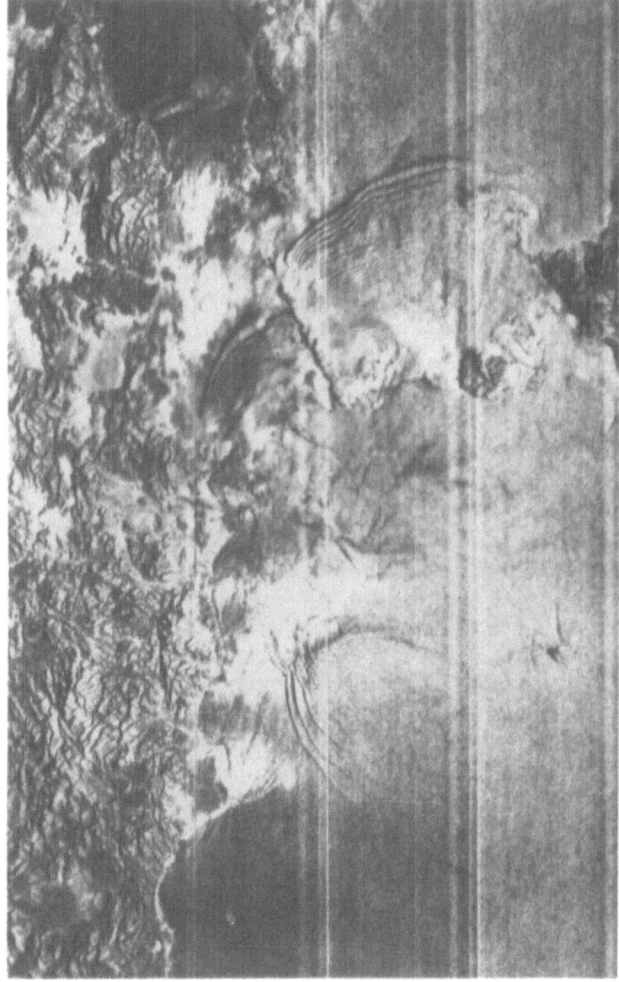

Figure 2.   SEASAT image of Baja California and the Gulf of
             California, showing three distinct groups of
             internal waves amidst an island archipelago. The
             waves are visible because of enhanced surface
             roughness existing over each of the oscillations
             within a group. Areas of light winds are dark in
             the image. Image is 100 km wide.

surface features to be seen readily. Such features include:
     * dominant length and direction of the ocean wave spec-
       trum for $125 \leqslant \lambda \leqslant 300$ m;
     * internal wave packets via roughness variations of the
       overlying surface;

* regions of low winds (or small roughness) which appear
  dark in the image;
* rainfall patterns;
* estuarine outflow patterns at sea;
* Gulf Strean boundaries;
* sea ice;
* ships, and probably oil spills, although the latter are
  not yet documented.

Figures 1 through 3 show some results obtained with the
SEASAT SAR imagery. The SEASAT SAR imagery shown in Figure 3
covers an area of approximately 100 by 100 square kilometers
over part of the southern California coast and the Pacific
Ocean. The upper left corner is the Santa-Anna region located
40 kilometers south-east of Los Angeles. The near vertical
streets mark the north/south direction of the ground surface.
Lake Elsinore and the Elsinore fault are in the top right part
of the image. The coast line extends from right east of Long
Beach to about 30 kilometers north of San Diego. Several ships
are visible in the ocean. They are clearly detected from their
wakes. The surface waves, observed in many parts of the image,
have a predominant wavelength of approximately 280 meters, and
are propagating in a south-north direction. Localized wave
trains of much longer wavelength are seen in the left center
part of the image. The abrupt change in the brightness of the
ocean could be due to wind effects. The near vertical line in
the center of the image is due to radar calibration codes
superimposed on the radar echo signals, and is parallel to the
satellite track. Radar illumination is from right to the left.

This particularly interesting image was acquired by the
SAR on September 25, 1978 (Rev 1291) at local time of 0:12 AM.
The image has a resolution of approximately 25 meters and has
4 looks. It is the first 100 x 100 km image frame processed
digitally at the Jet Propulsion Laboratory, using the Interim
Digital Processor (IDP) developed under NASA sponsorship.

Current throughout capability of the IDP is approximately
one SAESAT SAR image frame (100 x 100 km) per ten hours of
processing time.

From the characteristics of these images and from our
understanding of the imaging mechanisms at work on the sea
surface, it is felt that S-band SAR may be a prime detector
of oil spilled on the sea whenever the oil suppresses short
gravity waves appreciably.

Figure 3.   SEASAT SAR imagery of the Southern California coast
            and the Pacific Ocean (Santa-Anna region)
                     Image courtesy of
                     Radar Science and Engineering Section
                     Jet Propulsion Laboratory
                     California Institute of Technology
                     National Aeronautics and Space Administration
                     Pasadena, California, United States.

A LARGE SCALE MONITORING OF THE HYDROCARBONS POLLUTION

FROM THE LANDSAT SATELLITE

L. Wald[1], J.M. Monget[1], M. Albuisson[1],
and H.M. Byrne[2].

[1]Centre de Télédétection et d'Analyse des Milieux
Naturels
Ecole Nationale Supérieure des Mines de Paris
Sophia Antipolis, France
[2]National Oceanic and Atmospheric Administration
Pacific Marine Environmental Laboratory
Seattle, Washington, D.C., United States

ABSTRACT

The LANDSAT satellites have already been used in certain
instances for the remote sensing of oil spills at sea. The
detection of oil is mainly due to the variations of the
glitter radiance between the sea and the oil spill. This property has systematically been used in the framework of the
European "Archimedes" project, managed by the Joint Research
Centre (Ispra, Italy) in order to study the pollution in the
Mediterranean Sea. Eight hundred LANDSAT images were examined
for the period 1972 to 1975. After various corrections, mainly
allowing for the solar illumination angle, the cumulative area
covered by the hydrocarbons spread each year in the Mediterranean Sea is crudely estimated to be 175,000 km^2.

The usefulness of LANDSAT for the large scale monitoring
of oil pollution at sea as well as for the determination of a
pollution reference level is demonstrated.

For the time being, LANDSAT is the only satellite allowing the large scale inventory of oil spills and it could be
usefully used as a reference for evaluating the effects of
future conservation plans of our marine environment.

INTRODUCTION

Several authors have already reported on oil spills sensed by LANDSAT (Stumpf and Strong, 1974; Otterman et al., 1974; Deutsch et al., 1977; Albuisson and Monget, 1980; Deutsch and Estes, 1980). LANDSAT was in particular used for the tracking of oil slicks during the Campeche Bay oil-well blow out in 1979 (Hayes, 1980; Deutsch et al., 1980).

All these authors have demonstrated that LANDSAT is able to sense major oil spills at sea. This detection ability is explained by the difference in the reflection factor between an oil spill and the surrounding sea surface. However, LANDSAT has limitations when refined properties of oil spills are sought. The multispectral scanner MSS aboard LANDSAT does not provide information on the oil type, its thickness and its thermodynamic state. Moreover, LANDSAT cannot separate an oil spill from a monomolecular film of organic origin (Hühnerfuss, Garrett, 1981). Wald and Monget (1982) have also stressed the influences of the sea-state and of solar illumination angle on LANDSAT signatures.

Despite all these drawbacks, LANDSAT is presently the only sensor providing routine knowledge on statistics and location of the oil pollution level at sea. This is why it has been systematically used over the Mediterranean sea for a cartographic assessment of the most polluted areas.

VIEWING OF OIL SPILLS BY LANDSAT

The difference in the reflection coefficients between the spill and the surrounding sea allows the detection of the spill.

For detecting such a phenomena, the LANDSAT-MSS 7 channels is most often used, due to its near infrared spectral band (0.8 to 1.1 micron). In this spectral window, the measured reflectance can be considered as the sum of the atmospheric diffuse reflectance and the sun glitter reflectance after attenuation by the atmospheric absorption. The glitter reflectance is a function of the reflection coefficient of the sea surface, which can be expressed in terms of sea-state. In the MSS 7 LANDSAT imagery, the glitter reflectance of a rough sea is greater than that of a calm sea. Wald and Monget (1982) have shown that the reflectance contrast between a calm and a rough sea as measured by the LANDSAT-MSS 7 depends strongly on local time of observation with a maximum during the summer solstice (June-July) and a minimum during the winter solstice (December January).

The optical properties of the oil spill at sea are very difficult to study. They vary with the chemical composition of the spill, with its thickness, age, and more generally with its thermodynamic state. These parameters are a function of meteorological and oceanic conditions.

Goldman and Horvath, in 1975, outlined that the reflection factor of a thin oil film is the same as water and that it increases with thickness. On the other hand, Deutsch et al.

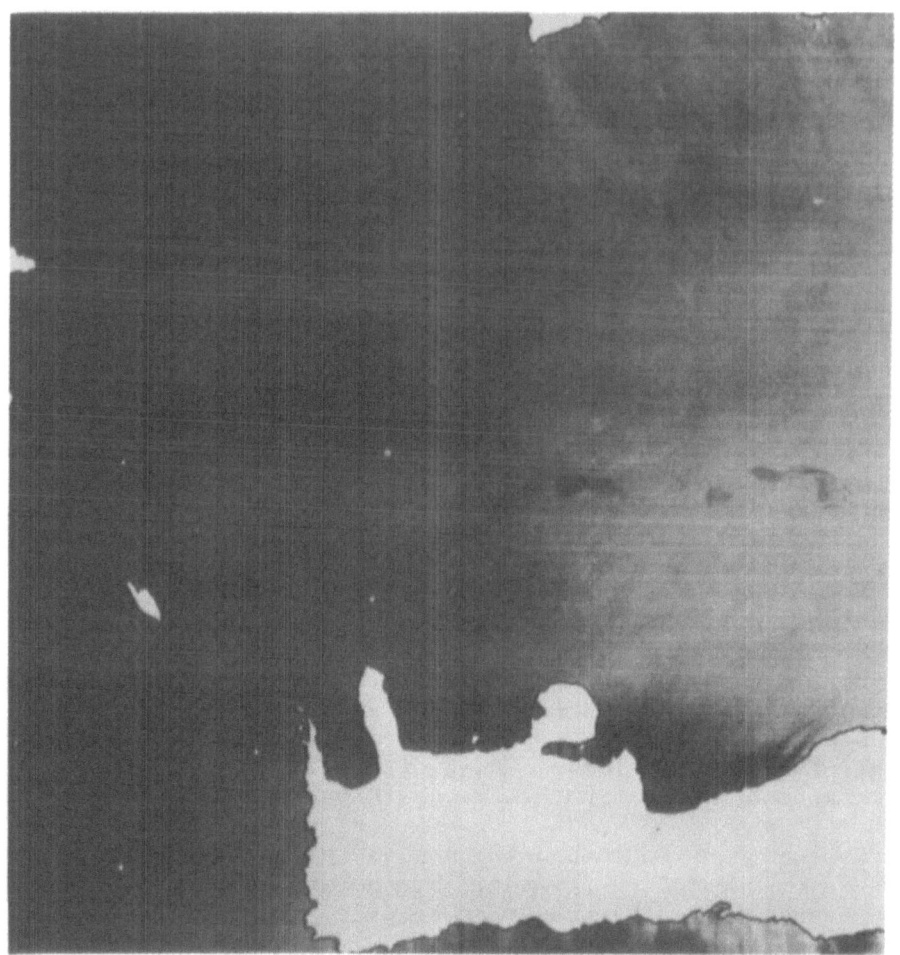

Figure 1.   Oil spills mapped by LANDSAT in east Mediterranean Sea. This MSS 7 image is centered on 36-00 N and 24-00 E and obtained on July 8, 1975; Crete is at bottom. Three major spills are clearly visible. Wind is blowing from the northwest quadrant.

(1980) suggest that oil-water emulsions of more than 50% of
water have a reflection factor greater than that of water.
Such emulsions are called "chocolate mousses", and have a solid
or semi-solid grease-like state. On contrary, emulsions with
less than 50% of water have a reflection factor which is equal
to that of water.

The thickness of a new oil spill is not constant.
Hollinger and Mennella have shown in 1973 that in this case 90%
of the oil is contained in less than 10 percent of the area
covered by the slick, the thicker region being in the lee-side
portion of the spill. An example of such a spill is presented
in Figure 1, on which the higher radiances are in white and the
lowest in black. The rightmost spill in the photograph shows
a bright region, because of the high reflection of the thick
oil. As the thickness decreases windward, the radiance decreases along side the wind direction and the thinnest part of
the spill displays the same radiance as a calm sea area, and
thus appears in black. Depending of the chemical composition
of oil, this part is composed of a very thin oil film, or of
an oil-water emulsion with less than 50 percent of water
(Berridge et al., 1968). Either way, the reflection factor is
very close to that of pure water.

Apart from its optical properties, oil changes the reflection of light because of its influence on the sea-state.
Surface films dampen the capillary waves, which are responsible
for the diffuse reflection of the light, because of the following processes: decrease of the water surface tension, modification of the wind stress and of wave-wave interaction. The thin
oil films and the oil-water emulsions other than mousses can
thus be considered equivalent to calm sea areas; they are
darker on the LANDSAT image than the rougher surrounding waters.

During the weathering of the spill, its thickness becomes
constant, and reaches an equilibrium value of about 1 micron
(Pilon and Purves, 1973). The radiance reflected by the spill
is uniform and equivalent to that of a calm water. Thus, oil
spills appear in dark in the images (Figures 1 and 2).

Some authors recommend the use of the multispectral capabilities of the MSS radiometer, in order to measure the reflectance backscattered by the upper layer of the sea (Deutsch and
Estes, 1980; Goldman and Horvath, 1975). Because the reflectance backscattered by natural waters exceeds that of oil
spills, the use of the four channels of the MSS allows to
better detect the spills and to minimize the limitations due
to the geometry of reflection induced by the use of the only
MSS 7 channel. However, such a method needs computer data processing capabilities which have not been used in this project,

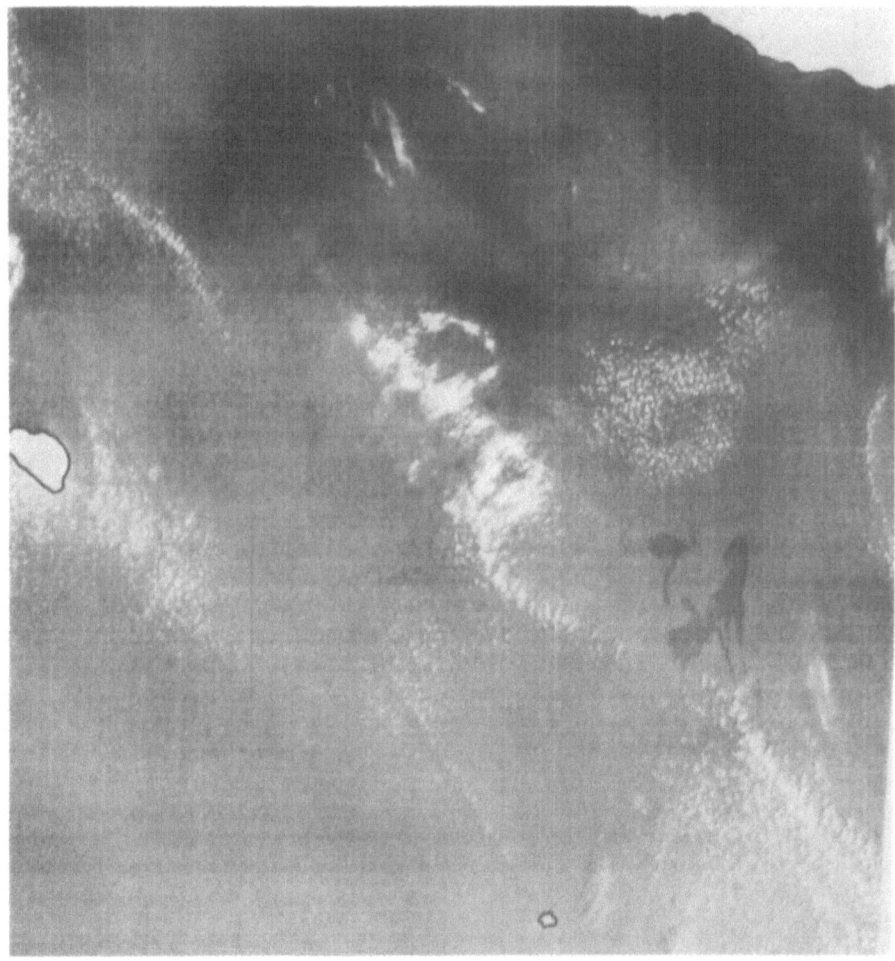

Figure 2.   Oil spills mapped by LANDSAT in central Mediterranean
             Sea. This MSS 7 image is centered on 36-30 N and
             13-00 E, and obtained on August 27, 1972. Sicilia is
             visible on upper right, Pantellaria island on middle
             left and Linosa island at bottom. The cumulative
             surface of the two oil spills amounts to 150 km^2.
             This image has also been presented by Deutsch et al.
             (1977).

due to reasons of costs because 800 images had to be interpreted
(Albuisson et al., 1981).

THE LANDSAT IMAGERY PROCESSING

      Three low-reflectance objects are distinguished in the MSS

images of the sea: cloud shadows; wind shadows; thin surface films.

Cloud shadows are easily interpreted because of the systematic association between clouds and their shadows in heliosynchronous imagery.

The second type of low-reflectance areas are due to local variability of the sea-state induced by the wind with most often an orographic origin. The resulting black signatures are usually connected to the coastlines and their outer limits are defined by diffuse boundaries.

The third type comprises both man-made and organic films. To separate these two kinds of films, a size criterion was used. Surface films of large size were interpreted as petroleum films while films of narrow transverse dimension (a few pixels) are associated to organic films. Although Hühnerfuss and Garrett (1981) reported on organic films of great size, this hypothesis is a posteriori justified by the statistical results obtained by this study.

In order to assess the large-scale oil pollution in the Mediterranean Sea, 768 LANDSAT images were examined, covering the periods August 11, 1972/November 18, 1973, and January 29, 1975/January 22, 1976. Albuisson et al. (1981) showed that this data set was a good estimation of the state of the oil pollution. Examples of such images are presented in Figures 1 and 2.

The data set reveals a great number of spills, most often of small thickness and which are mainly due to discharges of oil residues by tankers. This is in agreement with the results of Bertrand (1979, 1981), who outlined that large accidental pollution by tankers or platforms is rare in the Mediterranean Sea.

INVENTORY RESULTS

For the whole Mediterranean Sea, the number of available images per month is presented in Figure 3. The horizontal dotted line represents a statistical significance level computed by Albuisson et al., 1981. The monthly number of available images varies widely from 4 in April to 100 in June and November. It is too small during the first part of the year, so that significant statistical results can only be derived during the summer and fall seasons.

For each month, the percentage of images displaying pollution evidence is computed. An estimate of the pollution level is given by the measurement of the mean surface of the observed

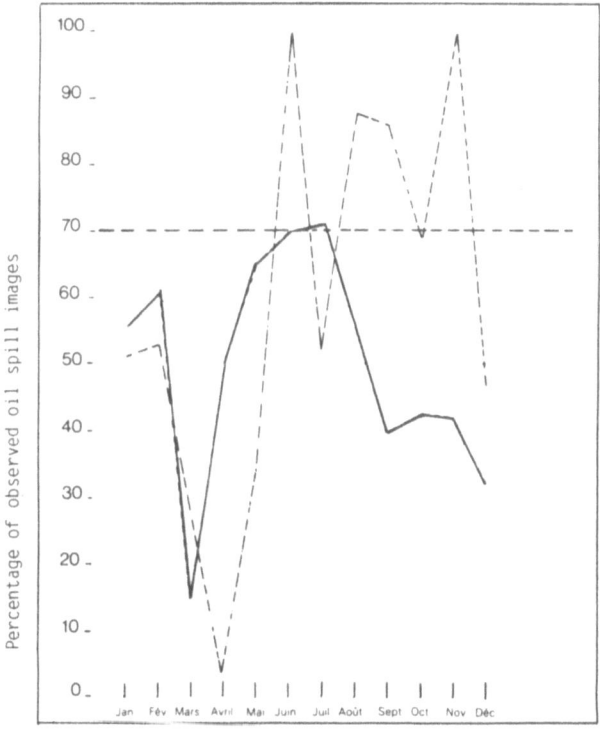

Figure 3.   Global statistical results on level of oil pollution
             variation in time. Number of processed images per
             month (dotted line) and monthly percentage of images
             displaying oil spills (full line) for the whole Med-
             iterranean Sea.

spills (Figure 3). The percentage of pollution evident images
is highly variable during the first five months, because of the
improper statistical significance of the number of available
images during this period. It reaches 70% in June-July, and
then drops to 40% in the fall. One should note that while the
monthly number of images remains relatively constant from June
to November, the percentage decreases during the same period.

    This decrease is in good agreement with the decrease of
the reflectance contrast between a calm and rough sea, induced
by the variations of the solar illumination angles along the
year (Wald and Monget, 1982). It illustrates how the use of
near infrared radiometer for the sensing of oil pollution at
sea leads to an important bias in the estimation of the number
of spills effectively present.

From the average cumulated surface of the films observed
in one image (100 km^2) and taking as a reference the percentage
of June-July, one obtains a rough estimation of the cumulative
area covered by the oil spread each year of the order of
175,000 km^2. This corresponds to 7% of the total area of the
Mediterranean Sea.

This cumulative surface cannot be linked in a satisfactory
manner to the mass of hydrocarbons spread each year. Indeed,
the quantities of hydrocarbons lost by evaporation or precipi-
tation account for more than 50% of the initial mass of the oil
spills, and these quantities cannot be estimated by LANDSAT
alone. Thus, the estimated cumulative area we propose is only
a qualitative indicator of the importance of the discharge of
oil residues by tankers.

However, it is interesting to compute a mass associated to
the selected spills, interpreted as aged petroleum spills.
Assuming a density of 0.9 and an oil thickness of 1 micron,
this aged spills mass reaches 160,000 tons.

It compares very well to the 150,000 tons of hydrocarbons
which U.S. National Academy of Sciences (NAS) estimates are
spread each year (In Goldberg, 1979).

Our estimation is computed from old oil spills, therefore

Figure 4.   Synoptic map of oil spills frequency.
            Map of the annual percentage of images
            displaying oil spills.

without taking into account the processes of evaporation, pre-
cipitation, dissolution, oxydation, bio-degradation and so on.
These processes have been considered by NAS; thus, it is most
probable that their evaluation must be increased in order to
be better compared to satellite results.

The new information brought by this LANDSAT inventory is
certainly not in this global estimation figure, but rather in
the production of regional cartography of oil pollution density.
The evolution in time of the pollution state of a particular
geographic area can thus be studied, allowing a better decision
in protection scheme as well as providing a means of assessing
the efficiency of particular enforcement projects.

The annual percentage of LANDSAT images showing evidences
of pollution is presented as a map on Figure 4, where symbols
have been centered at nominal image center coordinates.

This document shows that the more heavily polluted coastal
waters are particularly located along the industrial countries
of the western Mediterranean basin. Highest pollution is well
correlated with areas of heavy tankers traffic (detroits,
import and export harbours).

Two areas are tolerated for oil tankers residue discharges
by the 1962 amendment to the 1954 International Agreement,
because they are located at more than 100 miles from coastlines.
The first one is situated off Sicily and Libya, and shows heavy
signs of pollution; but the second one, off Turky and Egypt,
looks more clean. This may be caused by the non-operation of
the Suez canal during the period of analysis.

This regional results are in good agreement with the maps
published by the Regional Centre for Pollution Control (ROCC)
situated in Malta, as well as the measurements done by Zsolnay
(1979), thus giving more confidence in the reliability of the
LANDSAT cartographic inventory.

CONCLUSION

Statistical evaluations of the level of oil pollution in
the Mediterranean Sea have been derived from the interpretation
of the LANDSAT imagery. The importance of the discharges at sea
have thus been assessed leading to an updating of the estima-
tions previously made by NAS. A map showing the areas of
chronic high pollution level has been produced. Future monitoring of these areas should demonstrate the level of efficiency
of protection laws and oil pollution surveillance strategy.

From a methodological point of view, it has been shown

that LANDSAT provides an useful detection of oil spills. For
the first time, these satellites have been used on large-time
and space scale for such studies.

However, the use of LANDSAT for pollution monitoring is a
much debated question. The ad-hoc Working Group of the European
project "Archimedes" and the Working Group II of the French
pilot-study for the NATO Committee on the Challenges of Modern
Society recently concluded that LANDSAT was inadapted to the
routine monitoring of oil pollution, because of cloud-free requirements and of the long repeat period.

In fact, these remarks are relevant only if LANDSAT is
only used for surveillance.

On contrary, we propose to use LANDSAT (and further SPOT)
for statistical studies of oil pollution. Both pre-cited
working groups have underlined the ignorance of a reference
level, and, a fortiori, the evolution of this level. They have
also stressed the importance of the knowledge of this level,
because it is the only way to accurate modelling and predicting
of oil pollution.

Now, this level can be reached by LANDSAT. Archives of
LANDSAT imagery exist since 1972, and from their systematic
study a data bank can be built for the synoptic and local
surveys of the pollution state.

At last.the authors think that until the development of
better adapted oil detection sensors, LANDSAT is a low-cost
effective tool which should be considered.

ACKNOWLEDGEMENTS

The authors thank Mr. Bonnaure, Mr. Fraysse and
Mr. Hühnerfuss from Joint Research Centre, Ispra, for their
criticisms and suggestions.

They are very grateful to Dr. John Apel, Director of the
Pacific Marine Environmental Laboratory of NOAA, Seattle, for
letting them use the basic LANDSAT data necessary for this work
and giving them the necessary expertise in space oceanography.

This work was supported by the Joint Research Centre of
the Commission of the European Communities in Ispra, Italy.

The opinions or assertions expressed herein are those of
the authors and are not officially endorsed by the Commission
of the European Communities.

REFERENCES

Albuisson, M., Monget, J. M., 1980, Détection de la pollution
de surface en Méditerranée par le satellite Landsat,
in: "Proc. 26th C.I.E.S.M. meeting, Antalya, Turkey".

Albuisson, M., Monget, J. M., Wald, L., 1981, "Etude de l'In-
ventaire par Télédétection de la Pollution de Surface
en Méditerranée", Contract C.C.R. Ispra No. 1389-80-10

Bertrand, A. R. V., 1979, Les principaux accidents de déversements pétroliers en mer et la banque de données de
l'Institut Français du Pétrole sur les accidents de na-
vires (1955-1979), Revue de l'Institut Français du Pétrole, Rueil-Malmaison.

Bertrand, A. R. V., 1981, La banque de données de l'Institut
Français du Pétrole sur les déversements accidentels de
petrole en mer (1955-1980): nouveaux résultats, Revue
de l'Institut Français du Pétrole, 36, 2:229

Berridge, S. A., Thew, M. T., Loriston-Clarke, A. G., 1968,
The formation and stability of emulsions of water in
crude petroleum and similar stocks, Journ. Institute
of Petroleum, 54, 539:335

Deutsch, M., Strong, A. E., Deutsch, J. E., 1977, Use of
Landsat data for the detection of marine oil slicks,
in: "Proc. Offshore Technology Conference".

Deutsch, M., Estes, J. E., 1980, Landsat detection of oil from
natural seeps, Journ. American Soc. Photogrammetry, 66,
10:1313.

Deutsch, M., Vollmers, R. R., Deutsch, J. P., 1980, Landsat
tracking of oil slicks from the 1979 Gulf of Mexico oil
well blow-out, in: "Proc. 14th Intl. Symposium on
Remote Sensing of the Environment, San Jose, Costa Rica,
23-30 April 1980".

Goldberg, E. D., 1979, "La Santé des Océans", UNESCO, Paris.

Goldman, G. C., Horvath, R., 1975, Oil pollution detection and
monitoring from space, I.E.E.E. Ocean' 75.

Hayes, R. M., 1980, Operational use of remote sensing during
the Campeche Bay oil well blow-out, in: "Proc. 14th
Intl. Symposium on Remote Sensing of Environment, San-
Jose, Coasta Rica, 23-30 April 1980".

Hollinger, J. P., Mennella, R. A., 1973, Oil spills: measurements of their distributions and volumes by multifrequency microwave radiometry, Science, 181:54.

Hühnerfuss, H., Garrett, W. D., 1981, Experimental sea slicks:
their practical applications and utilisations for basic
studies of air-sea interactions, J. Geophys. Res., 86,
C1:439.

Otterman, J., Ginzburg, A., Ohring, G. and Mekler, Y., 1974,
Observations of water, air and soil pollution in Israel
and vicinity from the ERTS-1 imagery, Water, Air and
Soil Pollution, Vol. 3, 1:53.

Pilon, R. O., Purves, C. G., 1973, Radar imagery of oil slicks, I.E.E.E. transactions on Aerospace and Electronic Systems, Vol. AES-9, 5:630.

Stumpf, H. G.,and Strong, R. E., 1974, ERTS-1 views on oil slick, Remote Sensing of Environment, 3:87.

Wald, L.,and Monget, J. M., 1982, Remote sensing of the sea state using the O.8-1.1 micron channel, Intl. Journ. of Remote Sensing, (in press).

Zsolnay, A., 1979, Hydrocarbons in the Mediterranean Sea, 1974-1975, Marine Chemistry, 7:343.

THE FIRST EUROPEAN REMOTE SENSING SATELLITE (ERS-1):

OVERALL DESCRIPTION, POTENTIAL APPLICATIONS AND USERS

G. Duchossois
ERS-1 Mission Manager

European Space Agency
Paris, France

ABSTRACT

On 28 October 1981, the Member States* of the European
Space Agency (except Ireland), joined by Norway and Canada,
decided to proceed with the detailed design phase of the first
ESA Remote Sensing satellite (known as ERS-1). This paper
describes the overall ERS-1 system, its present status and the
various potential applications which may benefit from ERS-1
data.

INTRODUCTION

On 28 October 1981, the Member States* of the European
Space Agency (except Ireland), joined by Norway and Canada,
decided to initiate the first ESA Remote Sensing satellite
(known as ERS-1) programme. The detailed design phase, called
phase B, is being managed by the Earth Observation Department
of the Agency's Applications Directorate and is being executed
by an industrial consortium of European and Canadian companies
led by Dornier System of the Federal Republic of Germany. This
phase B was started in August 1982 and will be completed in
June 1983. The construction phase (Phase C/D) will start at
the end of 1983 aiming for launch at the end of 1987. The
currently foreseen schedule is presented in Figure 1.

---

* ESA MEMBERS STATES include Belgium, Denmark, Germany, France,
Ireland, Italy, Netherlands, Spain, Sweden, Switzerland,
United Kingdom. Austria, Canada and Norway are associated
members.

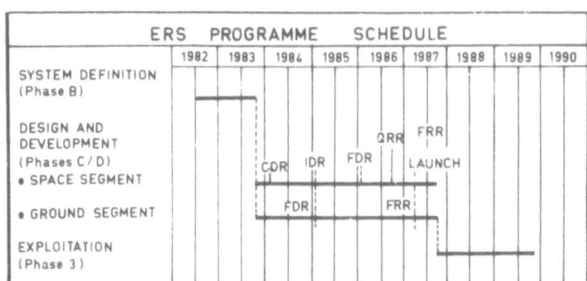

Figure 1. ERS programme schedule
(CDR: Critical design review; IDR: Intermediary design review;
FDR: Final design review; QRR: Qualification result review;
FRR: Flight readiness review).

The main objective of the programme is to give Europe the
ability to take part in both the management of the earth's resources and the monitoring of its environment.

ERS-1 aims primarily to:

* establish, develop and exploit the coastal, ocean and
ice applications of remote sensing data. Furthermore, all
weather high resolution imaging capability over land with the
synthetic aperture radar (SAR) will also provide useful data
as a complement to optical data provided by other satellites
such as Landsat and Spot.

* increase the scientific understanding of coastal zones
and global ocean processes which together with the monitoring
of polar regions, will provide a major contribution to the
World Climate Research Programme. ERS data used alone, or more
commonly in conjunction with complementary data from buoys,
radio-sondes, research vessels, other satellites in pre-arranged
global or regional experiments, will enable significant advances
to be made in physical oceanography, glaciology and climatology.

ERS-1 data will hopefully be utilised for a wide variety
of applications (scientific and economic) and by different communities of users.

The paper describes the overall ERS-1 system, its present
status, the requirements of potential users and the status of
various applications which may benefit from ERS-1 data. This
paper contains much information drawn from papers on specific
aspects of the project and the reader is invited to consult
these for a more complete picture (see references).

THE ERS-1 PAYLOAD

Priority in the payload has been given to a comprehensive set of radar instruments designed to observe the surface wind and wave structure over the oceans.

The instruments consist of:

* A C-band wind scatterometer designed to measure wind speed and direction;
* A Ku-band radar altimeter with the aim of measuring significant wave height and wind speed at nadir and of providing measurements over ice and major ocean currents;
* A C-band synthetic aperture radar to take all-weather high resolution images over polar caps, coastal zones and land areas. It will also be operated in a sampled mode with a reduced power level over oceans as a wave scatterometer with the aim of measuring the wave spectrum.

In addition to these priority instruments, the following elements are included:

* Laser retroreflectors for accurate tracking of the satellite and radar altimeter calibration;
* An along track scanning radiometer completed with a two-frequency microwave nadir sounder (ATSR-M) to measure sea surface temperature and to provide information for the "wet atmosphere" correction for the radar altimeter;
* A precise range and range rate experiment (PRARE) to provide high accuracy tracking information in support of radar altimetry for ocean circulation studies.

The last two items on the above list have been included as the result of an announcement of opportunity. A definite decision regarding their inclusion in the final payload complement will be taken at the end of the present design phase.

In order to reduce the required mass, volume and cost, the wind scatterometer and the SAR will be combined into one hardware package known as the C-band active microwave instrument (AMI).

Each of the above instruments is described below and parameters values are given in Tables 1 through 9.

The Active Microwave Instrument (AMI)

SAR Mode: The characteristics of the AMI-SAR mode are listed on Table 1.

In the SAR mode, the AMI will provide raw data for high resolution imagery, collected in C-band over a swath of at least 80 km. The system will consist of a fully-calibrated set of coherent RF electronics generating 300 W mean RF power which will be used to illuminate the earth via a planar antenna 10 m x 1 m constructed from metallised carbon-fibre waveguides.

The 37 μs transmitted pulse will be received and digitised to provide a signal coded to 5 bits I and 5 bits Q. This signal will then be transmitted to the ERS-1 ground segment where a digital processing facility will generate the final image products. The required pulse compression will be performed within this ground segment facility.

Wave Mode: In the wave mode, the SAR will be operated with a shorter pulse length (12.3 μs) to reduce the mean RF power to 100 W. Furthermore, the instrument will be operated intermittently to collect only sufficient data to provide an image of 5 km x 5 km size each 100 km along the track. This mode has been incorporated for the following reasons:

* to reduce the power consumption to enable global operation to be achieved;
* to collect imagery of sufficient size to facilitate study of ocean wave image spectra;

Table 1. ERS-1 AMI SAR mode characteristics.

Bandwidth	19 MHz
PRF (nominal)	1700 Hz
RF frequency	5.3 GHz
RF peak power	4.8 GHz
RF mean power at output of power amplifier	300 W[2]
Pulse length	37 μs
Compressed pulse length	64 μs
Spatial resolution	30 m x 30 m
Radiometric resolution at min of	2.5 dB − 18 dB
Orbital height (nominal)	777 km
Incidence angle (nominal)	23°
Swath width	80 km[1]
Raw data quantisation	5I + 5Q
Antenna length	10.0 m
Antenna height	1.0 m

[1]there is a goal to increase this to 100 km.

    * to reduce the data rate to enable on-board recording to
      be achieved;
    * to provide an intermittant mode which may be operated
      at the same time as the wind mode.

Wind Mode: In the wind mode, the AMI whose characteristics
are given in Table 2 acts as a 3-beam, C-band scatterometer
with geometry as shown in Figure 2.

The system will operate in vertical polarisation and
collect data from a 500 km swath. A special feature of the
ERS-1 scatterometer is that it has been designed to operate on
a yaw steering spacecraft, that is, the mid-beam has been designed to always point perpendicular to the satellite's ground
velocity factor. This feature ensures a minimum range of incidence angles (and hence minimum power requirement) needed to

Table 2.   ERS-1 AMI wind mode characteristics.

ERS-1 AMI Wind mode		
Frequency	5.3 GHz	
Polarisation	VV	
Peak power (RF)	4.8 kW	
Mean power (RF)	54 W	
Min. signal-to-noise ratio per cell	0 dB (for each pulse)	
No. of beams	3	
Orientation of beams[2]	0°	$\pm$ 45°
Pulse length	70 μs	130 μs
No. of pulses per 50 km	256	256
Pulse repetition interval	4.347 ms	4.878 ms
Spatial resolution	50 km	
Sample spacing (pixel size)	25 km	
Swath (one sided)	400 km (500 km)[1]	
Wind speed		
Range	4 ms^{-1} to 24 ms^{-1}	
Accuracy	2 ms^{-1} or 10%	
Wind direction		
Range	0-360°	
Accuracy	20°	
Antenna size	2.5 m x 0.3 m	3.6 m x 0.3 m

[1] data conforms to full specification only over 400 km swath
[2] 0° = perpendicular to the satellite ground track.

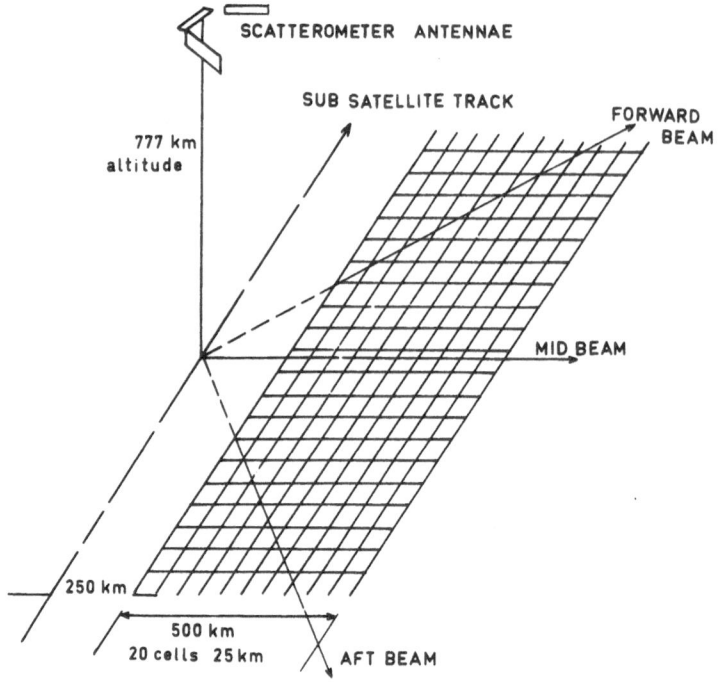

Figure 2.   Scatterometer geometry.

cover the 500 km swath, and additionally ensures that the two
45° beams operate under identical geometric conditions, and
conditions which do not vary with orbit position.

Each backscatter measurement is made with a spatial resolution of approximately 50 km, and the spacing between measurements is 25 km in each direction (along track and cross track).

Interleaved Wind/Wave Mode: In this mode, the wind mode
and the wave mode are operated simultaneously. Natural gaps in
the timing cycle of each mode enable this to be achieved without changing the format or quality of data from either mode.

The Radar Altimeter

The characteristics of the ERS-1 radar altimeter are given
in Table 3.

This altimeter will operate in Ku-band with a pulse of
20 µs which will be processed using the full de-ramp principle
as used in SEASAT. A derivative of the split-gate tracker will
provide a tracking precision of 10 cm over the oceans and will

Table 3. ERS-1 Altimeter parameters.

Frequency	13.5 GHz
Bandwidth	400 MHz
Pulse length	20 μs
Peak RF power	50 W
PRF	1.0 kHz (approx)
Altitude measurement accuracy	10 cm (1σ, 1 sec)
Backscatter coefficient accuracy	$\pm$ 1 dB (1σ)
Significant wave height Measurement range	1 m to 20 m
Measurement accuracy	0.5 m or 10%
Tracking window (ocean mode)	64 gates x 2.5 ms each
Tracking window (ice mode)	64 gates x 10 ms each
Antenna diameter	1.2 m

have a mode specifically adapted to use over ice. The ice mode
will degrade the altimetric precision to 40 cm but will increase the tracking window to 25 m to reduce the probability
of losing lock over uneven ice targets. A further aspect of the
design specifically adapted to ice is the dynamic range and
linearity of the signal handling; these features have been
assigned special importance in order to increase the utility
of this instrument for ice observation.

The Along Track Scanning Radiometer

This element has been included for study as a result of
an announcement of opportunity. The instrument was proposed
by Rutherford and Appleton Laboratory in association with other
UK and French laboratories. Its characteristics are shown in
Table 4.

The ATSR-M has been conceived and designed to provide the
following types of data and observations:

* Sea surface temperature with an absolute accuracy of
  better than $\pm$ 0.5°K with a spatial resolution of 50 km
  and in conditions of up to 80% cloud cover;
* Images of sea surface temperature with 1 km resolution
  and 500 km swath;
* Radar altimeter neutral atmosphere range correction to
  better than 5 cm;

Table 4. Along Track Scanning Radiometer-M characteristics.

Radiometer	
Spectral channels	3.7, 11 and 12 μm
Spatial resolution	1 km x 1 km square
Radiometric resolution	0.1°K
Absolute accuracy predicted	0.5°K over 50 km x 50 km square in 80% cloud cover conditions
Swath width	500 km
Data rate	205 kbps
Mass	33 kg
Power consumption	48 W
Microwave Sounder	
Channels (2)	23.8 and 36.5 GHz
Instantaneous FOV	22 km
Predicted accuracy	2 cm
Mass	21.5 kg
Power consumption	30 W

* Observations of clouds, aerosols haze and total water
  vapour content of the atmosphere.

Schematically, the instrument consists of two-principle
components, the infrared radiometer and the microwave radiometer, which share common interface facilities to the ERS-1
payload assembly.

The infrared instrument is an imaging radiometer having
3 co-registered channels. The channels are at 3.7 μm, 11 μm
and 12 μm wavelength, defined by beam splitters and multilayer
interference filters. The instantaneous field of view (IFOV)
at the nadir on the earth's surface is a 1 km x 1 km square
which is imaged onto the detector element via an F/2.3 off-axis
paraboloid mirror. The IFOV is scanned over the earth's surface
by a rotating plane mirror in such a way that it gives two
earth views (0° and 57° to the nadir), and also incorporates
views of on-board black bodies between the earth views in each
scan. One of the black body views is periodically replaced by
a space view via a hinged mirror. The mirror rotation rate is
chosen so that each scan is contiguous with the previous. The
final data product is average SST, with better than 0.5°K absolute uncertainty under most conditions, over 50 km x 50 km
areas each containing 2,500 pixels within the 500 km wide swath.

For the cloud histogram technique to work successfully, at least 500 pixels must be cloud-free (i.e. up to 80% cloud cover is allowed).

For the atmospheric correction to be determined to the desired accuracy, the uncertainty in the average brightness temperature over the cloud-corrected 50 km x 50 km areas must be <0.05°K. To obtain the necessary signal resolution performance, the detectors chosen are single-element photo-conductive HgCdTe cooled by a Stirling cycle mechanical cooler which is currently under development for space applications at Oxford University.

The microwave sounder consists of two channels at 23.8 and 36.5 GHz. The antenna, a Cassegrain, is limited to 50 cm for reasons of instrument accomodation, which results in a 3 dB beam width and footprint size of approximately 22 km. Calibration of the radiometer is achieved by an ambient load within the instrument, consisting of a terminated waveguide and a set of sky-horns viewing space.

The Precise Range and Range Rate Experiment

This instrument whose characteristics are given in Table 5 has also been included for study as the result of an announcement of opportunity.

The precise range and range rate experiment (PRARE) was conceived and proposed by the Technical University of Berlin and is sponsored by the German Federal Minister for Research and Technology.

The satellite equipment is to be compatible with the passenger payload facility to the ERS-1 platform and hence does not show up in the ERS-1 payload configuration.

The objectives of this equipment are to extend the ERS-1 mission towards geodetic and geodynamic applications, the most important target being to improve the radar altimeter mission

Table 5.   PRARE characteristics.

Up-link	10 MHz in the band 7.19-7.235 GHz
Down-link	10 MHz in the band 8.35-8.50 GHz
S-band down-link	1 MHz in the band 2.20-2.29 GHz
Predicted ranging accuracy	5-10 cm
Power consumption	35 W

towards global ocean circulation monitoring. It is recognised
that this requirement is very ambitious and is unlikely to be
attainable for the first mission. Thus, for ERS-1 the PRARE is
viewed as an experiment to demonstrate the feasibility of determining the satellite orbit radial component to a precision
of the order of 10 cm.

Although the final design of this equipment is not selected
at this time, as various options are still under study, it is
expected that it will have the following features:

Table 6. Main characteristics of the Multi-mission platform.

* Compatibility with sun-synchronous, circular orbits at
  altitudes between 600 km and 1,200 km.

* Compatibility with local time of satellite passes
  (ascending and descending node: 0800 to 1600 hours).

* Available power: 1.9 kW (beginning of life) provided by
  a solar generator.

* Power storage capacity: 4 x 23 ampere hours.

* Altitude control performance:

    . Pointing towards earth centre with yaw steering
      capability;
    . Stability on yaw: $1.1 \ 10^{-3}$°/s
    . Pitch and roll: 7.10^{-4}°/s
    . Accuracy   0.15° for all axes

* Altitude measurement accuracy better than 0.15°.

* Orbit control:

    . Parallel and perpendicular to the orbit plane
    . Maximum capability 580,000 Ns (300 kg of hydrazine)

* Satellite management by on-board computer with 20 kwords
  (16 bits) memory available for the payload management.

* Communication with ground:

    . Telemetry and telecommand via an S-band transponder
      compatible with ESA and NASA networks;
    . Transmitted power up to 200 mW
    . Data rate: 2 kbits/s.

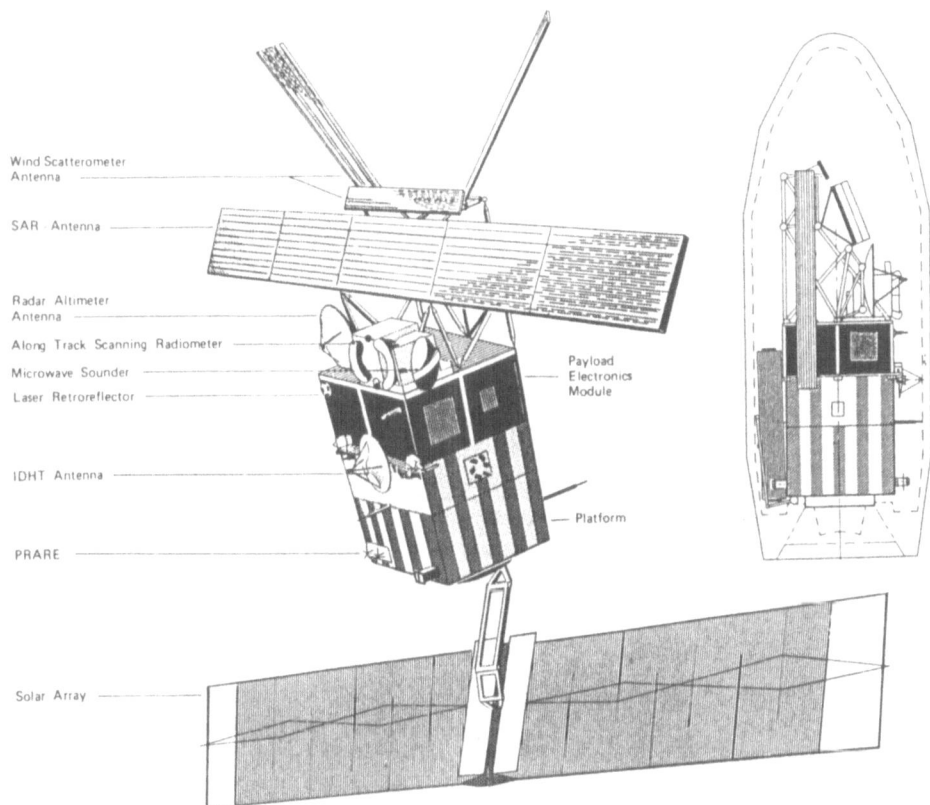

Figure 3. ERS-1 payload configuration.

* the satellite equipment will have low mass and power
  requirements;
* it will be a two-way X-band RF tracking system using
  pseudo-noise coding techniques. S-band one-way link be
  included to correct for the ionospheric errors;
* the ground-stations will be automatic with the necessary
  control commands being relayed via the satellite;
* all measurement data plus uplinked meteorological data
  for tropospheric corrections will be stored on-board to
  enable world-wide data to be assembled at a single processing centre.

SPACECRAFT

The ERS-1 spacecraft will use the multi-mission platform
(PFM) developed under the French national SPOT programme. The
principle characteristics of this platform are given in Table 6.

The platform will be configured to operate in a sun-synchronous orbit with a local hour of 10:15 (descending, equator
crossing). The altitude control will orient the yaw axis to
towards the earth's centre and the pitch axis will be oriented
along the instantaneous ground velocity vector (not, as has
been the case with previous earth observation satellites, along
the instantaneous orbital velocity vector). The use of an align-
ment along the ground velocity vector (a technique known as yaw
steering) has been selected to provide more favourable conditions for the operation of the AMI in wind mode.

The spacecraft configuration is shown in Figure 3.

ORBIT

ERS-1 will be launched from Kourou (French Guiana) by the
ARIANE launcher into a sun-synchronous, circular orbit with a
local time (descending node) of 10:15. This local time will be
maintained within 15 minutes. The main orbit parameters are
given in Table 7. The baseline repeat cycle will be three days,
but there will be sufficient fuel to enable the repeat cycle to
be changed several times within the three-year mission by means
of a small change of the orbit altitude. Other candidate repeat
cycles include 14 days and 26 days. In all cases, the stability
of ground track repeat cycles will be better than $\pm$ 1 km. The
nominal orbit altitude of 777 km has been selected as a compromise between the higher air drag of a lower orbit and the need
for greater power for the radar instruments at a higher orbit.

DOWNLINK AND GROUND SEGMENT

Three logical data streams are generated from the satellite
payload. They are:

* Format A: Direct read-out of the low data rate instruments (i.e. all instruments except the SAR in full imaging mode)
    Data rate: 800 kbits s^{-1}.

Table 7. Main characteristics of the ERS-1 nominal orbit.

Semi-major axis	a	7,153 km
Inclination	i	98.52°
Eccentricity	e	1.105 x 10^{-3}
Perigee Argument		90°
Nodal period	T	100.465 minutes
Repeat cycle		3 days
Orbital period per day		14 + 1/3

* Format B: Direct read-out of the raw SAR data (imaging mode of the AMI).
     Data rate: 100 Mbits s^{-1}
* Format C: Playback of the recorded data, consisting of the output of the low data rate instruments.
     Data rate: 15 Mbits s^{-1}

In each case, auxiliary data will be merged with the data stream to provide all data needed for ground processing. This auxiliary data will include parameters, such as orbit information, which have been provided to the spacecraft from the ground thus avoiding delays in ground data handling due to any need to merge data from different sources.

The downlink will be achieved in X-band using two distinct channels. Channel 1 will be for the format B, and channel 2 for formats A and C (simultaneously when needed). Details of the downlink are provided in Table 8. Integral with the ERS-1 project will be one ground station at Kiruna (Sweden) which will perform the following functions:

* Execution of Mission Management and Control functions;
* Reception of payload telemetry and other satellite telemetry, including the verification of the integrity of these telemetry links;
* Generation of near real-time products (see section " Product Definition" for definition) and their distribution to the user community;
* Onward transmission of all data to separate archiving facility.

It is planned that the Mission Management will be performed at the European Space Operations Centre (ESOC) at Darmstadt in Federal Republic of Germany. Commands and information will be routed between ESOC and ERS-1 via the Kiruna station.

Table 8.   ERS-1 downlink characteristics.

	Link 1	Link 2
Frequency band[1]	8.025-8.400	8.025-8.200
Modulation	QPSK	UQPSK
RF output power	20 W	20 W
Bit rate	100 MBs^{-1}	15 MBs^{-1}
Data margin[2] at 5°	2.7 dB	3.0 dB
Data margin[2] at 90°	4.0 dB	3.2 dB

[1]to be selected from within this band
[2]assuming ground stations G/T 31 dB/K

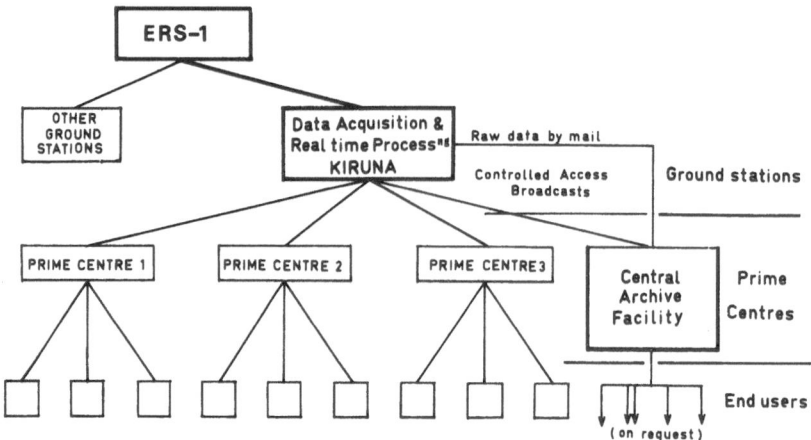

Figure 4.   ERS-1 data dissemination model.

     With a complex and power-limited satellite, none of the
above functions can be regarded as trivial, but the generation
of products and their dissemination to the users within three
hours of data reception represents a specific challenge. It is
intended to achieve this three-hour delivery time for the following products:

     * SAR "fast delivery" images;
     * wind speed and direction over a 500 km swath;
     * wind speed at nadir;
     * wave height at nadir;
     * wave image spectra.

     Clearly, this objective not only requires a suitable set
of processing facilities at Kiruna, but also a dissemination
method capable of handling large quantities of data, continuously and on an operational basis. Additionnally, a carefully
prepared operational method must be developed to ensure that
manual operations do not lead to unacceptable data delivery
delays.

     Current thoughts on data dissemination from Kiruna are
concentrated on a two-level system as shown in Figure 4. Examples of end users are: shipping companies, fishing companies,
oil industry (near real-time users). Examples of Prime Centres
are: national meteorological services, off-shore oil industry,
ESA central archive facility.

     The heart of this system is Controlled Access Broadcasting
via satellite to prime centres in member states with end users

being supplied via the prime centres. Advantages of this plan include:

* most end users already have links to prime centres (e.g. oil companies to meteorological services), and these links should be used, not bypassed,
* any end user with sufficiently urgent and important requirements may gain access to the controlled access broadcast, thus effectively becoming a prime centre.

Clearly, this plan will not satisfy those users which need, for example, raw SAR data, in real time. For those near real-time requirements not satisfied by the Kiruna station, participating nations must arrange direct reception from ERS-1.

Functions omitted from the list of tasks for Kiruna for less fundamental reason are an archiving facility and generation of "precision products", i.e. products which will not be generated in near real time, either because of the processing load involved, or because of the need for a posteriori data (e.g. refined orbit data for some altimeter products).

These two categories of facilities will be provided to the project in a manner as yet undefined. It is probable that one or more member state will wish to provide some precision-processing facilities, whereas it is also probable that ESA-EARTHNET (the ESA element responsible for the reception, processing and dissemination of remotely sensed data) will also conduct precision processing. Data archiving may be combined with such a precision-processing facility.

PRODUCT DEFINITION

At the present stage of the project, product definition is not complete. However, the following descriptions represent the current thinking for data product definition.

As previously stated, the ERS-1 project responsibility will only extent to generation and delivery of near real-time products and the delivery of all raw data to a separate processing and archiving facility. The subsequent (delayed time) processing of this raw data will not be covered by the ERS-1 project.

Near Real-time Products

This category of products will be generated at the Kiruna ground station. It will include:

Fast Delivery SAR Data: The facility will be capable of

Table 9. ERS-1 SAR fast delivery product.

Range resolution[1] r	40 m
Azimuth resolution[1] a·	40 m
Pixel size	/2
Image size (minimum)	80 x 80 km
Integrated sidelobes	- 9 dB
Radiometer resolution	(3 looks)
Intra image calibration }	System correction
Absolute calibration }	only
Location	2 km
	prediction only
Delivery time	3 hours
Production rate	12 frames/pass
	i.e. 8 mins/frame

[1]exact value to be optimised for the processor (e.g. convenient
FFI lengths).

producing 10 scenes of SAR data within 100 minutes of raw data
reception at Kiruna. These will conform to the description
given in Table 9.

Wind Scatterometer Data: The facility will produce wind
field maps; 500 km swath, 25 km grid, and deliver them to prime
centres within three hours of raw data reception in Kiruna.
Maps will be constructed without directional ambiguities, although it may be necessary to delivery two maps, one "correct"
and the other containing all of the 180° ambiguous measurements.
The customer then select the "correct" one based on existing
knowledge.

Radar Altimeter Data: Significant wave height measurements
and surface wind-speed measurements will be delivered within
three hours of collection of the raw data at Kiruna.

Wave Mode Data: The exact form of the data products from
this mode of the AMI are yet to be defined. The following
working assumptions have been made:

All wave mode data will be processed and distributed to
prime centres within 3 hours of reception of the raw data at
Kiruna. The processing will consist of the three main stages:

  * generation of 5 km x 5 km SAR images to the highest
    possible quality,
  * transformation of these images to spectra using a
    Fourier Transform,

* correction of these spectra to account for known modulation transfer function of the AMI and processor.

ATSR-M Data: At present, the processing of this data is
not included in the plans for the Kiruna facility. The present
agreement is to deliver tapes of raw data, annotated with satellite housekeeping and orbit data, to the sponsor of this
payload element. It is however, possible that this situation
will be revised, in which case sea surface temperature maps
500 km wide could perhaps be produced and disseminated within
three hours of data reception at Kiruna.

UTILISATION OF ERS-1 DATA: ERS-1 POTENTIAL USERS

ERS-1 satellite data will be utilised by two different
communities, one scientifically oriented and the other applications oriented, each having different requirements and characteristics.

Scientific Community

This community is at the origin of the development of
remote sensing techniques, sensors, basic physics understanding,
etc. It is research motivated and has no major operational requirements in terms of satellite data. It usually prefers to
handle annotated raw data used as inputs to models and algorithms, and has no requirement for short delivery times and
shipping by mail of tapes, so that computer compatible tapes
(CCT) would meet its needs.

Another important feature of this community is that it is
well organised through various structures (national and international) and used to meeting. It has a high educational level
with the aim to make progress in the scientific understanding
of remote sensing physics.

Applications Community

In principle, this is the end users community which would
use ERS-1 data on a continuous and operational basis for
various applications such as offshore petroleum activities,
ship routing, fishing, sea-ice monitoring, etc. In practice,
the community is extremely heterogeneous for the following
reasons.

Very diverse applications may use ERS-1 data and existing
(or potential) sub-communities do not mostly inter-communicate
(problems of compartmentalization).

The level of usefulness of remote sensing data to the

various applications is very unequal and is in a different
status (e.g. operational for weather and sea-state forecast,
and experimental for fisheries and pollution monitoring), although recent missions (Seasat, Nimbus-7) have demonstrated the
capability to measure from space a number of geophysical parameters with an adequate accuracy.

The sub-communities have a different level of resources
and education, e.g. offshore and fisheries.

They are not very well structured compared to the scientific community, particularly on the international level, in
view of aspects of economic competitiveness.

From the point of view of the ERS-1 system performance,
there is a fundamental difference from the scientific community
in that all the applications community requires timely (about
3-6 hours) disseminated thematic products, which is a very
challenging objective and requires a ground segment accordingly
sized that covers both processing and dissemination aspects.

POTENTIAL APPLICATIONS OF ERS-1 DATA

This section will only concentrate on applications requiring near real-time processing and dissemination, which are
both the driving and sizing requirements for the ERS-1 end-to-
end system configuration. ERS-1's capability to generate with
the SAR all-weather high resolution images over land should
also be borne in mind as they may be used to complement optical
data provided by other satellites (LANDSAT-D, SPOT) for a number
of land applications (geology, land use, etc.).

With the exception of meteorology, for which operational
satellite system exist, ocean surveillance systems have been
experimental (SEASAT, NIMBUS) or, as with ERS-1 and MOS-1, have
not yet passed the definition phase.

This is possibly one reason for which the application
users have adopted a cautious position vis-à-vis the commercial
utilisation of data coming from these satellites. Other difficulties encountered, such as the timely processing and dissemination of data, the lack of adequate models to accept asynoptic
data as inputs, etc., have also contributed to this situation.

Consequently, it is important to prepare the user community well in advance, and to size the ERS-1 ground segment
correspondingly in order that users may be ready and able to
utilise the data generated when the satellite is in orbit.

Data generated by the ERS-1 instruments can be categorised

as a function of their "maturity" for various applications and their ability to fulfil present and future requirements. This does not lead necessarily to establishing priorities but only to give indications to satellite designers in case of conflicting requirements in terms of platform resources (e.g. power allocation, storage capacity, etc.). Furthermore, this "categorisation" may evolve with time, taking into account the rapid progress made in some application sectors linked to the development of new industrial activities. For instance oil off-shore activities in North Canada (Beaufort Sea, Melville Straits) may trigger off and increase the development of microwave remote sensing techniques for ice monitoring applications which is currently considered to be at an early stage.

Category A: Operational (or quasi-operational) Data

These are typically data coming from the wind scatterometer, radar altimeter, and the along track scanning radiometer (ATSR). Wind field, wave height and sea surface temperature (SST) data are used today by weather and sea-state forecast centres on an operational basis. Some of these data (e.g. SST) are measured by satellite, and some are derived from non-satellite measurements (weather ships, radio sondes, etc.). A major problem identified here is the need for development of models accepting asynoptic data as generated by near-polar satellites, but work is already in progress on this subject, both in Europe and the United States.

Forecast activity includes a number of steps following the acquisition of satellite data and the transformation of these data into physical quantities. These are:

* merging of satellite data with data from other sources (i.e. buoys, radio sondes, research vessels, other satellite data, etc.);
* implementation of a numerical model;
* interpretation of model predictions in terms of the special operational conditions of the industrial activity;
* communication of refined information to field operators;
* monitoring of accuracy and usefulness of forecasts and feedback into operational systems.

The major limitation of current ocean and atmospheric forecasts is the lack of data over the open ocean. This is particularly severe in the vast oceanic areas of the southern hemisphere but is also apparent in the northern hemisphere due to the reduction of the number of weather ships in the Atlantic. Numerical experiments in the United States using SEASAT data, and in Europe (by the European Centre for Medium Range Weather Forecasts), have shown that the inclusion of such data in the

models provides more accurate forecasts, valid for longer
periods of time.

Weather Forecast: The basic information required to initiate meteorological models is wind field, temperature and
humidity in three dimensions and boundary conditions. This
information is used on three different levels:

* as initial conditions to numerical models,
* to validate and tune models, and
* to monitor climate parameters.

Improvements are expected for local forecast up to 36/48
hours requiring fine mesh models (100 km) and medium to long
range forecasts (up to 10 days) using coarse mesh (300 km) and
requiring hemispherical or even global data.

Sea-state Forecast: Most wave models use a wind field
derived from the atmospheric forecast model. ERS-1 capability
to generate directly wind information will result in improved
forecast. Various models exist in the United States and Europe
for local and regional forecast up to 36/48 hours.

An important additional benefit from ERS-1 will be the
capability to check and tune the models with direct measurements of wave height (Altimeter) and wave spectrum (Wave scatterometer).

The industrial activities which will benefit mostly from
forecast improvement and accurate measurements of sea surface
temperatures include:

* Offshore activities by providing:

. synoptic data for short term forecasting for planning
of operations during the construction of oil platforms and the
exploitation phase of the oil field;
. a continuous monitoring of ocean parameters for the
establishment of statistics on wave and wind field for engineering design of oil platforms (There is no real-time requirement for this activity).

* Ship routing by:

. reducing the time on trade routes and, consequently,
reducing total fuel consumption,
. improving the safety by reduction in hull damage,
cargo damage, marine insurance costs, catastrophic ship loses,
catastrophic hazardous substances pollution.

* Fisheries by:

. accurate measurements of sea-surface temperatures
which will allow the determination of temperature fronts. This
information, combined with meteorological and oceanic data
(e.g. wind field, wave field), will allow the evolution of the
thermal fronts to be monitored, which is of prime importance
to locate pelagic fish species living in the vicinity of the
sea surface, e.g. tuna. This will lead mainly to a more efficient management of the fish resources and to an improvement
in the efficiency of fishing fleets (navigation and routing
aspects).

Category B: Preoperational Data - SAR Imagery over Ice and Sea-ice

Experiments conducted, mostly in Canada, with airborne
SAR and SEASAT SAR data have demonstrated the capability of
SAR to identify some ice features (ice extent, ice type, ice
roughness, ice movement) that are of importance for operations
and activities in the high latitude areas infested by ice.

Today, airborne SAR systems are operationally used over
the Beaufort Sea in support of oil and gas exploration activities. It is expected that the C-band SAR on ERS-1 will satisfy requirements from "ice users". This implies a very fast
process time and delivery (about 3 hours from acquisition). A
daily (or less) coverage of the area to be monitored would be
required for an operational system.

Category C: Experimental Data - SAR Imagery over Oceans

Sea surface and wave imaging mechanisms are still in an
early stage of understanding, although progress have been made
since SEASAT imagery over oceans has been available. There is
still much controversy on the basic physics of the phenomena,
and also on the interpretation of data, partly due to lack of
simultaneous in-situ measurements.

Potential applications such as surface pollution detection
would require fast processing and delivery of data (about 3
hours) and this is again very challenging, in particular for
pollution where complete scenes have to be processed. Converse-
ly, for wave (image) spectrum, only small 5 km x 5 km scenes
will have to be processed.

It should be noted that some data classed in category A
will be of use for applications using data put in category B
or C. For example, ice drift forecast, pollution trajectory
prediction, etc., will make use of category A data (wind field,
wave height, etc.).

SCIENTIFIC VALUE OF ERS-1

Although the above list primarily addressed the economic
and commercial application aspects of ERS-1, it is clear that
ERS-1 will also benefit the scientific understanding of coastal
zones and global ocean process which, together with the monitoring of polar regions, will provide a major contribution to
the World Climate Research Programme. ERS-1 data used alone.or,
more commonly, in conjunction with complementary data from
buoys, radio-sondes, research vessels, other near-surface
platforms and other satellites in pre-arranged or regional experiments, will enable significant advances to be made in:

Physical Oceanography and Glaciology

The measurements of wind-field, wave spectra and wave
height that are used in near real time for forecasting ocean
conditions will also be used to improve the understanding of
wave generation and propagation, and to produce more refined
models of these dynamic processes. The little understood meso-
scale eddies are detectable by altimetry, and a precise knowledge (decimeter level) of the satellite orbit would allow sea
surface topography analysis.

A combination of altimetry measurements and microwave
imagery can be used for the study of ice-sheet profiles and the
behaviour of sea-ice in the polar regions.

Climatology

A priority research topic envisaged with the World Climate
Research Programme is the "controlling effect of the physics
and dynamics of the oceans on the global cycles of heat, water
and chemicals (specially carbon) in the climate system".

The wind stress on the sea surface is one of the primary
variables in climate research. Wind stress exerted upon the
surface of the ocean is the major driving force that maintains
the ocean currents and generates surface waves that are responsible for downward mixing of heat, thereby increasing the thickness of the heat storage layer. It affects the amount of thermal
energy in the ocean and is fundamental to model air-sea interactions.

Monitoring of the sea ice in the polar oceans is also
important because of the strong thermal effects resulting from
changing ice cover.

Table 10 illustrates the relevance of the various instruments to mission capabilities.

Table 10.   Relevance of instruments to mission objectives.

	Altimet. + PRARE	ATSR IR	C-band Active Microwave Instrum.		
			Imag. Mode	Wave Mode	Wind Mode
Weather forecast		▓			▓
Sea-state forecast	▓			▓	▓
Offshore activity	▓	▓		▓	
Ship routing	▓	▓		▓	
Fisheries (fish location)	▓	▓			
Sea and iceberg monitoring	▓	▓	▓		
Oil & pollution detection		▓			
Coastal processes		▓			
Ocean tides	2				
Ocean circulation	1		▓		▓
Wind field	3 ▓		▓	▓	▓
Wave field	3 ▓		▓	▓	▓
Sea surface temperature		▓			
Polar oceans	▓		▓		▓
Land ice	▓		▓		
Marine biology		▓			

1 For large-scale circulation, an accurate orbit determination over short arcs is required.
2 For solar tides, measurements from other satellites in complementary orbits is required.
3 The Altimeter and the C-band Active Microwave Instrumentation are mutually supporting in deriving the wind and wave fields.

NEED FOR DEMONSTRATION MISSIONS

    It is essential when designing the ERS-1 mission to involve the user community in the development of the requirements, the development of the system and the evaluation of the performance of the system. ESA has made a special effort to involve these communities in the definition of the mission requirements and intends to pursue this effort through the de-

development and exploitation phases of ERS-1, considered as
both an experimental and preoperational system.

It will be experimental since, as the first ESA mission,
it will have to demonstrate that the concept and technology for
both the space and ground segments are right for the application
envisaged, and that the users are ready and able to use the data
generated. This will require a number of activities before and
after launch which are very important for the success of the
mission such as:

* Simulation and optimisation of sensor performances (air-
borne testing)
* Development and testing of algorithms and models
* Setting up and testing of data products
* Definition of distribution networks to meet user require-
ments
* Develoment of pilot projects (small scale) and demonstra-
tion missions (large scale)
* Continuation of research and development to optimise the
use and value of the data provided by the satellite system.

On the other hand, ERS-1 will have to demonstrate, for some
appropriate applications and a limited scale, an operational
capability. This will, of course, require that data or products
be delivered in quasi real time to corresponding existing operational services or end users.

To this effect, the Agency has set up teams of experts,
called Instrument and Data Teams, to advise ESA on all-user-
related/scientific/technical aspects of each instrument and,
for the overall system, on the operation aspects, data hand-
ling/dissemination and the promotion aspects of the programme.

Scientific and application-oriented experts from the participating states will contribute to these teams throughout the
development and exploitation phases of the ERS-1 programme.

PREPARATION OF FOLLOW-ON OPERATIONAL MISSIONS

With ERS-1, it is expected that a gradual transfer of
applications from experimental to operational users will take
place, preparing the future users to later operational satellite
systems.

Although not yet approved, as a first step it is planned
to propose to the participating states in ERS-1 to consider the
launch of a second flight unit, nearly identical to ERS-1, two
or three years after the launch of ERS-1, i.e. in the 1989-1990
time frame, thereby providing the user community with five to
six years of continuous data.

Lastly, the ultimate goal would be to set up or to contribute to an operational multi-satellite system for global
rather than regional monitoring.

CONCLUSION

Within Europe, there is particular interest in a satellite
monitoring programme over such important economic areas as the
oceans, the coastal zones and ice caps. Equally, there exists
an active scientific community in Europe which recognises the
potential value of modern spaceborne sensors to longer-term
studies of ocean processes and their relevance to climatolog-
ical problems.

In conclusion, it is clear that if Europe wants to be
present on the world scene with operational remote sensing satellite systems in the 1990's, it is necessary to develop now
an experimental/preoperational satellite programme with a first
launch around 1987. All necessary capabilities, whether industrial, scientific or economic, exist in Europe and the positive
decision which has been taken recently by ESA member states,
plus Canada and Norway, to develop the ERS-1 programme should
allow Europe not only to guarantee its presence in the world
competition but also to provide a contribution to future world-
wide systems aiming at global monitoring of coastal zones and
oceans.

REFERENCES

Duchossois, G., 1982, ERS-1: First European remote sensing sat-
        ellite: User aspects and potential applications, in:
        "Proc. 33rd meeting of the IAF, Paris, 26 September -
        3 October".982".
Honvault, C., 1982, The first ESA remote sensing satellite,
        ERS-1: the programme and the system, in: "Proc. 33rd
        meeting of the IAF, Paris, 26 September - 3 October".
Jones, M., 1982, "The Ground Segment for an European Ocean
        Monitoring Satellite ERS-1".
Paci, G. and de Leffe, A., 1982, "Status and Future Plans for
        the First European Remote Sensing Satellite ERS-1",
        Post-graduate Summer school on remote sensing applica-
        tions in marine science and technology, University of
        Dundee, 1-21 August.
Reynolds, M. L. and Llewellyn-Jones, D. T., 1982, ERS-1 exper-
        imental payload package, in: "Proc. 33rd meeting of the
        IAF, Paris, 26 September - 3 October".
Vandeput, L., 1982, ERS-1 user benefits, in: "Proc. 33rd
        meeting of the IAF, Paris, 26 September - 3 October".

THE INTERNATIONAL STANDARDIZED

OIL WAKE EXPERIMENTS

(ISOWAKE)

THE PROCEDURES INTERNATIONAL STANDARDIZED OIL WAKE EXPERIMENT

(ISOWAKE)

R. Vollmers[1] and P.R. Morris[2]

[1]United States Coast Guard Headquarters
Washington, D.C., United States
[2]Warren Spring Laboratory
Stevenage, Hertfordshire, United Kingdom

ABSTRACT

ISOWAKE develops a standardized oil detection target which is used by each participating nation to examine the performance of various electro-optic and electro-magnetic remote airborne sensors. These sensors are flown over controlled discharges of oil which are at, above and below the levels called for by MARPOL 73/78 Annex I. Output sensor data would be nationally analyzed and subsequently reported internationally.

BACKGROUND

The question of necessary evidence of illegal discharges of oil in the marine environment has been addressed by the International Maritime Organization (IMO) at the thirteenth and fourteenth sessions of the Marine Environment Protection Committee (MEPC XIII and MEPC XIV) and guidelines (Document MEPC XIII/WP.3 as amended by document MEPC XIV/WP.1) were established. These guidelines included the use of conventional photography as well as other forms of remote sensing imagery. Amplifying information on the use of photographic evidence was included in these guidelines, however, details were not presented on the other forms of remote sensing.

Many different types of electronic sensors have detected illegal oil discharges, however, the degree and conditions under which these sensors can detect the levels specified in MARPOL 73/78 are uncertain. Therefore the interpretation of this imagery could be seriously questioned as legal evidence.

Within the framework of the present pilot-study, nations
participating to the pilot-study agreed at their meeting in
March 1980 to conduct experiments to establish the appearance
of wakes of ships containing known quantities of petroleum
oils when viewed by various remote sensing devices, and to
focus their efforts on determining the capability of existing
electronic sensors to detect illegal discharges of oil. To
accomplish this, the ideal approach would initiate a series of
international co-operative oil spill experiments where each
nation's oil detection sensors are flown simultaneously over
the same discharge. These experiments would continue until a
wide range of environmental conditions and oil discharge rates
were recorded. Unfortunately, the logistics complexities of
such an effort are beyond the economic means of Working Group I.

As an alternative, the United Kingdom has proposed a
simple experiment that can be performed individually by each of
the participating countries to test the performance of their
remote sensing systems.

SCOPE

In the subsequent paragraph, a standardized oil wake discharge is defined to such a degree that it can be duplicated
at any time and place convenient to each participating nation.
In addition, the depth to which each nation wishes to explore
the oil detection phenomena is left to their individual needs
and finances. Each nation is requested, however, to submit an
Isowake test plan to Dr. P.R. Morris (Warren Spring Laboratory,
Stevenage, Herts, UK) for information, prior to accomplishment.
In addition, Isowake results will be reported to the IMO/Mari-
time Environment Protection Committee.

The experiment includes:
* Requirements for the international standardized oil
  wake discharge;
* Recommended conduct for the airborne survey;
* Requirements for national test plans and final reports.

INTERNATIONAL STANDARDIZED OIL WAKE DISCHARGE

The oil wake discharge experiment is designed to simulate
real ballast-water discharges. The generation of a standardized
oil wake discharge shall be accomplished by:

(1) using a vessel of preferably greater than 100 tons,
    gross tonnage;
(2) conducting the tests at a speed between 8 and 12 knots,

preferably at 8 knots, into the wind and waves, in
water deep enough so no sediment is drawn up into the
wake;

(3) discharging the oil on the port side, approximately
0.5 m above the average sea level through a 100 mm
pipe facing vertically downwards between 0.2 m and
0.5 m from the side of the ship.

For simulated discharges, No. 2 Diesel Oil (whose physical
and chemical characteristics are described in Appendix I), and
if desired one or more of the following oils are discharged at
rates of 30, 60, 120, 240 litres/mile and mixed with 3,000 l/mile
of sea water. These oil flows become 240 litres/hour, 480 l/h
960 l/h and 1,920 l/h in a sea water flow of 24 cubic meters
per hour:

No.6 fuel oil (Physical and chemical characteristics are
described in Appendix I)
Bachaquero 17
Minas crude
Nigerian Medium
Sahara blend
Arabian light crude

CONDUCT OF THE AIRBORNE SURVEY

The success of Isowake is based on the standardization of
the oil wake discharge, documentation of various uncontrollable
factors and the conduct of the airborne survey. The following
guidelines are recommended:

* Conduct a trial run with no oil discharge;
* Fly principal flight lines either towards or away from
  the ship along the ship's track;
* Secondary flight lines perpendicular to the ship's track
  400 meters astern of the ship during discharge;
* Tertiary flight lines so as to vary sun-oil-sensor angle
  and the wind-oil-sensor angle;
* Preferred flight altitudes are 500 and 2,000 feet;
* Additional flight lines may be flown at other altitudes
  in order to exploit specific sensor performance characte-
  ristics;
* Preferred overflight times are: noon local time
                                  16:00-18:00 hr local time
                                  midnight local time

Consequently, survey flights have to be run as follows:

(1) Normal survey to be run from 2 miles astern of ship up
    the slick towards the ship or directly away from ship;

(2) One run at approximately 90° to the ship's track,
    approximately 400 m astern of ship;
(3) One run to obtain the best results from the equipment
    in the prevailing conditions.

REQUIREMENTS FOR NATIONAL TESTS PLANS AND FINAL REPORTS

     Each nation planning to participate in Isowake is invited
to submit a detailed test plan and schedule to Dr. P.R. Morris,
Warren Spring Laboratory (Stevenage, Hertfordshire, U.K.).

     This plan shall contain as a minimum a list of the sensors
to be tested and their specifications; vessel and aircraft
deployment strategy; aircraft, vessel and environmental factors
to be measured; the time and place of accomplishment and the
approach to be used for data analysis and reporting.

     The following measurements are recommended:

Measurements to Be Made at Sea Level

     The following measurements are required to give the sea
level conditions so that data can be interpreted:

* Exact position and time of release;
* Ship's heading (into wind standard);
* Ship's speed (8 knots standard and up to 12 knots)
  (Constant during test period);
* Rate of oil flow (standardized);
* Type of oil;
* Temperature of discharge;
* Rate of seawater flow (standardized);
* Wind speed (preferably less than 15 knots);
* Wind direction;
* Wind temperature;
* Water temperature;
* Wave height and visual description (in option, wave-
  rider buoy chart);
* Visibility, cloud type, cover;
* General weather description;
* Sun bearing from ship (preferably not due ahead or
  astern). In option, sky spectral irradiance, sea surface
  spectral irradiance, water attenuation coefficient;
* Water depth (preferably greater than 20 m for majority
  of tests);
* Presence of any visual sediment.

Measurements to Be Made from Survey Aircraft

     Airborne observations should be carried out in a manner

consistent with the safe operation of the one or more aircraft
involved. The following information should be recorded.

* Airspeed
* Altitude
* Track angle
* Drift angle
* Time of flight lines
* Data relevant to sensor systems on board (in option,
  sea surface temperature (radiometric temperature) and
  photographic record of wake).

Relevant data for the sensor systems and exact track flight
times should be noted together with any other relevant
observations.

Appendix 1:   Physical and Chemical Requirements.

(Source   Milspec 4300-1164)

GRADE OF FUEL OIL			N° 2 ①	N° 6 ②
Flash point °C (oF)		Min	38 (100)	60 (140)
Pour point °C (oF)		Max	−6 (20) ③ ④	– ⑥
Water Vol. %		Max	0.05	1.50 ⑦
Sediment Weight %		Max	0.02	0.50
Carbon Residue on 10 % Btms, %		Max	0.35	–
Sulfated Ash weight %		Max	–	1.00
Distillation Temperature °C (°F)	10 % point	Max	–	–
		Min	–	–
		Min	282 (540) ④	–
	90 % point	Max	338 (640) ⑤	–
Saybolt Universal 38 °C		Min	(32.6)	(> 900)
Viscosity, S ⑧ (100 °F)		Max	(37.9)	(9000)
Furol 50 °C		Min	–	(> 45)
Kinematic ⑧ Viscosity, cSt	(122 °F)	Max	–	(300)
	38 °C	Min	2.0 ④	–
	(100 °F)	Max	3.6	–
Specific gravity	50 °C	Min	–	> 92
	(122 °F)	Max	–	638 ⑨
	60/60 °F (°API)	Max	0.8762 (30 min)	–

①   A heavier distillate than grade No 1. It is intented for use in atomizing-type burners which spray the oil into a combustion chamber where the tiny droplets burn while in suspension. This grade of oil is used in most domestic burners and in many medium-capacity commercial-industrial burners where its ease of handling and ready availability sometimes justify its higher cost over the residual grades.

②   A high-viscosity oil sometimes referred to as "Bunker C", and used mostly in commercial and industrial heating. It requires preheating in the storage tank to permit pumping and additional preheating at the burner to permit atomizing. The extra equipment and maintenance required to handle this fuel usually preclude its in small installations.

③   Pour points other than those shown may be specified whenever required by conditions of storage and use, agreed upon by purchaser and seller.

④   When pour point less than - 18° is specified, the minimum viscosity for Grade No. 2 shall be 1.8 cst (32 s.) and the minimum 90% point shall be waived.

⑤   On occasion, it may be impossible to obtain 90% distillation values because of cracking. When this occurs, the percents recovery, residue and loss should be noted.

⑥   Where low sulfur fuel oil is required, Grade No. 6 will be classified as low pour (+ 15°C Max) or high pour (no Max) or high pour (no Max). Low pour fuel oil should be used unless all tanks and lines are heated.

⑦   A deduction in quantity shall be made if the sum for all water and sediment is in excess of 1.0%.

⑧   Viscosity values in parenthesis are for information only not limiting.

⑨   Where low sulfur fuel oil is required, fuel oil falling in the viscosity range of lower numbered grade down to and including No. 4 may be supplied by agreement between purchaser and supplier. This viscosity range of the initial shipment shall be identified and advance notice shall be required when changing from one viscosity range to another. This notice shall be in sufficient time to permit the user to make the necessary adjustments.

THE DUTCH ISOWAKE EXPERIMENT OF NOVEMBER 1980

W. Wijmans[1] and R. Spanhoff[2]

[1] Rijkswaterstaat
North Sea Directorate
Rijswijk, The Netherlands
[2] Rijkswaterstaat
Directorate of Watermanagement and
Hydraulic Research
The Hague, The Netherlands

ABSTRACT

A first Dutch Isowake experiment was held in November 1980. The Dutch experimental SLAR and the Swedish operational SLAR took part in this exercise. The first behaved poorly, but the last one did well. The water temperature was too low to use an Isowake-recommended type of oil, therefore a Nigerian light crude was chosen instead.

INTRODUCTION

In 1975, the Rijkswaterstaat of the Dutch Government commenced the airborne surveillance of the North Sea in the area in which the Netherlands are responsible for the prevention of oil according to the Agreement for Co-operation in Dealing with Pollution of the North Sea by Oil (Bonn Agreement).

Remote sensing techniques are required to be able to control the illegal oil spills at night.

The recommendation of the NATO/CCMS pilot study on remote sensing for the control of marine pollution was adopted to participate in the so-called Isowake-project, in which emphasis is laid upon the observation of oil spills made under controlled conditions. A Dutch experiment was conducted in November 1980. It served a dual purpose:

- to contribute to the Isowake-project;
- to gain experience with SLAR-systems in order to make a decision on the type of SLAR for an airborne oil detection system.

A Dutch Rijkswaterstaat report[1] and an experiment report of the Swedish Space Corporation[2] have appeared concerning this experiment.

DESCRIPTION OF THE EXPERIMENT

This first Isowake experiment of the Netherlands was used to investigate the sensitivity of the oil slick formation on the sea surface to the way the oil was pumped overboard.

Two different types of ship were used, while one ship applied two different discharge techniques, one with the oil/water mixture outlet-tube close to the ship hull, thus in the turbulent zone and consequently in the ship wake, and the other one with the outlet-tube at ca. 5 m from the ships hull, where the oil is less affected by the turbulence in the water caused by the ships motion.

The recommended oil types for Isowake were not useable due to the cold conditions in this time of the year. The viscosity appeared to be too high. To overcome this problem, Nigerian Bonny Light Crude was chosen instead.

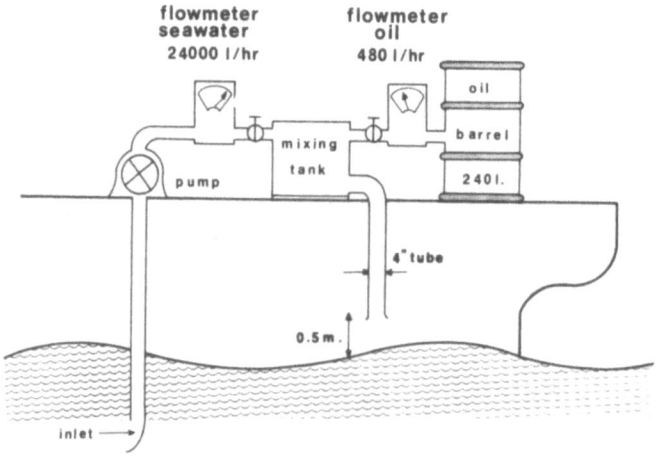

Figure 1.   Oil discharge set up according to CCMS standards as used on both ships.

Both ships were equipped with an oil-water mixing installation as sketched in Figure 1. This resulted in an oil
spill of 60 litres of oil per nautical mile.

The weather and sea conditions were marked down on the
ships as well as on the measurement platform operated by
Rijkswaterstaat 10 km offshore from Noordwijk and 60 km away
from the experimental area (Table 1). The movements of the
ship and the discharge actions are registered on board, and
are displayed in Figure 2.

Two SLAR-systems were available at the time: the Dutch
experimental SLAR and the Ericsson SLAR of the Swedish Coast
Guard. The Dutch SLAR is primarily intended for research
purposes and is as such equipped with a digitised output and
a digital recorder.

In order to use it for this special quasi-operational
purpose, a special analog interface was made to make the first
10 km of SLAR swath visible on a Visicorder. The most interesting parts of the SLAR registrations were recorded on the

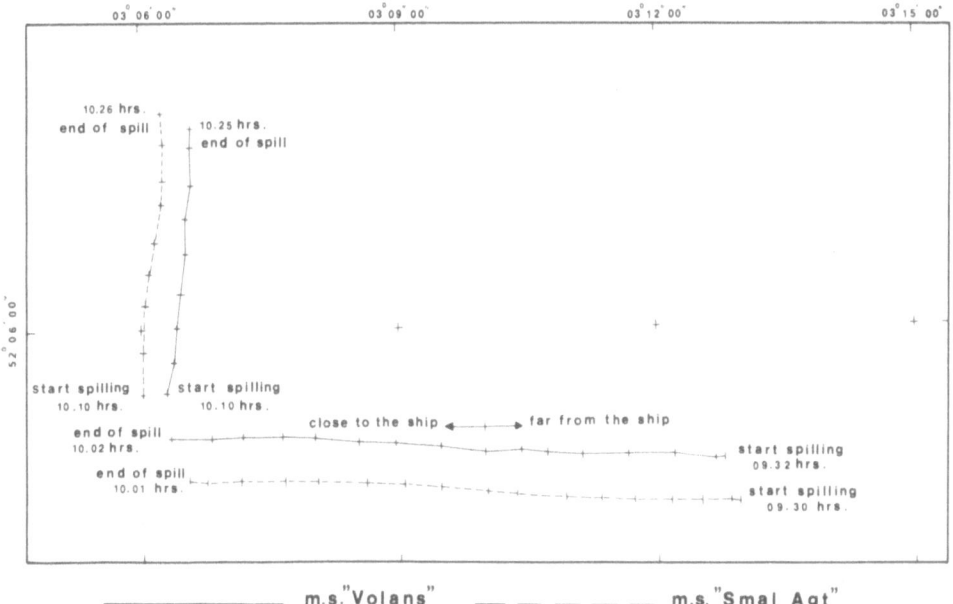

Figure 2.   Actual discharge pattern of both ships as measured
            with on board positioning system. During the first
            part of her first track, the mv. VOLANS discharged
            far from the hull and during the remainder close
            to it.

Table 1. Measurements made at sea level.

	Smal Agt	Volans
Type of ship	Coast tanker	Supply ship
Propellors	2	2
Length	54 m	49 m
Width	9.4 m	10 m
Draught	2.4 m	3 m
Size	660 tons	459 tons
Time of oil discharge	08:30-09:15 GMT	
Rate of oil flow	480 ltr/h	400 ltr/h
Type of oil	Bonny Light	Nigeria
Rate of water flow	24 m³/h	40 m³/h
Ships heading	280°/10°	280°/10°
Ships speed	8 knots	8 knots
Wind direction	280°	
Wind speed	10 m s⁻¹	
Wind temperature	10°C	
Sun position	Azim: 151° - Elev: 23°	
Water temperature surface	9°C	
Water temperature 2 m subsurface	9°C	
Wave height and description	H 1/3 = 1.5 m	
Water depth, presence of any visual sediment	30 m, turbid water	
Air humidity	71%	
Visibility, cloud type, cover	Vis. 15 NM, alto cumul. 5/8	
General weather description	Instable after passage cold front	
Sky irradiance	-	
Sea surface irradiance	-	
Visibility of standardized wake from stern	Barely visible during short period	

digital tape recorder as well for subsequent computer processing in the laboratory. The Dutch SLAR, flown in a METRO II
aircraft, is only left-looking.

The Ericsson SLAR, flown in a CESSNA 402, is part of an
operational maritime surveillance system. The equipment in
both aircraft is described in Table 2. In the same table,
mention is made of the Rijkswaterstaat PIPER NAVAJO surveillance
aircraft, with which visual observations and photographic
pictures were made according to normal surveillance routines

Table 2.   Equipment involved.

PIPER NAVAJO without remote sensing equipment doing routine
visual observations:

    * Position reference   Decca Mk 19
    * Operator seats   2

METRO II with digital SLAR:

    * Operating frequency   9.6 GHz
    * Antenna length   2.5 m
    * Peak output power   25 kW
    * Polarization   Horizontal
    * Pulse width   200 nanoseconds
    * Prf   100 Hz
    * Monitor   Visicorder with UV film
    * Recorder   Digital tape
    * Position reference   ISS with optional DME
    * Operator seats   2

CESSNA 402 C with SLAR:

    * Operating frequency   9.3 GHz
    * Antenna length   3.2 m
    * Peak output power   10 kW
    * Polarization   Vertical
    * Pulse width   500 nanoseconds
    * Prf   1,000 Hz
    * Monitor   Video display with polaroid film
    * Recorder   Analog video tape
    * Position reference   Decca Tans
    * Operator seats   2

and the subsequent actions when an oil spill is discovered.
This was done to make an honest comparison between the old,
conventional, technique and the modern, remote sensing, technique. The visual observations were carried out by trained
Rijkswaterstaat operators.

RESULTS OF THE EXPERIMENT

The weather conditions during the experiment were moderate
to calm. The weather and sea data can be found in Table 1
together with the type of oil and the way it was released.

The pattern of the oil spill is drawn in Figure 2. Visual
observations from the PIPER NAVAJO learned that the oil slick
was visible during 30 minutes. Since the whole spill operation

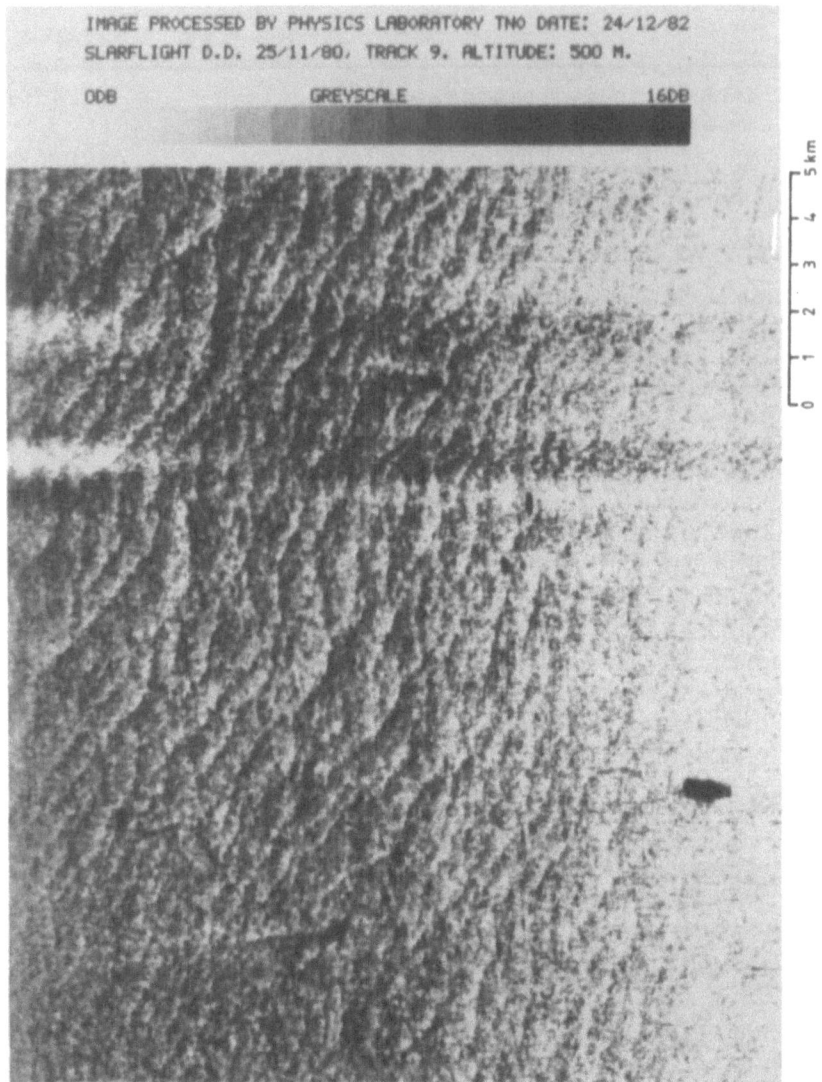

Figure 3. Computer processed image of the digital Dutch SLAR
recording of the part of the sea where the Isowake
standard oil discharge wake was supposed to be.
The oil wake is not discernable from other surface
features. This SLAR is very sensitive to sea surface
roughness variations influenced by the sea bottom
topography.

took 50 minutes, the beginning of the spill had faded away when
the last part of the spill was being made.

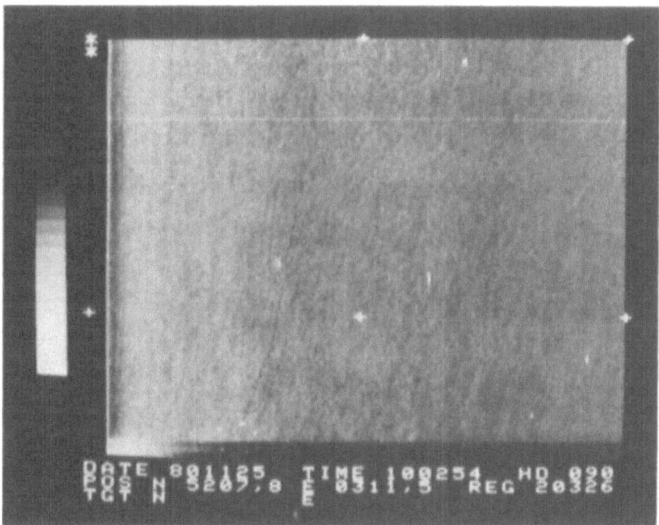

Figure 4. Radar image of the Isowake controlled discharge
          The first track of the L-shaped discharge is
          completed. The two ships with their wakes are
          visible at the lower left border of the picture.
          The first part of the port discharge is weaker.
          This could be due to the fact that the oil was
          discharged far from the hull during the first
          15 minutes. The whole of the port wake is weaker
          despite the fact that the rate of oil flow was
          higher.

With the Dutch SLAR it was impossible to detect the oil
slick except for the fresh oil spill. The flight pattern of the
METRO II was such that the oil slick was viewed from all possible directions with the left-side looking SLAR. These observations were also recorded on the digital tape recorder. Later
computer processing in the laboratory failed to give better
results (Figure 3). The Ericsson radar however was able to
detect the oil spill until 30 minutes after the time of the
actual release of the oil (Figure 4). The dependence on the way
the oil was released as well as on the shape of the ship turned
out to be very small.

CONCLUSIONS

Some conclusions may be drawn from this experiment. For
example that the detection capabilities of the Ericsson SLAR
for oil spills at sea is at least as good as detection by
visual means.

Another conclusion may be that, whenever the Ericsson radar detects an oil spill from a ship, the rate of discharge is at least 60 liters per nautical mile. The Dutch experimental SLAR was not capable to detect the oil spill for a number of reasons, the most important one being its limited sensitivity. The sensitivity of the Dutch SLAR is namely 10 to 12 dB lower than that of the Ericsson SLAR, which is mainly due to the difference in polarisation, HH vs. VV, which accounts for 8 to 10 dB.

To acquire an operational airborne oil detection system within a limited budget and within a limited time frame, the Dutch Government was advised to purchase a system similar to the maritime surveillance system used by the Swedish Coast Guard.

ACKNOWLEDGEMENT

The Rijkswaterstaat of the Dutch Government is grateful to Fast Airways, the Dutch National Aerospace Laboratory NLR, the Swedish Coast Guard and the Swedish Space Corporation for their fruitful co-operation in this experiment.

REFERENCES

1. W. Wijmans, Oliedetectie of de Noordzee met behulp van SLAR en IR/UV, North Sea Directorate report NZ-R-81.37.
2. Colliander, Oliedetectieproef, Rijkswaterstaat 1980, in Oil spill detection experiment with the Ericsson SLAR, Swedish Space Corporation Report FP1/3-1.

THE FRENCH ISOWAKE EXPERIMENT

(TOULON, SEPTEMBER 1980)

R. Burkhalter

Laboratoire National d'Essais
Paris, France

INTRODUCTION

Nations participating in the NATO/CCMS pilot study on remote sensing of marine pollution agreed to conduct experiments
to provide reliable information on the abilities of their remote sensing systems to detect oil discharges whether or not
conforming to the provisions of the 73/78 MARPOL Convention.
These experiments received the code-name of Isowake.

An Isowake experiment was carried out in September, 1980,
off the French Mediterranean coast, within the Protecmar campaign framework.

Protecmar purposes are mainly to measure the effectiveness
of oil dispersion at sea when using dispersants; for these purposes, small experimental oil slicks are spilled from a vessel.
Because the Protecmar trials are conducted with an important
ships and aerial support, it was thought that these experiments
could be used to test remote sensing techniques over controlled
oil spills conforming the provisions of the MARPOL Convention.

Unfortunately, because the Protecmar working programme was
too important, it was impossible to cover all the figures which
were planned in the Isowake standardized procedure, as established by the United Kingdom (Warren Spring Laboratory).

DESCRIPTION OF THE EQUIPMENT

Therefore, the test-programme was reduced and only four
controlled oil spills took place within the Protecmar test area.

401

Table 1. List of discharges.

No.	Date	Time (TU)	Oil type	Oil discharge rate
1	80/9/17	1407	Diesel oil	100 ppm
2	80/9/17	1428	Medium fuel oil	100 ppm
3	80/9/18	1225	Medium fuel oil	100 ppm
4	80/9/18	1235	Arabian Light	60.0 l/naut.mile

Table 1 lists all the discharges that were made, along which the remote sensing sensor was used.

For the 60.0 liters/nautical mile discharge rate, a light fuel oil, approximately equivalent to a partially evaporated crude oil (for example an Arabian Light crude oil without its light components up to a boiling point of 150°C) has been used.

The vessel used for the experiments was a 71 m overall tanker. The oil was discharged on the port-side, approximately 0.50 m above the average sea-level, through a 100 mm pipe facing vertically downwards from the side of the ship.

A micro-pump (1 liter per hour) was used to introduce oil in the water stream to simulate discharges at rates of 100 and 150 ppm. For the 60 liters per nautical mile discharge (simulating ballast waters), the system used the fire-circuit of the vessel. The oil flew at 480 liters per hour in a sea-water flow of 24 cubic meters per hour. The mixture oil/water occurred in a pipe situated on the starboard side about one third of the ship's length from the stern.

The measurements that were measured during the different trials are given in Table 2. .

SENSORS

Only a thermal infrared line scanner "SOSIE" manufactured by Société Anonyme de Télécommunications (SAT) was flown over the discharges. This scanner is working in the 8-10 micron detection band. Its main characteristics are listed below:

* Detection band   8-10 micron
* Detector   Hg Cd Te
* Total field of view   90°
* Instantaneous field of view   1.5 milliradian
* Scan speed   200 lines/second
* Thermal resolution   0.2°C

Table 2. Measurements during Isowake experiment.

Date	80/9/17	80/9/17	80/9/18	80/9/18
Time (UT)	14:07	14:28	12:25	12:35
Wind direction	240°	240°	140°	140°
Wind speed (knots)	10	10	10	10
Sea state	2	2	2	2
Visibility (km)	5	5	5	5
Cloud cover	2/8	2/8	3/8	3/8
Water depth	1000	1000	1000	1000
Ship speed (knots)	8	8	8	8
" heading	110°	110°	110°	110°
Flight altitude (feet)	1000	1000	1000	1000

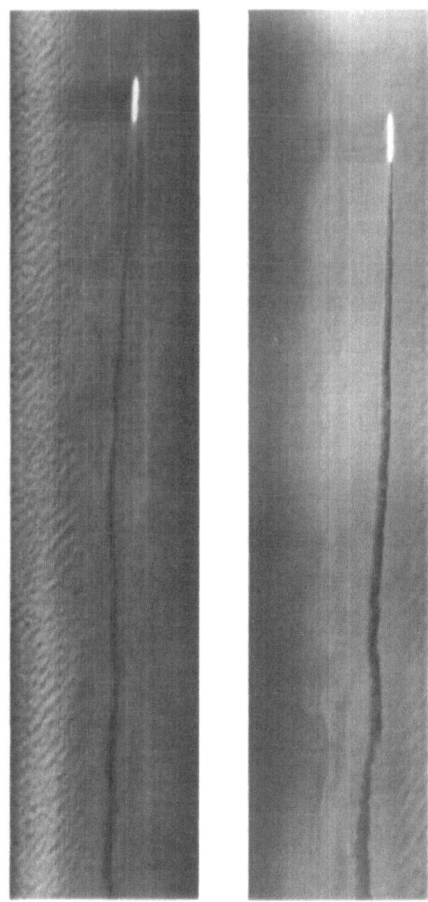

Figure 1. IR imagery of un-polluted (1) and polluted (2) wakes

RESULTS

The thermal imagery (Figure 1) shows the presence of a quite light superficial thermal gradient. A colder water up-welling can be. shown in the ship's wake, when there is no discharge.

The detection was possible only with the 60 liters per nautical mile discharge rate. Thermal imagery shows very clearly the presence of a negative anomaly in the ship's wake in comparison with the clean wake. No detection was observed with the 100 and 150 ppm discharge rates in the prevailing conditions.

CONCLUSIONS

This incomplete Isowake experiment shows that:

* the remote sensing system presently used in France is unable to detect oil discharged at rates of 100 and 150 ppm, as far as we know, considering the prevailing conditions during this Isowake experiment;

* the 60.0 liters per nautical mile oil discharge is clearly detected by the same system in the prevailing experimental conditions. This observation confirms an other observation made in 1979 during the 1979 Protecmar exercise: a controlled discharge corresponding to the rate of 120 liters per nautical mile had been seen in the same conditions;

* the lower detection limits of the present system is yet to be find.

USE OF AN INFRARED LINE SCANNER AND A SIDE LOOKING AIRBORNE
RADAR TO DETECT OIL DISCHARGES FROM SHIPS
(ISOWAKE EXPERIMENTS)*

N. Hurford, and F.N. Martinelli

Warren Spring Laboratory
Stevenage, Hertfordshire, United Kingdom

ABSTRACT

The report describes experiments to evaluate the use of
an infrared line scanner (IRLS) and a side-looking airborne
radar (SLAR) to detect oil discharges from tankers. Two oils
were used for the experiments, Kuwait Crude oil and No. 2 fuel
oil, and these oils were discharged into the sea at the
following rates: 30, 60 and 240 litres per nautical mile. The
oils were mixed with water prior to discharge to simulate
normal ballast water discharges.

The IRLS sensor was able to detect all the oil discharges,
but under the calm conditions in which the experiments were
carried out the SLAR sensor did not detect any oil.

The experiments were carried out as part of the UK contribution to the internationally agreed ISOWAKE experiments.

INTRODUCTION

Although accidental spillages of oil from tankers receive
widespread publicity, it has been estimated[1] that rather more
oil is discharged into the sea as a result of operational
activities, such as de-ballasting and tank cleaning. Regulations
for controlling the amount of oil discharged into the sea from

---

* Crown Copyright. Published by permission of the Director,
  Warren Spring Laboratory, Department of Trade and Industry.

405

ships have been agreed at the International Maritime Organisation (IMO), which is the UN specialist agency dealing with maritime affairs. The 1969 Amendments to the 1954 Oil Pollution Convention prohibit the discharge of oil from a tanker unless the following conditions are met:

* the tanker is proceeding en route;
* the instantaneous rate of oil discharge does not exceed 60 litres per mile (the units of litres per mile are obtained by dividing the oil discharge rate in litres per hour by the vessel's speed in knot.The 60 litres per mile criterion is based on work undertaken by Warren Spring Laboratory[2]);
* the total quantity of oil discharged on a ballast voyage does not exceed 1/15,000 of the total cargo carrying capacity;
* the tanker is more than 50.miles from land.

The above regulations are enforced in the UK by the Prevention of Oil Pollution Act, 1971.

In 1973 a Conference of IMO's member states drew up the International Convention for the Prevention of Pollution from Ships, which is more frequently known as MARPOL 73.

The regulations dealing with the discharge of oil were set out as Annex I to the Convention, and these regulations extended the 1969 Amendments in two important ways. Firstly, Annex I covers all oil products (including refined products), whereas the 1969 Amendments dealt only with black or persistent oils. Secondly, Annex I requires that monitoring equipment be fitted to record the oil content of ballast water discharges. All the discharge requirements of Annex I which apply to oil tankers are set out in Table 1. Although Annex I has been ratified by the UK, it is not expected to come into force until the beginning of October 1983.

THE ISOWAKE EXPERIMENT

Use of oil-water monitoring equipment, combined with inspection of the oil record book, should ensure that tanker operators comply with the requirements of Annex I. Nonetheless it would be unrealistic to expect that operators would never find it expedient to discharge oil illegally, and the regulations would be more effective if it were known that coastal states are carrying out policing activities. Indeed, it is a requirement of the Annex that whenever traces of oil are observed on the sea in the vicinity of a ship, Governments take reasonable steps to find out if there has been a violation of the Annex. In the UK there is a standing instruction to all

Table 1. Discharge requirements of Annex I of MARPOL 73
          (Oil tankers).

Criteria	Requirements
Max quantity allowed to be discharged	"Existing" ships: 1/15,000 cargo capacity "New ships": 1/30,000 cargo capacity
Vessel's position	50 miles from land
Vessels's speed	"en route"
Discharge rate	less than 60 litres per mile
Slop tank capacity	3% of cargo capacity
Location of discharge outlet	Above the water line
Control system	Oil content meter to be provided

aircraft that any sightings of oil should be reported to the
Department of Trade, and some prosecutions have been made on
the basis of such evidence. However, reliance on observation,
even when supported by photographs, is rather unsatisfactory,
particularly as it is not possible to detect ships which discharge oil at night, and it is extremely difficult to reliably
quantify the amount of oil which has been discharged.

To overcome these disadvantages it has been suggested that
electronic remote sensing devices be used to detect illegal
discharges of oil. The ideal aerial surveillance system would
meet the following requirements:

* all-weather, day/night capability;
* unambiguous identification of oil from long range with
  good resolution;
* quantitative information on the amount of oil discharged.

Not surprisingly, no single sensor utilising any one
region of the electromagnetic spectrum is capable of meeting
these requirements, and in practice it is necessary to find a
compromise using one or more sensors which will give sufficient
information.

Nations participating in the NATO/CCMS Pilot study on
remote sensing of marine pollution agreed to conduct experiments to provide reliable information on the abilities of
remote sensing devices to detect oil discharges. The experiments would enable a comparison to be made of the performance
of different sensors in detecting known quantities of oil in
the wake of a ship. It was recognised that the ideal procedure

would be to carry out a series of co-operative experiments in
which each nation's detection equipment would be flown over
the same discharge. However, the complexities and logistics
of such an operation made this impractical. As an alternative,
it was proposed that a simple experiment be defined in such a
way that it could be duplicated by any nation at any time; the
depth to which nations wished to explore the capabilities of
their detection equipment would be left to their needs and
resources. This international standardised oil wake experiment
was given the name ISOWAKE.

The Isowake experiment is described in detail in a
previous contribution; essentially it involves discharging
oil at, below and above the levels allowed by the discharge
regulations of Annex I, and then overflying with the airborne
sensors. In the experiments carried out in the UK, two oils,
No. 2 Fuel Oil (Gas Oil) and Kuwait crude oil were used for
the experiments. The physical properties of these oils are
given in Table 2. Each oil was discharged at the following
rates: 30, 60 and 240 litres per mile. The sensors used to
detect the discharges are described in the following chapter.

DESCRIPTION AND LIMITATIONS OF THE SENSORS

The UK has previously investigated the use of two airborne
sensors for the detection of oil slicks at sea: side-looking
airborne radar (SLAR) and an infrared line scanner (IRLS)[3].
The research has been carried out jointly by Warren Spring
Laboratory (WSL) and the Royal Signals and Radar Establishment
(RSRE) at Malvern, who fly a VICKERS VISCOUNT equipped with
remote sensing devices. The specifications of the two sensors
used for oil detection purposes are given in Table 3.

The SLAR is a sensor capable of detecting oil at a range
of 17 km and can be used either by day or by night[4]. The principle of operation is that a transceiver sends out a beam of
microwave radiation and picks up the backscatter from capillary
waves on the sea surface. Systems such as this which employ a
means of illumination are said to be active systems.

Table 2. Physical properties of oils.

Oil	Density Kg m^{-3} at 20°C	Viscosity cSt at 20°C
Kuwait crude oil	870	15
No. 2 fuel oil	830	3.5

Table 3.   Description of sensors.

```
SLAR (E.M.I.)

      * Frequency:              37 GHz (Q-band)
      * Polarisation:           Horizontal
      * Antenna:                Remotely controlled
      * Beam width:             0.1°
      * Transmitter peak power: 50 kw
      * Pulse length:           0.1 µs

IRLS (Texas Instrument AAS 18)

      * Detection band:         8-14 µm
      * Film:                   5 inch
      * V/H range:              0.1-2.6 rad s⁻¹
      * Spatial resolution:     1x1 mrad up to V/H 0.86 rad s⁻¹
                                1x3 mrad V/H 0.86-2.6 rad s⁻¹
                                ( 1x2 m at height of 700 m)
      * Thermal resolution:     0.25°K
      * Total field of view:    120°
```

Both systems employ film recorders and output is also stored
on video tape. The video output is processed through a
digital scan converter and provides a real-time TV display
facility.

When oil is present on the sea surface it damps the
capillary waves and this is detected by a reduction in backscatter. To obtain sufficient resolution between the clean sea
surface and the wave damped contaminated surface, the signal
from the clean sea surface must be sufficiently above the noise
level for the reduction to be evident. In practice, this means
that the sea must be roughened by winds of approximately four
knots or more.

The disadvantages of the SLAR sensors are they have poor
spatial resolution and image contrast, the imagery is difficult
to interpret and it is impossible to discriminate between thick
and thin oil on the basis of the imagery. Since oil thicknesses
down to monolayers are capable of damping capillary waves, the
detection of slicks with this sensor is not evidence of oil in
any significant quantities.

The IRLS does not employ a source of illumination and is
described as a passive system. It detects radiation in the
8-14 µm band, at an intensity which is dependent upon the
thermal temperature of the surface and its emissivity. IRLS is
more restricted in its use than SLAR, for although it can be

used both day and night, the signal is attenuated by cloud cover. IRLS does, however, give some information on slick thickness. The radiance of oil can appear either greater or less than the surrounding water. Thin oil films in thermal equilibrium with the underlying water have a lower emissivity and are detected as being cooler than surrounding water. These regions appear darker in the imagery. Thick oil, which absorbs solar radiation and is hotter than the surrounding water, appears as bright spots in the imagery. These two regions are quite distinct and there appears to be little or no transition stage between them.

In order to assign oil thicknesses to the two regions in the imagery WSL undertook a series of trials[3] using a pad sampling technique for determining the thickness of oil resulting from small experimental spills, and correlating the samples with the corresponding imagery taken during overflights at the time of sampling. The technique was necessarily approximate due to the difficulty in taking representative samples at a particular region of the slick and then pinpointing the position of the sampling dinghy on the imagery. However, the conclusions were that the lower limit of detection of IRLS was of oil in the region of 10-50 µm thick, and that the lower limit for the hot spots was around 1,000 µm, although this threshold is likely to vary with ambient conditions. No further resolution of oil thickness is possible, since there is no continuous intensity variation noticeable in the imagery (Cf. passive microwave sensors).

The use of SLAR and IRLS extends materially the amount of information that can be gathered from aerial surveillance, over and above that obtained by observers and photographic cameras. Other sensors for use in detecting oil spills have been described[5], but in general these are still in the development stage. They include laser fluorosensors, which detect the emitted radiation after excitation from a laser pulse, positively identifying that the slick is indeed oil and giving a rough classification of the oil type. Another sensor detects passive microwave radiation at an intensity dependent upon the oil thickness, thereby providing a means of quantifying the amount of oil spilled.

In Europe, several countries have aircraft equipped with IRLS and SLAR sensors, and routinely fly these to search for oil spills and to furnish evidence for prosecutions.

DESCRIPTION OF THE EXPERIMENTS

The vessel used for the experiments was the WSL research vessel SEASPRING; the principal dimensions of the vessel are given in Table 4.

Table 4.   RV SEASPRING.

Length	56.7 m
Beam	11.3 m
Depth (moulded)	4.2 m
Normal economic speed	8 knots

Description of the Discharge Equipment

Figure 1 shows a schematic diagram of the equipment used to discharge the oil-water mixtures. By adjusting the flow control valves at the Y-piece it was possible to select a

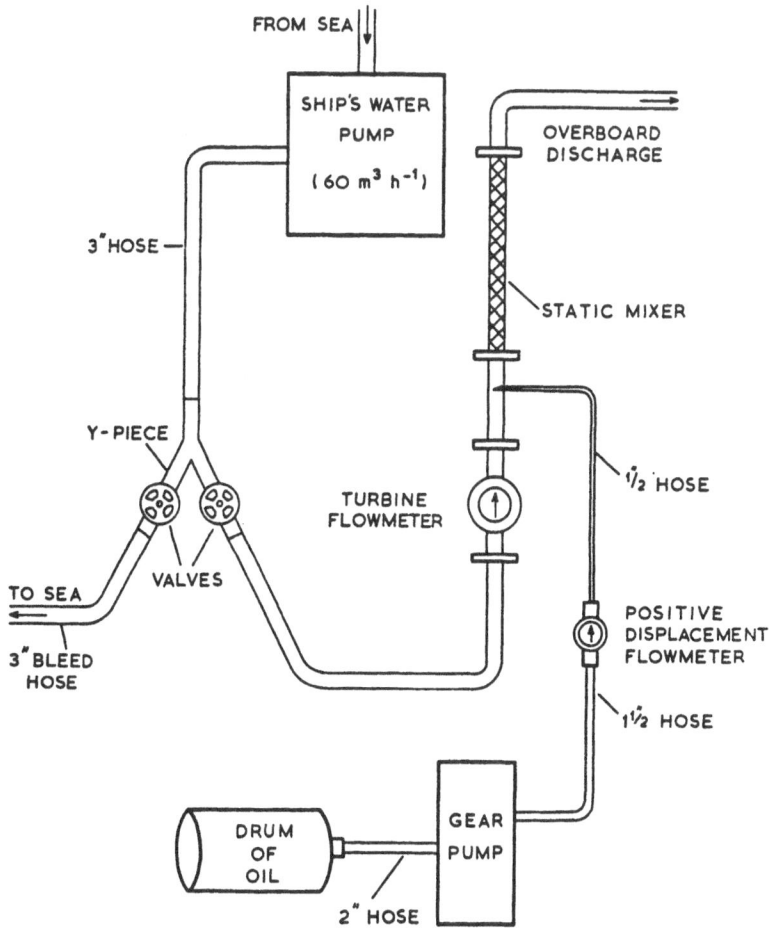

Figure 1.   Schematic layout of discharge apparatus.

Table 5.   Measurements taken during Isowake Experiment.

	09:00 GMT	13:00 GMT
Wind direction	300 degrees	300 degrees
Wind speed	10-12 knots	5 knots
Air temperature	16°C	21°C
Sea temperature	14°C	14°C
Wave height	0.5 m, sea slight	0.3 m, sea smooth
Visibility	Good, 10 km	Good, 10 km
Cloud type	Cirrus, cirrostratus	Cirrus, cirrostratus
and cover	3/8 cover	3/8 cover
Water depth	30-40 m	30-40 m

flowrate of water of 22 m^3 h^{-1}. A variable speed gear pump was
used to introduce oil into the water stream, and a positive
displacement flowmeter was used to measure the rate of oil discharge. The use of a static mixer ensured that the oil and water
were intimately mixed together.

Description of the Discharge Procedure

All the discharges took place within the area 02° 20' E -
02° 40' E, 52° 25' N - 52° 30' N, where Warren Spring Laboratory
has permission to discharge oil for research purposes. Throughout the discharges RV SEASPRING proceeded at 8 knots in a north-
westerly direction. The measurements that were taken during the
trial are given in Table 5.

The following procedure was followed when the infrared
line scanner sensor was used. The oil-water mixture was discharged into the wake to produce a slick about 1 km long. The
aircraft then flew directly over the wake at a height of about
150 m, and as soon as it reached SEASPRING the oil pump was
switched off. The next discharge was not made until SEASPRING
was clear of the previous slick.

A similar procedure was followed when the SLAR was used,
but for these discharges the aircraft flew in a box pattern
around the track of the vessel so that imagery was obtained with
the aircraft flying first perpendicular and then parallel to the
ship's track. In general, the aircraft was about 3 km away from
the ship when the SLAR imagery was obtained.

Table 6 lists all the discharges that were made, along with
the sensors that were used. In all cases, except discharge No. 8,
the flowrate of water was 22 m^3 h^{-1}. In the case of discharge
No. 8, no water was discharged.

Table 6.   List of discharges.

No	Time GMT	Oil type	Oil discharge rate l/min	Oil discharge rate l/mile	Sensor
1	0947	Kuwait Crude	5	37.5	IRLS
2	0954	Kuwait Crude	8	60.0	IRLS
3	1001	Kuwait Crude	30	225	IRLS
4	1008	–	0		IRLS
5	1018	No. 2 Fuel Oil	5	37.5	IRLS
6	1026	No. 2 Fuel Oil	9	67.5	IRLS
7	1032	No. 2 Fuel Oil	33	248	IRLS
8	1042	–	0		IRLS
9	1310	No. 2 Fuel Oil	9	67.5	SLAR
10	1319	No. 2 Fuel Oil	35	263	SLAR
11	1331	Kuwait Crude	30	225	SLAR
12	1344	–	0		SLAR
13	1357	Kuwait Crude	8	60.0	SLAR

It should be noted that three "blanks" were included; these are essential in ensuring unambiguous interpretation of the results.

RESULTS

No oil could be detected in the imagery of the SLAR runs in these experiments. This was the expected result because, as can be seen from the ground truth data, the sea was comparatively smooth during the experiments, and there was insufficient backscatter above background noise to produce a contrast with the oil-smoothed water. It has been demonstrated, however, that in somewhat rougher conditions, similar discharges to those used in the present experiments can be detected from a range of several kilometres by SLAR scanners[4]. The SLAR which successfully detected oil discharges used an X-band transceiver with vertical polarization, and it is probable that this system is more suitable for detecting oil than the Q-band horizontally polarized system used in the present experiments.

The imagery obtained in the infrared line scanner runs is shown in Figures 2-9. Oil was detected unambiguously by IRLS under these conditions, and neither the ship underway discharging clean water (run 4) nor the ship underway with no discharge (run 8) gave rise to wakes which could be mistaken for oil.

The lower limit of detection for IRLS was not determined

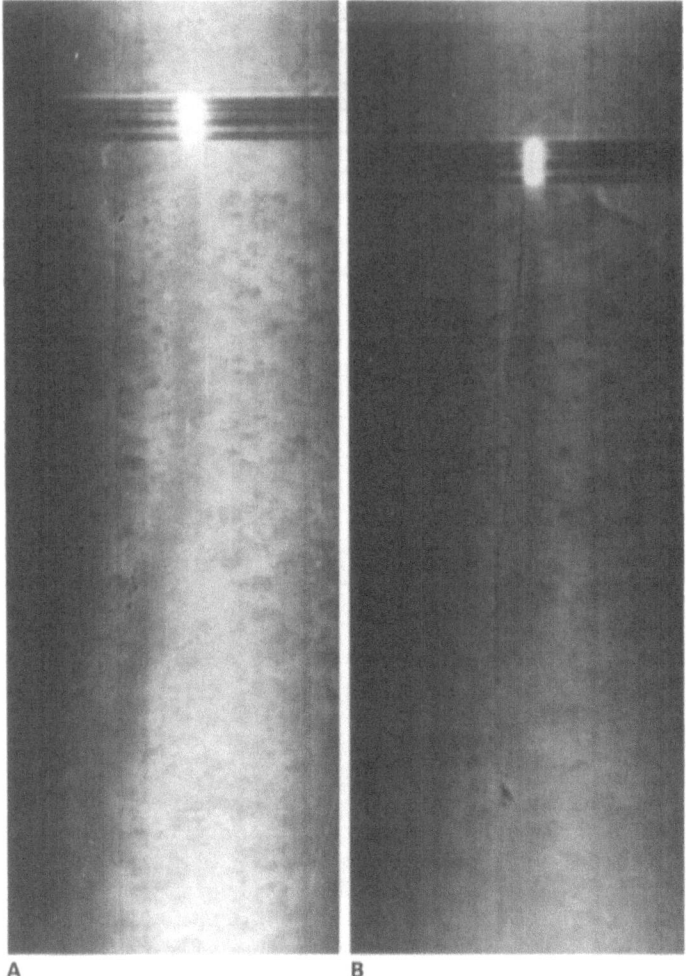

A   B

Figure 2.   IRLS imagery of Kuwait Crude Oil discharges.
             (A): 30 litres per mile
             (B): 60 litres per mile.

although under these conditions it is obviously quite close to
the lower discharge rate of 5 l min^{-1} (37.5 litres per mile).

    The No. 2 fuel oil discharges were more easily visible in
the imagery than the Kuwait Crude discharges at the corresponding rate, despite the latter being a thicker, blacker oil.
There are two possible explanations for this. Either the No. 2
fuel oil has a lower emissivity than the Kuwait Crude, and
therefore presents a better contrast with the surrounding water,

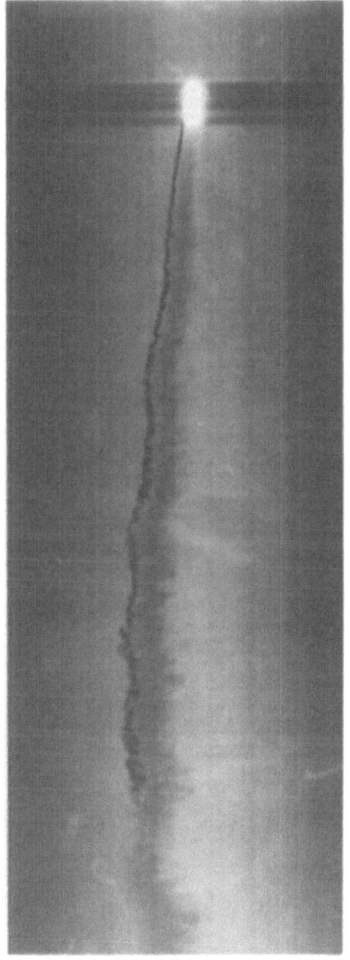

Figure 3.   IRLS imagery of Kuwait Crude Oil discharge
             (240 litres per mile).

or the No. 2 fuel oil spreads more rapidly and more evenly than
the Kuwait crude oil and presents a larger area to be detected.

DISCUSSION

     The use of IRLS for imaging oil in the wake of a ship
gives a discrimination which is not available using visual observation or photographic cameras. The limit of detection for
the IRLS is close to 30 litres per mile; by contrast, oil discharged at a rate of about 0.1 litre per mile is clearly
visible to the naked eye, but it is extremely difficult to

distinguish visually between oil discharged at a low rate and '
oil discharged at a much higher rate[2].

 In order to discuss the usefulness of the IRLS in detecting
illegal discharges, the various limits imposed on shipping can
be discussed separately.

Tankers within 50 Miles of Land

 No oil can be discharged within 50 miles of the UK coast,
and it is the enforcement of this regulation which is the first

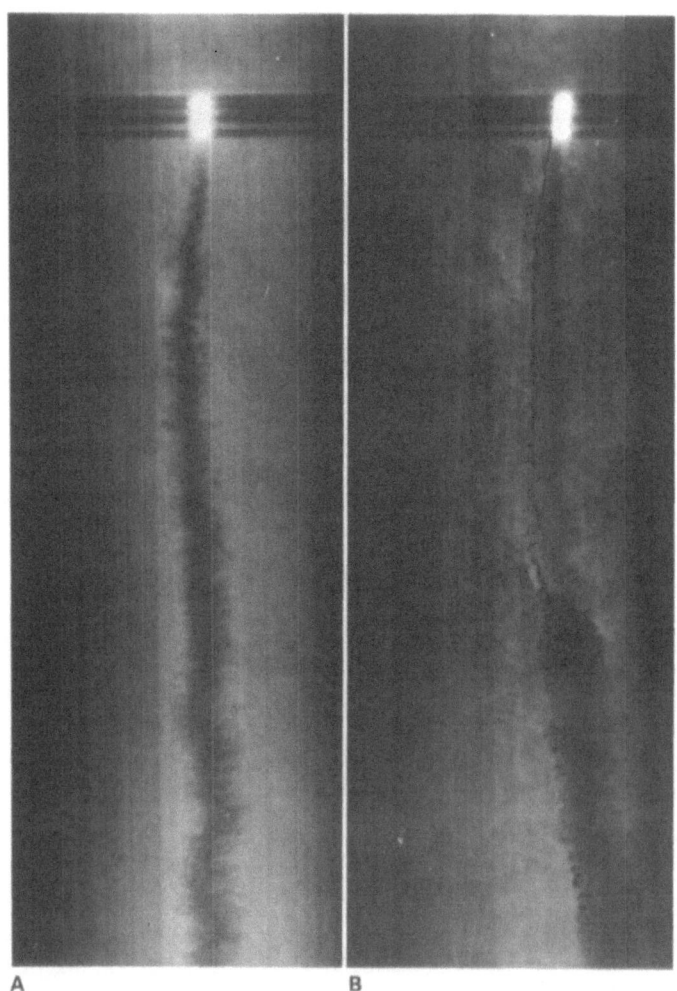

Figure 4. IRLS imagery of Gas Oil discharge (30 litres per mile)
    (A): Blank, water 22 m^3 h^{-1} discharge rate
    (B): Gas oil.

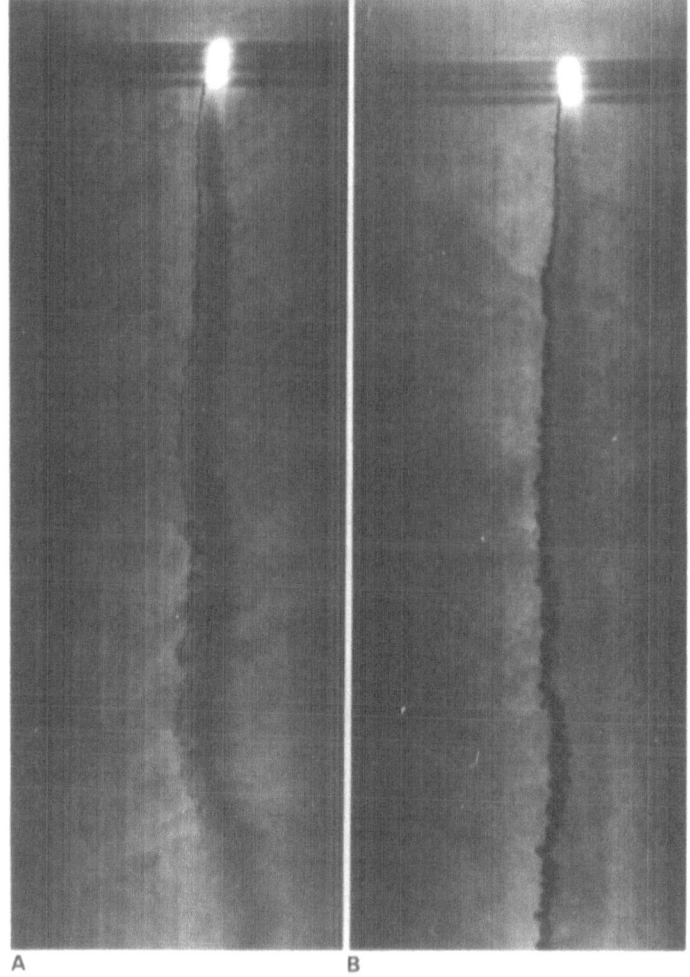

A   B

Figure 5.   IRLS imagery of Gas Oil discharges
            (A): 60 litres per mile
            (B): 240 litres per mile.

priority, since oil discharges in coastal regions are likely to
cause most damage, both to coastal amenities and to marine flora
and fauna.

    The IRLS is ideally suited to detect illegal discharges
within this region since it detects the oil when it exists in
amounts beginning to be significant in terms of pollution.

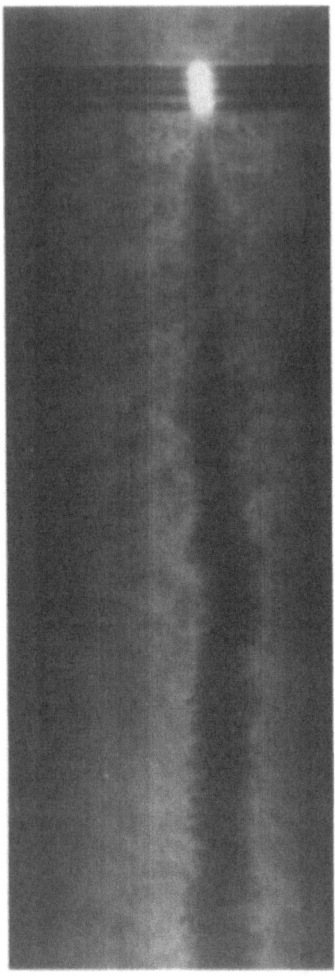

Figure 6.   Clean ship's wake as seen by the IRLS.
(blank, no discharge).

Tankers Further than 50 miles from Land

Outside the 50-mile limit oil may be discharged into the
sea, but the discharge rate must no exceed 60 litres per mile.
The IRLS will detect oil at this level but it cannot be used to
determine the precise discharge rate and, except in the case of
a gross discharge, it is unlikely that the imagery could be
used to show that the discharge rate had been exceeded. The
most useful procedure would be to take note of any ships discharging oil and then refer to the oil content meter to provide

evidence as to whether the discharge had been carried out
legally or illegally.

Outside the 50-mile limit the SLAR sensor becomes more
useful, owing to the large area of sea to be surveyed. A sensor
for giving quantitative data on the rate of oil discharge would
be the passive microwave imager, development of which is currently being funded by the UK Department of Trade.

Other Ships

This report has concentrated on discharges of oil from
tankers, however, all ships have the potential to discharge oil
when cleaning machinery space bilges. Such discharges are also
regulated by Annex I and may only take place if:

* The vessel is at least 12 miles from land and proceeding
  en route,
* The oil content of the effluent is less than 100 ppm.

However, discharges may take place when the ship is less
than 12 miles from land if the oil content is below 15 ppm.

The amount of oil discharged from bilges should be very
small, typically in the region of 0.1 litres per mile. It is
therefore very unlikely that an IRLS sensor would detect a
bilge water discharge unless it was massively in excess of the
permitted limit.

FURTHER WORK

An airborne surveillance system will only act to deter
ships for contravening the regulations of Marpol 73, if it is
capable of furnishing good evidence for prosecutions. Further
work is required on the IRLS sensor to resolve ambiguities
which may bring the evidence into question.

Detection of Oil at Night

It is quite likely that when oil is discharged illegally,
it will be discharged at night. It is understood that the IRLS
sensor will detect oil at night in exactly the same way as
during the day, but this has not been unequivocally demonstrated.

Effect of Changing the Location of the Discharge Outlet

Annex I requires that oil is discharged above the waterline. However, when discharging oil illegally, tanker operators
are likely to by-pass the normal discharge system and discharge

the oil below the waterline in the hope that this will further
dilute the slops and render the oil less visible. It is therefore necessary to check that oil discharged below the waterline and well mixed in the wake can also be detected.

Response to Substances other than Oil

Marpol 73/78 not only regulates discharges of oil but also
those of chemicals. Although many chemicals are components of
refined oil products, the discharge regulations for chemicals
are quite different from those which deal with oil, and perhaps
the most important difference is that chemicals may be discharged into the sea when a vessel is only 12 miles from land
(cf. 50 miles for oil). Since modern tankers may carry both oil
and chemical products at the same time, an operator might
attempt to claim that his ship was discharging chemicals
legally rather than oil illegally. Since most chemicals will
rapidly dissolve or evaporate on discharge into the sea, few
substances will form persistent slicks and it should be possible
to distinguish between oil and chemical discharges on this
basis. It is recommended that work be undertaken to determine
how this might be achieved in practice.

Positive Identification of Vessels

In addition to proving that a discharge contravenes the
Annex I regulations, it is essential to be able to positively
identify the offending vessel. Oblique angled cameras can be
used to photograph a ship during the day when its name will be
visible but this approach cannot be used at night. A low light
level television camera could possibly be used on such occasions but it might be more suitable to take low altitude
imagery of the ship using the IRLS sensor and use this for
identification if enough discriminating features are evident.

In addition to photographing/imaging the vessel, navigation data (i.e. ship's position, heading, etc.) should be obtained and ideally should be superimposed on the imagery. It
is recommended that flight trials on the identification of
ships be undertaken in conjunction with the above mentioned
work.

CONCLUSIONS

1. The feasibility of using an IRLS sensor to detect
   illegal discharges from oil tankers within the 50-mile
   prohibited zone has been demonstrated.

2. The SLAR sensor did not detect oil under the calm conditions of these tests.

3. Because the IRLS sensor does not provide thickness data, it can only be used to give a broad estimate of the amount of oil discharged. The possibility of using other sensors, such as the passive microwave imager and the laser fluorosensor, for obtaining more precise information on quantities of oil discharged should be investigated.

4. When the oil is discharged above the waterline it is possible to distinguish between oil and the wake of the ship.

5. Further work is required to ensure that the IRLS sensor provides unambiguous evidence which can be used in the courts. A system for positive identification of vessels both day and night should be developed for use in conjunction with the infrared line scanner.

ACKNOWLEDGEMENTS

We would like to thank the Royal Signals and Radar Establishment, Malvern, for their help in carrying out the work described in this report.

REFERENCES

1. K. G. Thomas, How significant is marine pollution, Paper given to the Conference on Reduction of Pollution from Shipping, London, Inst. Mar. Engrs., Dec. 1977
2. Establishment of criteria for determining pollution of the sea by oil (Note submitted by the United Kingdom Government), IMO Doc. OP/WG III/WP. 2, 4 December 1968
3. H. D. Parker and D. Cormack, "Evaluation of Infrared Line Scan (IRLS) and Side-looking Airborne Radar (SLAR) over Controlled Oil Spills in the North Sea", Warren Spring Laboratory report No. LR 315 (OP) (1979).
4. C. Colliander, "Aerial Surveillance Trial in the English Channel", Swedish Space Corporation report (1981).

The ISOWAKE-MED EXPERIMENT

S. Galli de Paratesi and B.M. Sørensen

Commission of the European Communities
Joint Research Centre
Ispra, Varese, Italy

ABSTRACT

Isowake-Med was an experiment of monitoring of standard oil spills which was performed in 1981 in the North Sea, some 40-50 miles west of The Hague (The Netherlands).

EXPERIMENT OBJECTIVES

The goal of the Isowake is the detection of oil films when spilled according to a legal procedure.

It is important to determine the limits of sensitivity in the detection of films by automatizable techniques (which could serve as a basic for an operational surveillance system): reflectometry, UV induced fluorescence, short/long level absorption spectrometry, microwaves and chromatographic reference techniques. It has to be shown to what extent these measurements techniques will be able to be adapted to the subsequent construction of operational detection and measurement networks. In particular, it is important to study the influence on detection thresholds, of atmospheric conditions (optical transmission, wind), of the state of the seas (waves and currents).

In the case of Isowake the potentiality of radar was investigated. This initiative was taken as a consequence of the SAR-580 availibility. SAR-580 is a Canadian aircraft equipped with X-, L- and C-bands SAR which flew in Europe under contract JRC/ESA in May/June 1981.

The spills were overflown with the following objectives:

* evaluating the capability of detecting a few distinctly
  different types of oil by SLAR (SAR and RAR) when spilled
  within legal limits[2.3].

* making sequential measurements of the oil spills with
  SLAR and UV/IR line scanner until oil can no longer be
  detected.

* investigating the differences between HH and VV polarization in SLAR detection of clutter suppression resulting
  from oil on the sea surface.

* evaluating the differences in SAR backscattering from
  oil covered and non-covered sea surface resulting from
  the use of steep and shallow depression angles.

* investigating on the C-band capability of oil detection
  by SAR.

* evaluating the detection differences between SAR and
  RAR when flown simultaneously and/or sequentially over
  the same spill with the same polarization.

ISOWAKE STANDARDS

The following concepts stand behind the experimental oil
spills:

Till 1981, no report had clearly defined the smallest detectable concentration of oil for any of the available remote
sensing systems. Nor did any report clearly describe the
difference in detectability of the various types of oil.

Investigating on these problems with an arbitrary or even
iterative approach is both impractical and prohibitively expensive. Therefore in dealing with the detectability of various
types of oils, Isowake referred to the following documents of
the International Maritime Organization (IMO):

A. Recommendation on International Performance and Test
   Specification for Oily-Water Separating Equipment and
   Oil Content Meters, Resolution A.393 (X), Nov. 14, 1977,
   Part II, Paragraph 2.1.6.:

   "Changing the feed to the separating equipment from
   oily water to oil, or from oil and or water to air
   should not result in the discharge overboard of any
   mixture containing more than 100 ppm of oil".

B. International Convention for the Prevention of Pollution
   from Ships, 1973, Regulation 9, Paragraph 1.a, Section
   IV:

".... any discharge into the sea of oil or oily mixtures from ships to which this Annex applies shall be prohibited except when the following conditions are satisfied:
..... the instantaneous rate of discharge of oil content does not exceed 60 litres per nautical mile".

In 1978 a group of experts from Warren Spring Laboratory (United Kingdom) proposed a standard method for spilling of oil for experimental remote sensing purposes, to be executed within the legal limits as referred to the above A and B points. It called for a series of tests which were designated as International Standardized Oil Wake Experiment (ISOWAKE). They were agreed by the NATO/CCMS working group 1.

As experimental guidelines it was proposed as follows:

"Fishing vessels will be employed to spill Diesel, Crude and Bunker C oils. Vessels shall be of 100 tons or more to reduce rapid movements and hence steady the discharge. They will steam into wind and waves to reduce disturbances to ship's wake, and have a speed of 8 knots. The oil discharge will take at port side approximately 0.5 m above average sea level.

The following oil quantities will be spilled from each vessel:

* 30 l/mile + sea water to make 3,000 l/mile
* 60 l/mile + sea water to make 3,000 l/mile
* 150 l/mile + same volume sea water
* 1,000 l/mile + same volume sea water
* 100 ppm oil in 10 m^3 sea-water/hour (bilge water simulation .

The main advantage of this experimental procedure is that spilling itself can be carried out in some standard and reproducible way everywhere all over the ocean within a wide range of environmental conditions.

EXECUTION OF THE FIELD PROGRAMME

Isowake-Med was developed as an international co-operative oil spill experiment co-ordinated by the Joint Research Centre, Ispra (JRC) (Table 1).

This initiative was taken as a consequence of the SAR-580 availability in Europe. SAR-580 is a Canadian airborne SAR equipped with X-, L- and C-bands, which flew over the European test-sites in May/June 1981 under a contract JRC/ European Space Agency.

Table 1. Participation in the Isowake-Med Experiment 1981.

---

* Joint Research Centre of the European Communities,
  Ispra, Italy.

* Rijkswaterstaat, North Sea Directorate, Rijswijk,
  The Netherlands.

* Ente Nazionale Idrocarburi, Milan, Italy.

* Swedish Space Corporation, Solna, Sweden.

* Swedish Coast Guard, Sweden.

* Technical University of Denmark, Lingby, Denmark.

* Nordan Pollution Control, Kertemunde, Denmark.

* Terma, Electronic, Aarhus, Denmark.

* Water Quality Institute, Horsholm, Denmark.

Experiment Plan

        Originally the test-site was chosen in the Mediterranean
Sea. Three SLARs (SAR-580), RARs from the Swedish Coast Guard
and from the Technical University of Denmark and one UV/IR from
Swedish Coast Guard were considered to be flown over the
spillage area.

        The experiment had to be moved to the North Sea, some
40-50 miles west of the Hague (The Netherlands) (approximately
52° N, 3°20' E) on short notice because of difficulties in obtaining a permission from Italian authorities to spill the
400-500 litres of test oil in the Mediterranean Sea. On request
of Rijkswaterstaat, the Netherlands provided two vessels, the
permission to spill and the necessary oil.

        Fortunate circumstances allowed for rescheduling of the
SAR-580 flight plan so that it could fly over the spill area
on July 1st, 1981 contemporaneously with the two RARs from
Swedish Coast Guard and Technical University of Denmark. On
July 2nd a new series of spills were performed with SCG and TUD
overflights. Upon completion on this experiment, the aircraft
was made available to fly over a 200 lt controlled oil spill
arranged by Rijkswaterstaat to test a newly constructed Dutch
oil combatting vessel "COSMOS".

Description of the Campaign

Day 1: On July 1st (Day 1) the weather conditions were
very rough. Strong wind (force 4) and swell made operation at
sea difficult. This made utilization of sea-truth's instrumentation impossible. Oceanographic and radiometric measurements
were cancelled but an attempt was made to carry on with the
planned aerial activities.

Four discharges of two types of oil (Crude oil and Diesel
fuel oil No.2) for a total amount of a 400 liters were spilled
in quantities of 60 and 150 l/mile (Table 2). The spilling
equipment showed to run satisfactorily with oil of low and
middle viscosity only. One was obliged to cancel also oil sampling owing to sea conditions.

Totally overcast sky with low cloud ceiling severely impaired aircraft operations. In fact, the Swedish Coast Guard
aircraft did not find the oil spills and missed half of them
due to navigation difficulties. The Technical University of
Denmark aircraft only passed it once and covered oil by the
radar.

The SAR-580 had some problems with its Inertial Navigation
System (INS) and arrived one hour late on the scene. They made
an attempt to up-date the INS by descending below the cloud
cover and identifying the oil visually before flying the to-
and-fro scheduled lines at 10,000 ft altitude. The SAR-580

Table 2. Oil spills and some significant atmospheric
          and sea conditions.

Date	Time*	Oil type	Quantity l/nm	Wind Speed m/s	Significant wave height m
July 1 Day 1	10.00	Crude oil	60	7	0.78
		Crude oil	150		
		Diesel Fuel oil No.2	150		
	11.00	Diesel Fuel oil No.2	60	7.5	0.85
July 2 Day 2	08.00	Fuel oil No.6	30	6	1.05
		Diesel Fuel oil No.2	150		
		Fuel oil No.6	30		
	09.00	Fuel oil No.6	60	5.5	1.13

*Local time (GMT + 2H)

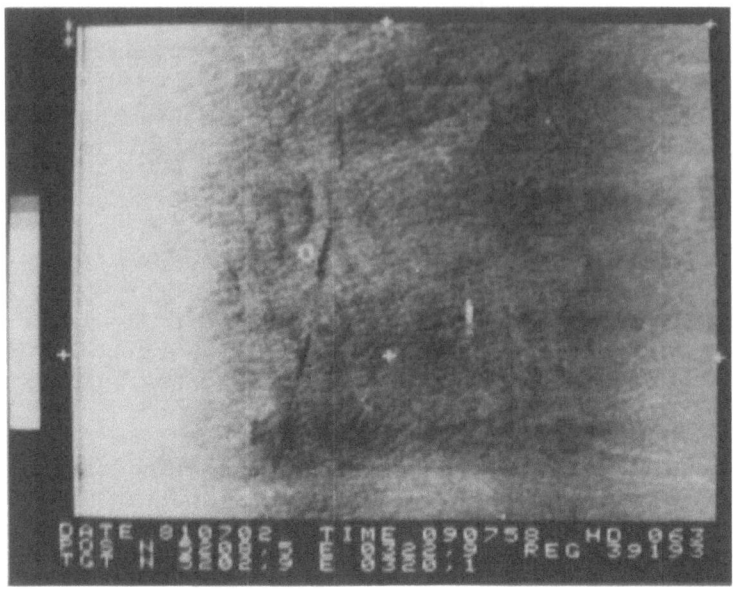

Figure 1.   RAR image of the four 1-nautical mile long spills
            of different types and quantities of oil.

real-time product (quick look) did not show the oil wakes and
neither the optically processed data did.

    The Diesel fuel oil No. 2 disappeared within one and half
hour from spilling, while the fuel oil No. 6 remained visible
considerably longer.

    Day 2 (contingency): In view of the failure to acquire
useful aerial data, the experiment was repeated, even though
it was not possible to reschedule the SAR-580 participation.

    Weather had improved considerably. The complete overcast
and low cloud ceiling, however, remained unchanged from the day
before. In spite of the reduced wind speed, the significant
wave height had increased due a strong swell.

    Four oil spills were discharged for a total amount of
about 300 litres (Figure 1). Spilling of fuel oil No. 6 was
made in quantities of 30 and 60 litres/mile, while Diesel oil
No. 2 was only discharged as a 150 litres/mile spill. The
Diesel fuel oil No. 2 wake vanished again within hours in spite
of the reduced windspeed. Oil sampling was cancelled again.
Same difficulty as for the day before to take local ship-obser-
vation due to sea conditions.

Both RARs and IR/UV line scanner covered oil spills. The
RARs, (VV and HH polarizations) were able to detect all the
wakes at least until one hour after the spilling (Figure 1).
VV polarization appeared more suitable for oil slick reconnaissance.

In the afternoon the new Dutch oil combatting vessel
"COSMOS" was tested, and both Swedish Coast Guard RAR and UV/IR
line scanner flew a number of successful sorties over the 200
ton oil spill discharges for the experiment.

In all three days some floating corner reflectors were
lowered into the sea surface as reference points in the scene
for radar intensity calibration (Figure 2).

CONCLUSIONS AND RECOMMENDATIONS

During the Isowake experiment 1981 in the North Sea, the
oil spilling followed strictly the international guidelines
agreed by NATO/CCMS.

The relatively rough weather-and-sea conditions allow the
experimental campaign to be partially successful only. On the
other hand, they provided a unique opportunity to carry on the
experiment under realistic oil spill surveillance conditions.

During both days, the high sea impaired sampling and
radiometric measurements, but it had no significant impact on
the spilling.

Failure of the INS mounted on SAR-580 was a significant
drawback which did not allow to reach all appointed goals. In

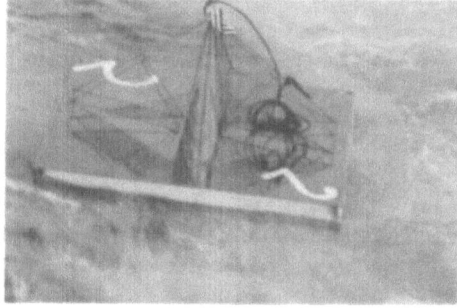

Figure 2.   A radar reflector being lowered into the sea. It
            was used as a position reference in the radar
            image for oil spill identification when carrying
            out visual inspection.

particular, no comparison between SAR and RAR could be carried out simultaneously applied to the same standard oil spill.

Some useful conclusions and recommendations can however be draught:

1. (First day): Diesel fuel oil No. 2 was visible after spilling at 60 1/n mile. It began to disintegrate one hour later and was no more detectable within two hours for both 60 and 150 1/n mile discharge.

Crude oil remained clearly visible 4-5 hours after discharge at 60 1/n mile. Traces remained the following day when spilled at 150 1/n mile.

2. (2nd day): Diesel Fuel oil No. 2 vanished again within 2 hours when discharged at 150 1/n mile.

Fuel oil No. 6 remained visible considerably longer even at 30 1/n mile.

3. Diesel fuel oil No. 2 suggested by NATO/CCMS demonstrated to be critical for remote detection. It disappears so fast from the experimental scene that time delay between spilling and overflight becomes critical to the point at which it becomes impractical. This suggests another type of oil to be chosen for the Isowake standard, e.g. U.S. standard fuel oil No. 6.

4. Using such a different type of oil, Isowake experiments can still be conducted at wind force 4 (7 m/s) or higher.

5. Crude oil demonstrated to be remotely detectable and suitable for spilling experiments.

6. Both airborne real aperture radars (VV and HH polarization) detected all the oil spills at least until one hour after the spilling. VV polarization appears more suitable for oil slick reconnaissance.

7. The UV/IR scanner system confirmed the usefulness of such system to obtain high resolution imagery at chosen range. IR channel can give information on the spreading of oil also indicating the relative oil thickness within the oil spill. The UV channel adds confidence to IR registrations (Figure 3).

8. Oil sampling was not possible to perform. The smallest remotely detectable concentration of oil by remote sensing systems (i.e. UV/IR and microwave) does remain to be single out through future experiments.

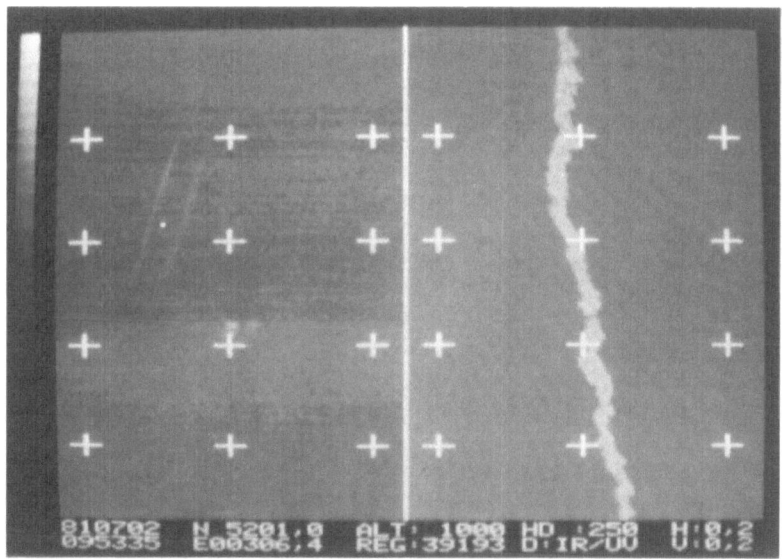

Figure 3. Ultraviolet image of a 150 litres/nautical mile spill
(on the right side of the photograph).

9. The joint campaign managed by the Joint Research Centre
required experienced participation in a number of scientific
and engineering fields. It was necessary to construct new equipment such as the "spill-kit" and the floating corner reflectors
(Figure 2). Some oil sampling schemes as for taking oil thickness samples by means of absorbent pads were designed. Such a
technical, logistic and organisation experience may be usefully
available to some future similar initiatives.

10. Obtaining the permission to spill a few hundred litres
of oil in the Mediterranean Sea was the most time consuming part
of the preparation of the experiment, and all in vain. It is
expected that a standard procedure will be set up to obtaining
permission of discharging small quantities of oil into the sea
for scientific purposes.

One can conclude that, in spite of the adverse meteorological conditions and some technical drawbacks, the Isowake-Med
has provided a set of useful informations to develop a standardized oil detection target for the examination of various remote
airborne sensors performance

REFERENCES

Sørensen, B. and Sturm, B., 1982, Some results from experiments

on remote sensing of water quality and oil pollution in
the Mediterranean Sea, in: "Remote Sensing of Arid and
Semi-arid Lands, ERIM, El Cairo, January 1982".
Galli de Paratesi, S., Sørensen, B. M., 1982, Report on the
Isowake-Med experiment, Programme Progress Report,
Remote Sensing from Space, Cat. 18, Nr. 3989, JRC, Ispra.
"International Convention for the Prevention of Pollution from
Ships", 1973, International Maritime Organisation,
London.
"Recommendation on International Performance and Test Specifi-
cations for Oily Water Separating Equipment and Oil
Content Meters", 1977, International Maritime Organisa-
tion, London.
"Isowake: Procedures of International Standardized Oil Wake
Experiment", 1980, NATO/CCMS, Brussels.

(Editor's note: This paper was sollicited by the Paris workshop
organisation committee but the authors were unable to present
it to the workshop. It is included as an addendum to the proceedings, because of its pertinence and interest to the pilot
study participants.)

CONTRIBUTORS

Mr. M. Albuisson
Centre de Télédétection et d'Analyse des Milieux Naturels
Ecole Nationale Supérieure des Mines de Paris
Rue Claude Daunesse
Sophia Antipolis
06565 Valbonne Cedex, France

Mr. R. Alderman
National Oceanic and Atmospheric Administration (NOAA)
Code RDI
Rockville, MD 20852, United States

Mr. U. Alvarado
General Electric Company
P.O. Box 8555
Philadelphia, PA 19101, United States

Mr. J. Apel
National Oceanic and Atmospheric Administration (NOAA)
Pacific Marine Environmental Laboratory
3711, 15th Avenue, N.E.
Seattle, WA 98105, United States

now with

The Johns Hopkins University
Applied Physics Laboratory
Johns Hopkins Road
Laurel, Maryland 20707, United States

Mr. L. Backlund
Swedish Space Corporation
Tritonwägen 27
S-17154 Solna, Sweden

Mr. W.E. Brown Jr.
Jet Propulsion Laboratory
4800 Oak Grove Drive - Bldg 264-420
Pasadena, California 91109, United States

Mr. H. Brunsveld van Hulten
Directorate for Water Management and Hydraulic Research
District Kust en zee
Rijkswaterstaat
400 Van Alkemadelaan
2597 AT 's Gravenhage, The Netherlands

Mr. M. Brussieux
Centre Océanologique de Bretagne (COB)
Centre National pour l'Exploitation des Océans (CNEXO)
BP 337
29273 Brest Cedex, France

Ms. L. Buja Bijunas
Canada Centre for Remote Sensing
2464 Sheffield Road
Ottawa, Ontario, K1A OY7 Canada

now with

Design and Development Division
Ontario Hydro
700 University Avenue
Toronto, Ontario, M5G 1X6 Canada

Mr. R. Burkhalter
Laboratoire National d'Essais
1 Rue Gaston Boissier
75015 Paris, France

Mr. H.M. Byrne
Pacific Marine Environmental Laboratory
National Oceanic and Atmospheric Administration (NOAA)
3711, 15th Avenue N.E.
Seattle, Washington 98105, Unites States

Mr. Y. Caltabiano
Department of Engineering
Machinery Institute
University of Catania
Viale Andrea Doria 6
95125 Catania, Italy

Mr. G. Chafin
U.S. Environmental Protection Agency (EPA)
Room 819, West Tower (A-106)
401, M street S.W.
Washington, D.C., 20460, United States

Dr. D. Cormack
Warren Spring Laboratory
Gunnels Wood Road
P.O. Box 20
Stevenage, Hertfordshire SG1 2BX, United Kingdom

now with

Department of Transport (Marine Division)
Marine Pollution Control Unit
Sunley House, 90/93 High Holborn
London WC1V 6LP, United Kingdom

Mr. W.F. Croswell
National Aeronautics and Space Administration (NASA)
United States

Mr. K. Dagg
The Genesys Group
1755 Courtwood Crescent
Ottawa, Ontario K2C 3J2, Canada

now with

Computer and Information Systems Branch
Department of Communications
Room 862,,Journal Tower North
300 Slater Street
Ottawa, Ontario K1A OC8, Canada

Mr. G.P. de Loor
Institute of Applied Physics TNO
Stieltjesweg 1
P.O. Box 155
Delft, The Netherlands

Mr. J. De Villiers
Canada Centre for Remote Sensing (CCRS)
DEMR
2464 Sheffield Road
Ottawa, Ontario, K1A OY7, Canada

Mr. G. Duchossois
Directorate of Application Programmes
European Space Agency (ESA)
8-10 Rue Mario Nikis
75015 Paris, France

Mr. O. Fäst
Swedish Space Corporation
Tritonwägen 27
S-17154 Solna, Sweden

Mr. A. Fontanel
Institut Français du Pétrole
1-4 Avenue de Bois Préau
92506 Rueil Malmaison Cedex, France

now

Directeur Général
SPOT Image (Société de Distribution des Images SPOT)
Centre Spatial de Toulouse
18 Avenue Edouard Belin
31055 Toulouse Cedex, France

LCDR D. Freezer
U.S. Coast Guard Headquarters
COMDT (G-DOE-2/TP54)
400 7th Street, S.W.
Washington, D.C., 20590

Dr. S. Galli de Paratesi
Commission of the European Communities
Joint Research Centre, Ispra Establishment
21020 Ispra (Varese), Italy

Mr. A. L. Gray
Canada Centre for Remote Sensing (CCRS)
2464 Sheffield Road
Ottawa, Ontario, K1A OY7, Canada

Mr. F. Günneberg
Bündesanstalt für Gewasserkunde
Postfach 309
D 54000 Koblenz, Federal Republic of Germany

Mr. R. K. Hawkins
Canada Centre for Remote Sensing (CCRS)
2464 Sheffield Road
Ottawa, Ontario, K1A OY7, Canada

Mr. J.P. Hollinger
Department of the Navy
Naval Research Laboratory
Washington, D.C., 20375, United States

Mr. W.A. Hovis
Director, Satellite Experimental Laboratory
National Oceanic and Atmospheric Administration (NOAA)
NOAA/NESS-S/32
Washington, D.C., 20233, United States

Mr. N. Hurford
Warren Spring Laboratory
Gunnels Wood Road
P.O. Box 20
Stevenage, Hertfordshire 2G1 2BX, United Kingdom

Mr J. D. Kingham
Director
Environment Canada
15th Floor, Place Vincent Massey
Ottawa, Ontario, K1A 1C8, Canada

Mr. J. S. Knoll
National Oceanic and Atmospheric Administration (NOAA)
National Environmental Satellite, Data, and Information
Service
Satellite Experimental Laboratory
Federal Building No. 4, room 0128
Washington, D.C., 20233, United States

Mr. F. Lacoste
Thomson CSF
Division Equipements Avioniques
Rue Toussaint Catros
33160 La Haillan, France

Mr. R. Landers
U.S. Environmental Protection Agency (EPA)
EMSL - LV/EPA
P.O. Box 15027
Las Vegas, Nevada 89114, United States

Capt. T. Lutton
U.S. Coast Guard Headquarters
COMDT (G-DOE/TP54)
400, 7th Street, S.W.
Washington, D.C., 20590

Mr. G. Massart
Centre National pour l'Exploitation des Océans
Centre Océanologique de Bretagne (COB)
B.P. 337
29273 Brest Cedex, France

Mr. J.M. Massin
Secrétariat d'Etat auprès du Premier Ministre,
Chargé de l'Environnement et de la Qualité de la Vie
Direction de la Prévention des Pollutions (SE/EM)
14 Boulevard du Général Leclerc
92524 Neuilly-sur-Seine Cedex, France

LCDR R. L. McFadden
U.S. Coast Guard Headquarters
COMDT (G-WEP65/73)
400, 7th Street, S.W.
Washington, D.C., 20590, United States

Mr. R. A. Mennella
Department of the Navy
Naval Research Laboratory
Washington, D.C., 20375, United States

Mr. Cl. Meyer
Laboratoire National d'Essais
1 Rue Gaston Boissier
75015 Paris, France

Mr. P. R. Morris
Warren Spring Laboratory
Gunnels Wood Road
P.O. Box 20
Stevenage, Hertfordshire SG1 2BX, United Kingdom

Mr. R. A. Neville
Intera Environmental Consultants Ltd.
785 Carling Avenue
Ottawa, Ontario, K1S 5H4, Canada

now with

    Canada Centre for Remote Sensing (CCRS)
    Sensor section, Data acquisition Division
    2464 Sheffield Road
    Ottawa, Ontario, K1A OY7, Canada

Mr. D. Nicholson
Royal Signals and Radar Establishment (RSRE)
St. Andrews Road
Great Malvern, Worcestershire WR14 3PS, United Kingdom

Mr. R. A. O'Neil
Canada Centre for Remote Sensing (CCRS)
2464 Sheffield Road
Ottawa, Ontario, K1A OY7, Canada

Mr. H. Parker
Warren Spring Laboratory
Gunnels Wood Road
P.O. Box 20
Stevenage, Hertfordshire SG1 2BX, United Kingdom

now with

International Tanker Owners Pollution Federation Ltd.
(ITOPF)
Staple Hall, Stonehouse Court,
87-90 Houndsditch
London EC3A 7AX, United Kingdom

LT. W. E. Plage
U.S. Coast Guard Headquarters
Washington, D.C., 20593, United States

Mr. D. Rayner
National Research Council of Canada
Division of Chemistry
Sussex Drive
Ottawa, Ontario, K1A OR6, Canada

LT. R. Schmidt
U.S. Coast Guard Headquarters
COMDT (G-DOE-2/TP54)
400, 7th Street, S.W.
Washington, D.C., 20590, United States

Mr. C. Smorenburg
Institute of Applied Physics TNO-TH
Stieltjesweg 1
P.O. Box 155
Delft, The Netherlands

Mr. B. M. Sørensen
President
Intradan, Environmental Consultants A/S
Vesterbrogade 2D
1620 Copenhagen V, Denmark

Mr. R. Spanhoff
Rijkswaterstaat
Directorate for Watermanagement and Hydraulic Research
Hooftskade 1
2526 KA 's Gravenhage, The Netherlands

Mr. W. Strome
Chief Application Division
Canada Centre for Remote Sensing (CCRS)
2464 Sheffield Road
Ottawa, Ontario, K1A OY7, Canada

Mr. V. Thomson
Intera Environmental Consultants Ltd.
785 Carling Avenue
Ottawa, Ontario, K1S 5H4, Canada

now with

National Research Council of Canada
Division of Mechanical Engineering
Building M-2, Montreal Road
Ottawa, Ontario, K1A OR6, Canada

Mr. H. Visser
Institute of Applied Physics TNO-TH
Stieltjesweg 1
P.O. Box 155
Delft, The Netherlands

CDR R. Vollmers
NATO/CCMS Working Group I Chairman
U.S. Coast Guard Headquarters
Office of Research and Development
400, 7th Street, S.W.
Washington, D.C., 20593, United States

LT. G. S. Voyik
U.S. Coast Guard Headquarters
Washington, D.C., 20590, United States

Mr. L. Wald
Centre de Télédétection et d'Analyse des Milieux Naturels
Ecole Nationale Supérieure des Mines de Paris
Rue Claude Daunesse
Sophia Antipolis
06565 Valbonne Cedex, France

LCDR J. R. White
U.S. Coast Guard Headquarters
COMDT (G-DOE-2/TP54)
400, 7th Street, S.W.
Washington, D.C., 20590, United States

Mr. Wijmans
Rijkswaterstaat
North Sea Directorate
P.O. Box 5807
2280 HV Rijswijk, The Netherlands

PARTICIPANTS

BELGIUM

    ✹Mr. T.G. Jacques
    Unité de Gestion du Modèle Mathématique Mer du Nord
    et Estuaire de l'Escaut
    Ministère de la Santé Publique et de la Famille
    Manhattan Center,
    3 Rue des Croisades
    1000 Brussels

    Mr. G. Pichot
    Unité de Gestion du Modèle Mathématique Mer du Nord
    et Estuaire de l'Escaut

    (same address as Mr. Jacques)

CANADA

    Ms. L. Buja Bijunas
    Canada Centre for Remote Sensing
    Ottawa, Ontario, K1A OY7

  now with

    Design and Development Division
    Ontario Hydro
    700 University Avenue
    Toronto, Ontario, M5G IX6

    Mr. J. De Villiers
    Canada Centre for Remote Sensing (DEMR)
    2464 Sheffield Road
    Ottawa, Ontario, K1A OY7

✹Principal national contact
✿Ex-national contact
★Pilot study Steering Committee member

Mr. R.K. Hawkins
Canada Centre for Remote Sensing
2464 Sheffield Road
Ottawa, Ontario, K1A OY7

Mr. J.S. Klenavic
Environmental Protection Service
Environment Canada
Place Vincent Massey
Ottawa, Ontario, K1A OH3

Mr. R.A. Neville
Intera Environmental Consultants Ltd.
785 Carling Avenue
Ottawa, Ontario, K1S 5H4

now with

Sensor Section, Data Acquisition Division
Canada Centre for Remote Sensing
2464 Sheffield Road
Ottawa, Ontario, K1A OY7

★✹ Mr. R.A. O'Neil
Canada Centre for Remote Sensing
2464 Sheffield Road
Ottawa, Ontario, K1A OY7

Mr. W. Strome
Chief, Application Division
Canada Centre for Remote Sensing
2464 Sheffield Road
Ottawa, Ontario, K1A OY7

Mr. V. Thomson
Intera Environmental Consultants Ltd.
785 Carling Avenue
Ottawa, Ontario, K1S 5H4

now with

Division of Mechanical Engineering
National Research Council of Canada
Building M-2, Montreal Road
Ottawa, Ontario, K1A OR6

DENMARK

✹ Mr. B.M. Sørensen
Intradan Environmental Consultants Ltd.
Vesterbrogade 2D
DK 1620 Copenhagen V

FRANCE

Mr. R. Burkhalter
Laboratoire National d'Essais
1 Rue Gaston Boissier
75015 Paris

Mr. R. Camps
Direction Environnement
Société Nationale Elf Aquitaine
Les Miroirs, 18 Avenue d'Alsace
92400 Courbevoie

Mr. J.Y. Comby
Centre de Documentation, de Recherche et d'Expérimentations
sur les Pollutions Accidentelles des Eaux (CEDRE)
C.O.B. Plouzané
B.P. 308 F
29274 Brest Cedex

Mr. A. Fontanel
Institut Français du Pétrole
1-4 Avenue de Bois Préau
B.P. 311
92506 Rueil Malmaison Cedex

now

Directeur Général
SPOT Image (Société de Distribution des Images SPOT)
Centre Spatial de Toulouse
Avenue Edouard Belin
31055 Toulouse Cedex

Mr. M. Houdart
Direction des Programmes
Centre National pour l'Exploitation des Océans (CNEXO)
66 Avenue d'Iéna
B.P. 107.16
75763 Paris Cedex 16

Mr. F. Lacoste
Division Equipements Avioniques
Thomson C.S.F.
Rue Toussaint Catros
33160 Le Haillan

✡Mr. L. Laidet
Groupement pour le Développement de la Télédétection
Aérospatiale (G.D.T.A.)
18 Avenue Edouard Belin
31055 Toulouse Cedex

now

    Scientific Attaché
    French Embassy
    2011 "I" (Eye) Street NW
    Washington, D.C., 20006, United States

    Mr. P. Lefebvre du Prey
    Service Recherches Commerciales
    Division Equipements Avioniques
    Thomson C.S.F.
    178 Boulevard Gabriel Péri
    92242 Malakoff Cedex

    Mr. J.P. Le Gorgeu
    Administrateur Adjoint, Chargé de la Recherche
    Groupement pour le Développement de la Télédétection
    Aérospatiale (GDTA)
    18 Avenue Edouard Belin
    31055 Toulouse Cedex

now with

    SPOT Image (Société de Distribution des Images SPOT)
    Centre Spatial de Toulouse
    18 Avenue Edouard Belin
    31055 Toulouse Cedex

    Mr. G. Massart
    Centre Océanologique de Bretagne (COB)
    Centre National pour l'Exploitation des Océans (CNEXO)
    Le Plouzané
    B.P. 337
    29273 Brest Cedex

★✹Mr. J.M. Massin
    Service de l'Eau, Sous-direction des Eaux Marines
    Direction de la Prévention des Pollutions
    Secrétariat d'Etat auprès du Premier Ministre, chargé de
    l'Environnement et de la Qualité de la Vie
    14 Boulevard du Général Leclerc
    92524 Neuilly sur Seine Cedex

    Mr. A. Prouvost
    Mission Interministérielle de la Mer
    Ministère de la Mer
    9-11 Rue Georges Pitard
    75015 Paris

    Mr. A. Roussel
    Centre National pour l'Exploitation des Océans
    66 Avenue d'Iéna, Paris

Ms. C. Valerio
Centre d'Etudes Techniques de l'Equipement
Division Labo
Rue Einstein
13762 Les Milles

Mr. L. Wald
Centre de Télédétection et d'Analyse des Milieux Naturels
Ecole Nationale Supérieure des Mines de Paris
Rue Claude Daunesse
Sophia Antipolis
06565 Valbonne Cedex

GERMANY (FEDERAL REPUBLIC OF)

★✳Mr. F. Günneberg
Bundesanstalt für Gewâsserkunde
Postfach 309
D 5400 Koblenz

GREECE

★✳Mr. M. Moutsoulas
National Centre for Space Research
Z Road, Hellinikon
Athens

Mr. T. Xanthopoulos
Ecole Polytechnique
Université d'Aristote
Thessaloniki

ITALY

Ms. G. D'Agostino
Direzione Generale Demanio e Porti
Ministero Marine Mercantile
Via Le Asia
Roma 00100

Mr. E. Carloni
Centre for Study and Applications of Advanced Technology
(C.S.A.T.A.)
Via Amendola 173
70126 Bari

Mr. G.P. Francalanci
AGIP S.p.A.
Exploration et Production Hydrocarbures
P.O. Box 12069
Milan

Mr. A.L. Geraci
Machinery Institute, Department of Engineering
University of Catania
95125 Catania, Sicily

Mr. . Passino
Consiglio Nazionale delle Recerche
Instituto di Recerca sulle Acque
Via Reno 1 C.A.P.
00198 Roma

THE NETHERLANDS

✪ Mr. H.W. Brunsveld van Hulten
Directorate for Water Management and Hydraulic Research
District Kust en Zee
Rijkswaterstaat
400 Van Alkemadelaan
2597 AT 's-Gravenhage

Ms. J. Buzeman
Head of Bureau for Foreign Relations
Toegepast -Natuurwetenschappelljk Onderzoek (TNO)
148 Juliana ven Stolberglaan
2595 CL 's-Gravenhage

Mr. K. Smorenburg
Institute of Applied Physics TNO
1 Stieltjesweg
P.O. Box 155
Delft

✱ Mr. R. Spanhoff
Directorate for Water Management and Hydraulic Research
Rijkswaterstaat
Hooftskade 1
2526 KA 's-Gravenhage

Mr. H. Visser
Institute of Applied Physics TNO
1 Stieltjesweg
P.O. Box 155
Delft

Mr. P. Wijmans
North Sea Directorate
Rijkswaterstaat
Nijverheldsstraat 1
P.O. Box 5807
2280 HV Rijswijk (z.h.)

NORWAY

> Mr. H.C. Christensen
> Head of Division for Environmental Pollution
> The Royal Noewegian Council for Scientific and Industrial
> Research
> Gaustadalleen 30 C
> Oslo 3

> *Mr. A. Flikke
> The Norwegian State Pollution Control Authority
> P.O. Box 8100 DEP
> Oslo 1

> Mr. J.N. Langfeldt
> Head, Oil Pollution Control
> Research and Development Programme
> Ministry of Environment
> Postboks 70
> Tåsen, Oslo 8

PORTUGAL

> *Mr. C.N. Silva Sousa
> Direcçao-Geral Dos Servicos de Fomento Maritimo
> 1188 Lisboa Cedex

SWEDEN

> *Mr. L. Backlund
> Swedish Space Corporation
> Tritonvägen 27
> S 17154 Solna

> Mr. L. Holmstrom
> Swedish Space Corporation
> Tritonvägen 27
> S 17154 Solna

> Mr. B. Loostrom
> Swedish Coast Guard Headquarters
> Generaltullstyrelsen
> Box 2267
> S 103 16 Stockholm

> Mr. P. Ortendahl
> Swedish Coast Guard Headquarters
> Generaltullstyrelsen
> Box 2267
> S 103 16 Stockholm

TURKEY

★❋Mr. T. Guvenç
Institute of Marine Sciences and Technology
Mühendislik Bilimleri Fakültesi
Aegean University
178 Mithatpasa Caddesi Karatas
Bornova, Izmir

Mr. S. Basoglu
Institute of Marine Sciences and Technology
Mühendislik Bilimleri Fakültesi
Aegean University
178 Mithatpasa Caddesi Karatas
Bornova, Izmir

Mr. H. Ogelman
Temel Bilimler Fakültesi
Culcurova University
Adana

UNITED KINGDOM

Mr. D. Cormack
Warren Spring Laboratory
Department of Trade and Industry
P.O. Box 20
Gunnels Wood Road
Stevenage, Hertfordshire SG1 2BX

now with

Marine Pollution Control Unit
Marine Division
Department of Transport
Sunley House
90 High Holborn
London WC1V 6LP

Mr. B. Dowsett
Warren Spring Laboratory
Department of Trade and Industry
P.O. Box 20
Gunnels Wood Road
Stevenage, Hertfordshire SG1 2BX

★❋Mr. P.R. Morris
Warren Spring Laboratory
Department of Trade and Industry
Gunnels Wood Road
Stevenage, Hertfordshire SG1 2BX

Mr. R.J. Moulton
Marine Technology Support Unit
Atomic Energy Research Establishment (AERE), Harwell
Didcot, Oxfordshire OX11 ORA

Mr. D. Nicholson
Royal Signals and Radar Establishment (RSRE)
St. Andrews Road
Great Malvern, Worcestershire WR14 3PS

☆☼ Mr. H.D. Parker
Warren Spring Laboratory
Department of Trade and Industry
Gunnels Wood Road
Stevenage, Hertfordshire SG1 2BX

now with

International Tanker Owners Pollution Federation Ltd.
Staple Hall, Stonehouse Court
87-90 Houndsditch
London EC 3A 7AX

Mr. D.W.C. Rodda
Head of Computing Services
Water Data Unit
Reading Bridge House
Reading RG1 8PS

UNITED STATES

Mr. R. Alderman
National Oceanic and Atmospheric Administration (NOAA)
Code RDI
Rockville, MD 20852

Mr. U. Alvarado
Technology Centre
General Electric Company
Valley Forge Space
P.O. Box 8555
Philadelphia, PA 19101

Mr. J. Apel
National Oceanic and Atmospheric Administration (NOAA)
Pacific Marine Environmental Laboratory
3711, 15th Avenue N.E.
Seattle, WA 98105

now with

The Johns Hopkins University
Applied Physics Laboratory
Johns Hopkins Rd.
Laurel, Maryland 20707

Mr. G. Chafin
U.S. Environmental Protection Agency (EPA)
Room 819, West Tower (A-106)
401, M Street, S.W.
Washington, D.C., 20460

★✻CDR Ch. R. Corbett
Chief of Marine Environmental Protection
U.S. Coast Guard Headquarters
G-WEP Room 7305
400 7th Street S.W.
Washington, D.C., 20590

CDR R.E. Ettle
U.S. Coast Guard Headquarters (G-WEP)
Prevention and Enforcement Division
2100, Second Street S.W.
Washington, D.C., 29593

LCDR D. Freezer
U.S. Coast Guard Headquarters
COMDT (G-DOE-2/TP54)
400, 7th Street S.W.
Washington, D.C., 20590

Mr. F. A. Harris
Director
Environmental Protection Agency
Office of International Activities
101, M. Street S.W.
A 110 Washington D.C., 20460

Mr. J.P. Hollinger
Department of the Navy
Naval Research Laboratory
Washington, D.C., 20375

Mr. W.A. Hovis
National Oceanic and Atmospheric Administration (NOAA)
Director, Satellite Experimental Laboratory
NOAA/NESS-S/32
Washington, D.C., 20233

Mr. J.S. Knoll
National Oceanic and Atmospheric Administration (NOAA)
National Environmental Satellite, Data and Information
Service, Satellite Experimental Laboratory
Federal building No. 4, room 0128
Washington, D.C., 20233

Mr. R. Landers
U.S. Environmental Protection Agency (EPA)
EMSL - LV/EPA
P.O. Box 15027
Las Vegas, Nevada 89114

Mr. R. Livingstone
A-106 U.S. Environmental Protection Agency (EPA)
401, M Street, S.W.
Washington, D.C., 20460

Capt. T. Lutton
U.S. Coast Guard Headquarters
COMDT (G-DOE/TP54)
400, 7th Street S.W.
Washington, D.C., 20590

LCDR McFadden
U.S. Coast Guard Headquarters
COMDT (G-WEP-5/73)
400, 7th Street, S.W.
Washington, D.C., 20590

Mr. W.L. Pugh
National Oceanic and Atmospheric Administration (NOAA)
6010 Executive Bld.
Rockville MD 20852

LT. R. Schmidt
U.S. Coast Guard Headquarters
COMDT (G-DOE-2/TP54)
400, 7th Street, S.W.
Washington, D.C., 20590

LCDR J.D. Sipes
U.S. Coast Guard Headquarters (G-WEP-2)
Environmental Coordination Branch
400, 7th Street, S.W.
Washington, D.C., 20590

Ms. J. Snider
U.S. National Oceanic and Atmospheric Administration
6010 Executive Bld.
Rockville, MD 20652

☼ Mr. R.R. Vollmers
U.S. Coast Guard Headquarters
(G-DOE-2/TP54)
Marine Environmental Protection Division
400, 7th Street, S.W.
Washington, D.C., 20590

LT G.S. Voyik
U.S. Coast Guard Headquarters
400, 7th Street, S.W.
Washington, D.C., 20590

LCDR J.R. White
United States Coast Guard Headquarters
COMDT (G-DOE-2/TP54)
400, 7th Street, S.W.
Washington, D.C., 20590

EUROPEAN COMMUNITIES

Mr. R.H. Gillot
Commission of the European Communities
Joint Research Centre
Ispra Establishment
I-21020, Ispra (Varese), Italy

Mr. P. Guillot
Commission of the European Communities DG XII/G
200 Rue de la Loi
1049 Brussels, Belgium

EUROPEAN SPACE AGENCY (ESA)

Mr. G. Duchossois
European Space Agency
Directorate of Application Programmes
8-10 Rue Mario Nikis
75015 Paris, France

NATO (Observers)

Dr. M. di Lullo
NATO Scientific Affairs Division
Boulevard Leopold III
1110 Brussels, Belgium

Mr. M. Sudarskis
NATO Executive Secretariat
Boulevard Leopold III
1110 Brussels, Belgium